Theory of Fusion Plasmas: Joint Varenna-Lausanne International Workshop

Journal of Physics: Conference Series Volume 260

Varenna, Italy
30 August – 3 September 2010

Editors:

X. Garbet **O. Sauter**

ISBN: 978-1-61782-249-0
ISSN: 1742-6588

Printed from e-media with permission by:

Curran Associates, Inc.
57 Morehouse Lane
Red Hook, NY 12571

Some format issues inherent in the e-media version may also appear in this print version.

Printed by Curran Associates, Inc. (2011)

Institute of Physics
Dirac House, Temple Back
Bristol BS1 6BE UK

Phone: 44 1 17 929 7481
Fax: 44 1 17 920 0979

techtracking@iop.org

Additional copies of this publication are available from:

Curran Associates, Inc.
57 Morehouse Lane
Red Hook, NY 12571 USA
Phone: 845-758-0400
Fax: 845-758-2634
Email: curran@proceedings.com
Web: www.proceedings.com

TABLE OF CONTENTS

ORBIT-AVERAGED QUASILINEAR DIFFUSION OPERATOR IN THE ION-CYCLOTRON RANGE OF FREQUENCIES 1
V. Belmondo, R. Bilato, M. Brambilla, O. Maj

THE INFLUENCE OF THE EDGE DENSITY FLUCTUATIONS ON ELECTRON CYCLOTRON WAVE BEAM PROPAGATION IN TOKAMAKS 9
N. Bertelli, A. A. Balakin, E. Westerhof, O. E. Garcia, A. H. Nielsen, V. Naulin

MAGNETIC CHAOS HEALING IN THE HELICAL REVERSED-FIELD PINCH: INDICATIONS FROM THE VOLUME-PRESERVING FIELD LINE TRACING CODE NEMATO 16
D. Bonfiglio, M. Veranda, S. Cappello, L. Chacón, G. Spizzo

PFIRSCH–SCHLÜTER IMPURITY TRANSPORT IN STELLARATOR EDGE PLASMAS WITH LARGE RADIAL GRADIENTS 24
S. Braun, P. Helander

ONE-DIMENSIONAL PARTICLE MODELS FOR HEAT TRANSFER ANALYSIS 32
H. Bufferand, G. Ciraolo, Ph Ghendrih, P. Tamain, F. Bagnoli, S. Lepri, R. Livi

ON THE EFFECTS OF THE EQUILIBRIUM MODEL IN GYROKINETIC SIMULATIONS: FROM S-A TO DIVERTED MHD EQUILIBRIUM 40
A. Burckel, O. Sauter, C. Angioni, J. Candy, E. Fable, X. Lapillonne

ANALYTICAL AND NUMERICAL STUDIES OF THE COLD ELECTROMAGNETIC LH WAVE EQUATION IN THE MODE CONVERSION REGIME 50
A. Cardinali, V. Fusco

PARAMETRIC INSTABILITY AND LOWER HYBRID CURRENT DRIVE AT PLASMA DENSITIES REQUIRED FOR THERMONUCLEAR REACTORS 58
R. Cesario, L. Amicucci, D. De Arcangelis, M. Ferrari, F. Napoli, E. Pullara

TRANSPORT PHYSICS OF ASDEX UPGRADE CURRENT RAMPS: EXPERIMENT AND THEORY 65
E. Fable, C. Angioni, S. Fietz, J. Hobirk, G. Pereverzev

ENTROPY PRODUCTION RATE IN TOKAMAK PLASMAS WITH HELICAL MAGNETIC PERTURBATIONS 72
X. Garbet, J. Abiteboul, Y. Sarazin, A. Smolyakov, S. Allfrey, V. Grandgirard, P. Ghendrih, G. Latu, A. Strugarek

NONLOCAL EFFECTS IN GYROKINETIC TURBULENCE SIMULATIONS USING GENE 79
T. Görler, X. Lapillonne, S. Brunner, J. Chowdhury, T. Dannert, F. Jenko, B. F. McMillan, F. Merz, D. Told, L. Villard

BARRIERS TO FIELD LINE TRANSPORT IN 3D MAGNETIC CONFIGURATIONS 87
D. Grasso, D. Borgogno, F. Pegoraro, T. J. Schep

THE IMPACT OF ENERGETIC PARTICLES AND ROTATION ON TOKAMAK PLASMAS 94
M. Hole, K. G. McClements, G. Dennis, M. Fitzgerald, R. Akers

GLOBAL LINEAR GYROKINETIC SIMULATIONS FOR LHD INCLUDING COLLISIONS 102
K. Kauffmann, R. Kleiber, R. Hatzky, M. Borchardt

ELECTRON RESPONSE IN GYROFLUID SIMULATIONS OF MAGNETIC RECONNECTION 109
A. Perona, D. Grasso, D. Borgogno

DIFFUSION COEFFICIENT AND CONVECTIVE VELOCITY PROFILES OF NI-IMPURITY IN JET PLASMAS WITH ICRH 120
N. I. Pometescu

FLUX-DRIVEN GYROKINETIC SIMULATIONS OF ION TURBULENT TRANSPORT AT LOW MAGNETIC SHEAR 126
Y. Sarazin, A. Strugarek, G. Dif-Pradalier, J. Abiteboul, S. Allfrey, X. Garbet, Ph Ghendrih, V. Grandgirard, G. Latu

MICROTURBULENCE STUDIES IN RFX-MOD 135
F. Sattin, S. C. Guo, I. Predebon, S. F. Liu, X. Garbet, M. Veranda

GYROKINETIC INVESTIGATION OF MAGNETIC ISLANDS IN TOKAMAKS 143
M. Siccinio, E. Poli, W. A. Hornsby, A. G. Peeters

GYROFLUID SIMULATIONS OF COLLISIONLESS RECONNECTION IN THE PRESENCE OF DIAMAGNETIC EFFECTS 150
E. Tassi, F. L. Waelbroeck, D. Grasso
GLOBAL COLLISIONAL GYROKINETIC SIMULATIONS OF ITG MICROTURBULENCE STARTING FROM A NEOCLASSICAL EQUILIBRIUM 159
T. Vernay, S. Brunner, L. Villard, B. F. McMillan, O. Sauter, S. Jolliet, T. M. Tran, A. Bottino
KINETIC STRUCTURES OF SHEAR ALFVÉN AND ACOUSTIC WAVE SPECTRA IN BURNING PLASMAS 168
F. Zonca, A. Biancalani, I. Chavdarovski, L. Chen, C. Di Troia, X. Wang
Author Index

FOREWORD

The 2010 session of the joint Varenna–Lausanne workshop on the theory of fusion plasmas was doubtlessly a great success. The programme encompassed a wide variety of topics, namely turbulence, MHD, edge physics and RF wave heating. With 22 papers appearing in this volume of JPCS and another set of 19 papers to be published in PPCF – all refereed – it is certainly fair to say that the production was both abundant and of high scientific quality. This is important for fusion in the present scienctific climate. These are times of decreasing resources in many laboratories. This difficult situation increases the responsibilities of the community of theorists and modellers. Indeed, theory is often considered as a major indicator of the quality of science in many fields of research. This is why the Varenna–Lausanne meeting is somewhat special in the landscape of conferences, as there are few forums worldwide for the exchange of ideas on theory and modelling, and for substantiating the outstanding results which are being obtained in our field.

The tradition of the meeting is to provide a forum for numerical modelling. This tradition is clearly respected given the respectable number of papers on this area. This reflects the revolution we have been living through for some years now, with the emergence of high-performance computers. It also appears that analytical theory is flourishing. This is important for producing new ideas and also for providing guidance for numerical simulations. Finally, code validation and comparison to experiments were also well represented. I believe that this is good news given the complexity of the non-linear physics we are dealing with.

Another subject of satisfaction was the presence of many young scientists at the meeting. It is certainly crucial to train a new generation in the context of ITER operation. This is one of the missions of the meeting and once again the objective was met, thanks also to the valuable and active participation of senior scientists.

In conclusion, I anticipate great success for this special issue of JPCS and I wish readers will find guidance and inspiration herein.

Xavier Garbet
Cadarache, October 27 2010

CONFERENCE DESCRIPTION

The Joint Varenna–Lausanne international workshop on Theory of Fusion Plasmas takes place every other year in a place particularly favourable for informal and in-depth discussions. Invited and contributed papers present state-of-the-art research in theoretical plasma physics, covering all domains relevant to fusion plasmas. This workshop always welcomes a fruitful mix of experienced researchers and students, to allow a better understanding of the key theoretical physics models and applications.

- Theoretical issues related to burning plasmas
- Anomalous Transport (Turbulence, Coherent Structures, Microinstabilities)
- RF Heating and Current Drive
- Macroinstabilities
- Plasma-Edge Physics and Divertors
- Fast particles instabilities

The conference is organized by:

- Centre de Recherches en Physique des Plasmas, Ecole Polytechnique Fédérale de Lausanne, Association EURATOM - Confédération Suisse
- 'Piero Caldirola' International Centre for the Promotion of Science and International School of Plasma Physics
- Istituto di Fisica del Plasma del CNR, Milano

Editors: X Garbet (CEA, Cadarache, France) and O Sauter (CRPP-EPFL, Lausanne, Switzerland)

Theory of Fusion Plasmas: Joint Varenna–Lausanne International Workshop IOP Publishing
Journal of Physics: Conference Series **260** (2010) 012001 doi:10.1088/1742-6596/260/1/012001

Orbit-averaged quasilinear diffusion operator in the ion–cyclotron range of frequencies

V Belmondo[1], R Bilato[1], M Brambilla[1] and O Maj[1,2]

[1] Max–Planck Institut für Plasmaphysik – EURATOM Association – Garching, Germany
[2] Max-Planck Institut für Sonnensystemforschung, D-37191 Katlenburg-Lindau, Germany

E-mail: vtb@ipp.mpg.de

Abstract. The absorption of radio-frequency waves in the ion-cyclotron range of frequencies (ICRF) in fusion plasmas is adequately described in the frame of the quasilinear theory. A peculiarity of ICRF heating is to increase the fraction of trapped particles, whose the guiding-center trajectories are typically banana orbits with finite width. One possible way to properly account for the effects due to the finite-width orbits is provided by the orbit-averaged quasilinear theory. Here, we propose a routine for the evaluation of the orbit-averaged quasilinear diffusion tensor, given the the wave fields from the full wave code TORIC. Particular care is taken in the evaluation of the contribution of each wave-particle resonance as well as for the transition between the two regimes of correlated and uncorrelated resonances along the orbit. We discuss the quasilinear diffusion coefficient thus obtained and validate the result by comparing the power deposition profile computed from the orbit-averaged quasilinear operator with the one determined by TORIC. For future applications, a novel algorithm for the calculation of the corresponding Monte Carlo operators is also presented.

1. Introduction

The evolution of the ion distribution function in the presence of high frequency waves in the ion-cyclotron (IC) range of frequencies is usually described in the frame of the quasilinear (QL) theory [1, 2, 3, 4, 5]. A common approximation is to surface average the quasilinear operator, which is equivalent to assuming that all the resonant particles are tied to magnetic field lines. This is a good approximation for the description of waves interacting with the thermal bulk of particles. However, IC fast-wave heating typically increases the fraction of energetic deeply trapped ions characterized by large banana orbits. To describe properly the finite banana-width effects it is necessary to follow the particles along their guiding-center drift orbits [6]. The Doppler shift in the wave-particle resonance condition, moreover, is usually calculated by retaining in the wave fields description only one toroidal wavenumber (typically the dominant value of the antenna spectrum), and neglecting the contribution of the poloidal wavenumber to the parallel component of the wave vector.

Here we relax these constraints: we present a routine which evaluates the resonance kernel of the quasilinear operator by following the guiding-center orbits, and takes into account all Fourier components of the wave field calculated with the full-wave code TORIC [7]. For each Fourier component, the Doppler shift is evaluated at the resonant position corresponding to its poloidal and toroidal wave numbers and the local particle velocity.

The orbit-averaged quasilinear theory upon which our calculations are based is briefly reviewed in section 2. The algorithm and the first numerical results are then reported in section 3. In particular, the quasilinear diffusion coefficient is studied in relation to the different topologies of ion orbits. As a check, we have compared the heating rate predicted by the QL operator assuming a Maxwellian distribution function, with the power deposition profile evaluated by `TORIC` under the same assumption. In section 4 we present a novel algorithm to compute numerically the Monte Carlo operators in a general set of invariants starting from the orbit-averaged quasilinear operator. In the last section we draw preliminary conclusions.

2. The orbit-averaged quasilinear diffusion tensor

In axisymmetric tokamak plasmas, the unperturbed ion orbits are characterized by three adiabatic invariants [1, 8], i.e., the energy ε, the magnetic moment μ and the toroidal momentum P_φ, together with the poloidal and toroidal angles (ϑ, φ) of the ion guiding center and the phase ϕ_v of the fast ion gyration around the guiding center. Along with $\vec{I} = (\varepsilon, \mu, P_\varphi)$, it is useful to consider also arbitrary (curvilinear) coordinates $\vec{I} = (I^1, I^2, I^3)$ in the space of invariants; the value of $\vec{I}$ determines the guiding-center trajectory which evolves either with (trapped particle) or without (passing particle) change of the sign of the velocity component parallel to the local confining magnetic field. In both cases, the motion in ϑ is periodic, and the corresponding time period is referred to as the bounce time τ_b, which depends on the particle invariants $\vec{I}$. On a slower time scale the evolution of the particle distribution function F of a given ion species is governed by the balance between resonant wave-particle interactions and collisions. In the quasilinear picture, the kinetic equation describing this evolution is [1, 3, 4]

$$\frac{\partial F}{\partial t} = \mathcal{Q}_{\text{avg}}(F) + \mathcal{C}_{\text{avg}}(F) \quad \text{with :} \quad \mathcal{Q}_{\text{avg}}(F) = \frac{1}{\sqrt{|g|}}\frac{\partial}{\partial I^\alpha}\left(\sqrt{|g|}\,\bar{D}^{\alpha\beta}_{\text{rf}}\frac{\partial F}{\partial I^\beta}\right) \tag{1}$$

where $F = F(\vec{I})$, $\bar{D}^{\alpha\beta}_{\text{rf}}$ is the orbit-averaged quasilinear diffusion tensor, and $\mathcal{C}_{\text{avg}}(F)$ is the orbit-averaged collision operator.

The QL operator in (1) is in divergence form with g being the metric tensor associated with the generic set of invariants $\vec{I}$. For example, on making use of the action variables $I^\alpha = J^\alpha$, one finds that the metric is Euclidean and, thus, $|g| = 1$, while, in the case of $\vec{I} = (\varepsilon, \mu, P_\varphi)$, one can show $|g| \propto \tau_b^2$ [3]. The QL diffusion tensor can be written as

$$\bar{D}^{\alpha\beta}_{\text{rf}}(\vec{I}) = \frac{1}{2}\text{Re}\left\{\frac{1}{\tau_b}\int_0^{\tau_b} \text{d}t \int_0^{2\pi}\frac{d\phi_v}{2\pi}\Delta^\alpha(t,\phi_v)^* \int_{-\infty}^{t} dt'\;\Delta^\beta(t',\phi_v)\right\}. \tag{2}$$

The first integral is the average over the bounce period, the second integral comes from the gyro-average over the fast phase ϕ_v, and the third integral is the result of integrating the linearized Vlasov equation along the unperturbed particle orbits. As a consequence of axisymmetry [3, 4], the averages in (2) can be shown to be equivalent to the standard orbit average of reference [1]. The quantities $\Delta^\alpha = \vec{a}\cdot\nabla_v I^\alpha$ are evaluated on the particle trajectory corresponding to invariants $\vec{I}$, with $m\vec{a} = Ze(\vec{E} + \vec{v}\times\vec{B}/c)$ being the Lorentz force due to the radio-frequency fields. In our study, the wave fields are provided by the full-wave code `TORIC` [7] which makes use of the spectral representation $(\vec{E}(t,\vec{r}), \vec{B}(t,\vec{r})) = \sum_{(m,n)}(\vec{E}_{m,n}(\psi), \vec{B}_{m,n}(\psi))e^{i(m\vartheta+n\varphi)-i\omega t}$, ψ being a label of the magnetic surfaces. Due to the contravariant nature of the QL diffusion tensor, $\bar{D}^{\alpha\beta}_{\text{rf}}$ of (2) can be readily transformed into any other coordinate system.

Following the classical derivation [2], we obtain, for a single cyclotron harmonic p and neglecting the parallel component of the electric field,

$$\bar{D}^{\alpha\beta}_{\text{rf}} = \frac{Z^2e^2}{2m^2}\text{Re}\left\{\sum_{\nu_1,\nu_2}\frac{1}{\tau_b}\int_0^{\tau_b}\text{d}t\,\Pi^\alpha_{\nu_1}(t)^* e^{-i\chi_{\nu_1}(t)}\int_0^{+\infty} dt'\Pi^\beta_{\nu_2}(t-t')e^{i\chi_{\nu_2}(t-t')}\right\}, \tag{3}$$

where, for brevity, $\nu = (m, n)$, and

$$\Pi^{\alpha}_{\nu}(t) = \frac{v_\perp}{\sqrt{2}}\left[J_{p-1}(k_\perp\rho)E^{+}_{\nu}e^{-i\delta} + J_{p+1}(k_\perp\rho)E^{-}_{\nu}e^{i\delta}\right]X^{\alpha},$$
$$\chi_\nu(t) = \int_0^t (p\Omega_c - \omega + k_\parallel v_\parallel)dt'. \tag{4}$$

Here, Ω_c is the ion cyclotron frequency, J_p are Bessel functions, $E^{\pm}_{\nu}$ are the rotating components of the wave electric field, δ is a phase shift, and $\vec{X} = (X^{\alpha})$ is a vector field over the space of invariants. In equations (4) $v_\parallel$ and $v_\perp$ are to be expressed in terms of the invariants (hereafter, $\parallel$ and $\perp$ are defined with respect to the direction of the local magnetic field), $k_\parallel$ is obtained from m, n, and $k_\perp$ is then computed from the cold-plasma dispersion relation for the fast root. It is worth noting that the phase $\chi_\nu(t)$ is the integral of a periodic function (since the guiding center motion is periodic in ϑ), but is not periodic itself. The integrals in (3) are evaluated using the stationary phase method, the stationary-phase points being defined by the usual Doppler-shifted resonance condition $p\Omega_c - \omega + k_\parallel v_\parallel = 0$. Hereafter we denote $t_{\nu,j}$ the transit time of the guiding center at the resonance positions labelled by the index $j = 1, 2, \ldots$. It is implicitly assumed that collisions decorrelate the wave-particle phase over times longer than a bounce period, thereby effectively suppressing resonances with harmonics of the bounce frequency [9]. On applying the stationary phase method, particular care must be taken for those stationary-phase points that hit the boundary of the integration domain. Finally we obtain a relatively simple approximation of (3),

$$\bar{D}^{\alpha\beta}_{\rm rf} \approx \frac{Z^2e^2}{2m^2}\frac{1}{\tau_b}\mathrm{Re}\left\{\sum_{\nu_1}\left(\sum_{j_{\nu_1}}\mathcal{I}^{\alpha}_{\nu_1,j_{\nu_1}}\right)^{*}\sum_{\nu_2}\left(\sum_{j_{\nu_2}}\mathcal{I}^{\beta}_{\nu_2,j_{\nu_2}}\right)\right\}, \tag{5}$$

where for each ν the index j_ν runs over all resonances within one period. As well known [2, 10], when resonances are well separated in the sense of the stationary phase evaluation, there are two of them for passing ions, and four for trapped ions. In this case each contribution is given by the non-degenerate stationary-phase formula

$$\mathcal{I}^{\alpha}_{\nu,j} = \Pi^{\alpha}_{\nu}(t_{\nu,j})\,\mathcal{W}_\nu(t_{\nu,j}) = \Pi^{\alpha}_{\nu}(t_{\nu,j})\left|\frac{2\pi}{\chi^{(2)}_{\nu}(t_{\nu,j})}\right|^{\frac{1}{2}} e^{i\frac{\pi}{4}\mathrm{sgn}(\chi^{(2)}_{\nu}(t_{\nu,j}))}e^{i\chi_\nu(t_{\nu,j})}. \tag{6}$$

where $\chi^{(n)}_{\nu} = d^n\chi_\nu/dt^n$. Resonances can merge in pairs, however, for particles either on banana orbits with reflection point close to resonance, or on passing orbits nearly tangential to the cyclotron layer. In such cases the second-order derivative of the phase at resonance is small and the denominator in (6) blows up; for a correct evaluation the phase correlation between the two resonances cannot be neglected and their combined contributions take the form [2]

$$\mathcal{I}^{\alpha}_{\nu,j} = \Pi^{\alpha}_{\nu}(\Theta_j)\,\mathcal{W}_\nu(\Theta_j) = \Pi^{\alpha}_{\nu}(\Theta_j)\,2\pi e^{i\chi_\nu(\Theta_j)}[\chi^{(3)}_{\nu}(\Theta_j)/2]^{-\frac{1}{3}}\mathrm{Ai}(\chi^{(1)}_{\nu}(\Theta_j)[\chi^{(3)}_{\nu}(\Theta_j)/2]^{-\frac{1}{3}}), \tag{7}$$

where Ai is the Airy function, and the argument Θ_j is defined as the point between the two considered resonances where the second derivative vanishes, $\chi^{(2)}_{\nu}(\Theta_j) = 0$. We have investigated numerically the transition between the case of two isolated and two merging resonances (cf. section 3), and adopted a matching procedure similar, but not identical, to that proposed in [10].

For sake of simplicity, in this paper we limit our analysis to the dominant [Appendix B of [2]] diagonal terms in the double sum (5) (the significance of the off-diagonal terms will be addressed in the future). On making use of (6) and (7), the quasilinear diffusion tensor takes the form

$$\bar{D}_{\rm rf}^{\alpha\beta} \approx \sum_{\nu}\sum_{j_\nu} D_{\nu,j_\nu} X^{\alpha}_{\nu,j_\nu} X^{\beta}_{\nu,j_\nu}, \tag{8}$$

where $D_{\nu,j}$ are positive coefficients and $\vec{X} = (X^{\alpha}_{\nu,j})$ is the vector field $\vec{X} = (X^{\alpha})$ evaluated at either $t_{\nu,j}$ for well-separated resonances or Θ_j for merging resonances. The quasilinear diffusion tensor thus obtained (8) amounts to a sum of dyadics with positive coefficients, hence it is symmetric and positive definite [3].

The dyadic structure of $\bar{D}_{\rm rf}^{\alpha\beta}$ is crucial for the calculation of the Monte Carlo, cf. section 4. Let us recall that in a Monte Carlo (MC) code the invariants of a given particle are periodically (Δt) updated according to $I^{\alpha} \mapsto I^{\alpha} + \Delta I^{\alpha}$, with [3, 4, 11]

$$\Delta I^{\alpha} = \dot{\bar{I}}^{\alpha}\Delta t + \sqrt{2\Delta t}\sum_{a=1}^{3} A^{\alpha}_{a}\xi_a, \tag{9}$$

where ξ_a are random numbers sampled from a normal distribution of zero mean and unit variance; the time-derivatives $\dot{\bar{I}}^{\alpha}$ of the expectation values and the vectors $\vec{A}_a = (A^{\alpha}_a)$ should be evaluated from $\bar{D}_{\rm rf}^{\alpha\beta}$ according to [4]

$$\dot{\bar{I}}^{\alpha} = \frac{1}{\sqrt{|g|}}\frac{\partial}{\partial I^{\beta}}\left[\sqrt{|g|}\bar{D}_{\rm rf}^{\alpha\beta}\right], \qquad \bar{D}_{\rm rf}^{\alpha\beta} = \sum_{a=1}^{3} A^{\alpha}_a A^{\beta}_a. \tag{10}$$

By inspection of the second equation in (10) one can see that the coefficients A^{α}_a, regarded as a matrix, amount to the square root of $\bar{D}_{\rm rf}^{\alpha\beta}$. The analytical calculation of the derivative of the expectation value, on the other hand, can be rather cumbersome [4]. In section 4 we propose a new algorithm for the calculation of the Monte Carlo operators (10).

3. Algorithm implementation and numerical results

In this section, the algorithm for the evaluation of the quasilinear diffusion tensor (8) is briefly outlined. Then, we report about the first numerical results that provide a rather detailed study of the quasi-linear diffusion tensor and its relationship to the topology of guiding-center trajectories, together with an estimate of the heating rate.

3.1. Numerical evaluation of the quasilinear diffusion tensor from `TORIC` *wave fields*

The value of the quasilinear diffusion tensor (8) depends on the Fourier coefficients of the wave electric field, and on the specific guiding-center trajectory corresponding to the set of invariants $\vec{I}$. The numerical evaluation of the diffusion tensor is obtained by integrating the guiding-center drift equations [6] by means of an *ad hoc* routine based upon the explicit solver `LSODE` [12], then locating the resonances along the trajectory and evaluating their contributions to the sum (8). The guiding-center trajectories have been sampled according to a uniform mesh in energy ε and an adapted non-uniform mesh in the normalized magnetic moment $\Lambda = \mu B_{\rm eq}/\varepsilon$, where $B_{\rm eq}$ is the value of the magnetic field at the point of the external equatorial transit of the particle; instead of P_{φ}, we use the radial position $\psi_{\rm eq}$ of the equatorial transit, uniformly discretized, together with the sign σ_v of the the parallel velocity. For each point $(\varepsilon, \Lambda, \psi_{\rm eq}, \sigma_v)$ of the grid the value of $\bar{D}_{\rm rf}^{\alpha\beta}$ is computed and written onto a table, which can subsequently be read and interpolated by any Monte Carlo code. The frames (a) and (b) of figure 1 show an example of poloidal projections of the guiding-center trajectory of passing and trapped particles respectively. The bullets mark the

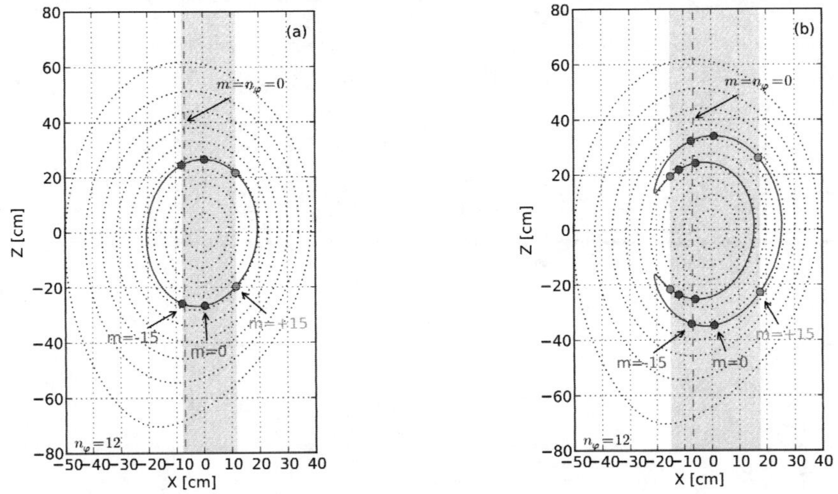

Figure 1. Poloidal projections of the guiding-center trajectory of passing (a) and trapped (b) particles. The bullets mark the wave-particle resonance positions for $m = -15$ (green), $m = 0$ (magenta), $m = +15$ (light blue) and the same $n = +12$. The vertical dashed line is the resonance position of the $(m = 0, n = 0)$ mode, i.e. without Doppler shift.

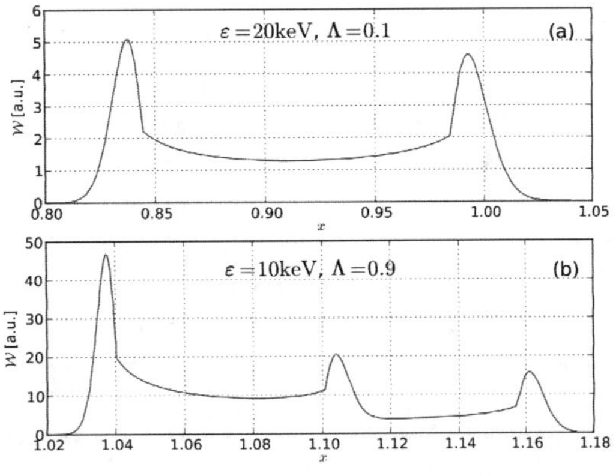

Figure 2. The Resonance kernel $\mathcal{W}_\nu = \sum_j |\mathcal{W}_{\nu,j}|^2$ for passing (a) and trapped (b) particles as function of $x = \omega/(\Omega_c + k_\| v_\|)$. By varying x the resonance layer moves from the high field side to the low field side of the guiding-center trajectories.

wave-particle resonance positions for the different values of m, all the resonances calculated for the same $n = +12$. As reference, the vertical dashed line corresponds to the cyclotron resonance in the absence of Doppler shift. For passing particles (frame (a)) the resonances with the same (m, n) values are almost vertically aligned. This is not the case for trapped particles (frame (b)) where the Doppler shift is influenced by the strong variations of $v_\|$ along the guiding-center trajectory. In particular, $v_\|$ changes sign when passing over the banana tips: for this reason the resonances in the internal and external part of the trajectory are Doppler-shifted in opposite directions.

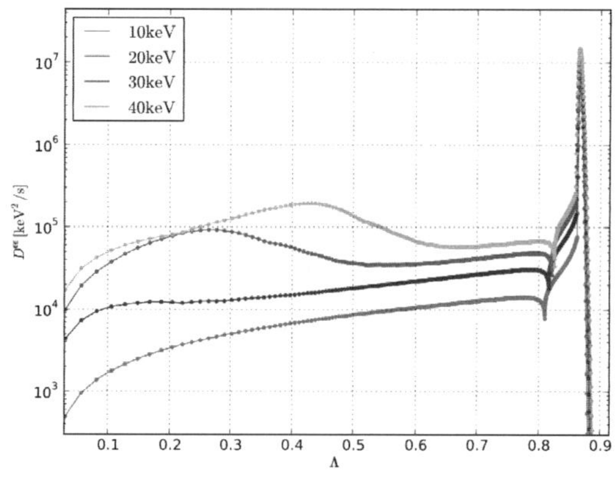

Figure 3. Diffusion coefficient $\bar{D}^{\varepsilon\varepsilon}$ as function of Λ for $\psi_{\rm eq} = 0.5$ and $\sigma_{\rm v} = +1$.

As an example of the transition from isolated resonances (6) to the phase-correlated regime (7), figure 2 shows the sum of the resonance kernels $\mathcal{W}_\nu = \sum_j |\mathcal{W}_{\nu,j}|^2$ with $\nu = (m = 10, n = 12)$ for passing and trapped particles, respectively frame (a) and (b). The quantity $\mathcal{W}_\nu$ is plotted as function of $x = \omega/(\Omega_c + k_\parallel v_\parallel)$; with increasing x the resonances move from the high (left) to the low (right) field side of the guiding-center trajectory. In the case of figure (2.a), for $0.8 < x < 0.84$ the resonances are phase-correlated and the resonance kernel is described by the Airy function, cf. (7). In the interval $0.84 < x < 0.98$ the resonances are well separated and the stationary phase formula (6) is applied. For $x > 0.98$ the resonances are phase-correlated again. In the case of trapped particles, figure (2.b), the picture is more complicated. The first peak on the left originates from the contribution of two pairs of correlated resonances; this happens when the resonance layer is approximately close to the banana tips. Then, a first plateau corresponding to four uncorrelated resonances develops into a second smaller peak which is due to a pair of correlated resonances together with two uncorrelated; this happens when the resonance layer, moving towards the low field side, leaves the inner branch of the banana orbit. At last, after a second plateau, where only two uncorrelated contributions are significant, a third peak appears when the two remaining resonances becomes correlated. The transition between the two regimes described by equations (6) and (7) is implemented by imposing a threshold on the value of the argument of the Airy function in equation (6); the specific value has been selected in order to ensure the continuity of the resonance kernel.

Figure 3 shows $\bar{D}^{\varepsilon\varepsilon}$ as function of the normalized magnetic moment for different values of energy and $\psi_{\rm eq} = 0.5$, $\sigma_{\rm v} = +1$, $n_\varphi = +12$ and summing over m in the range $-15 \le m \le +15$. The transition from passing to trapped particles is visible as a sharp feature of $\bar{D}^{\varepsilon\varepsilon}$ pattern around $\Lambda = \Lambda_c = B_{\rm eq}/B_{\rm max} \approx 0.8$. The $\bar{D}^{\varepsilon\varepsilon}$ peak is in correspondence of trapped particles which are in resonance with the wave close to their banana tips. If Λ is increased further, the particle is reflected before encountering the resonance and $\bar{D}^{\varepsilon\varepsilon}$ rapidly decreases to zero.

3.2. An estimate of the heating rate

As an order-of-magnitude validation we compare the power deposition profiles predicted by `TORIC` and the heating rate profile predicted by the quasilinear operator evaluated above. Without solving the kinetic equation (1), in both cases we assume that the distribution function $F(\vec{I})$ is a Maxwellian $F_M(\varepsilon)$. If we use as set of invariants $\vec{I} = (\varepsilon, \Lambda, \bar{\psi})$, with $\bar{\psi}$ the third

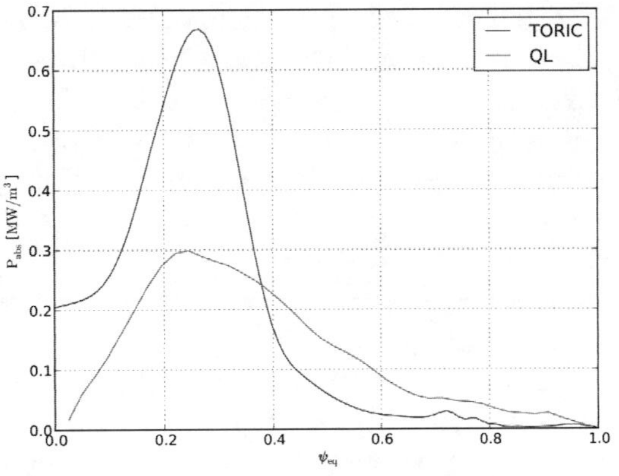

Figure 4. Radial power deposition profiles.

invariant [8] and $\sqrt{|g|} = \tau_b\, \varepsilon/(2\,\pi\, B_{\rm eq})$, we have

$$\mathcal{P}_{\rm abs}(\psi) = -\left(\frac{2\pi}{m}\right)^2 \frac{\mathcal{F}_p(\psi)}{V(\psi)} \int_0^{\varepsilon_{\rm max}} {\rm d}\varepsilon\, \frac{\varepsilon}{B_{\rm eq}} \frac{\partial F_{\rm M}}{\partial \varepsilon} \int_0^1 {\rm d}\Lambda\, \tau_b\, \bar{D}_{\rm rf}^{\varepsilon\varepsilon} \tag{11}$$

where $\mathcal{F}_p(\psi) = {\rm d}\Psi_p/{\rm d}\psi$, with Ψ_p the poloidal flux, $V(\psi)$ is the specific volume of the magnetic surface ψ, and $\bar{D}_{\rm rf}^{\varepsilon\varepsilon}$ is calculated according to (8). In order to be consistent with the model of TORIC, in equation (11) we consider the limit of well passing particles $\bar{\psi} \approx \psi$. Figure (4) shows the radial power deposition profiles: the red line has been obtained implementing equation (11) given the diffusion coefficients computed in section 3.1, whereas the blue line is obtained from `TORIC`. The profile predicted by QL coefficient is broader than the profile computed by `TORIC`. This difference can be explained by the finite orbit widths, which are not included in `TORIC` model. However, this broadening can also be due to the neglected off-diagonal terms in (5). This will be investigated in the future. Moreover, the two profiles in Figure (4) exhibit a different behavior near the magnetic axis. However, the power balance is not affected by this discrepancy since upon integrating over ψ the power density is weighted with a specific volume which is small for $\psi \to 0$. In fact, the total absorbed power computed from the two deposition profiles are in good agreement, the values being: 0.667 MW from `TORIC` and 0.704 MW from integrating equation (11) over ψ, both for 1 MW of absorbed ICRF power.

4. Numerical evaluation of Monte Carlo operators

In the evaluation of the Monte Carlo (MC) operator (10), one must perform the derivatives of the quasilinear diffusion tensor. To avoid to perform these derivatives analytically, we propose to calculate numerically $\dot{\bar{I}}^{\alpha}$ by using the divergence of the quasilinear diffusion tensor $\bar{D}_{\rm rf}^{\alpha\beta}$ written in the form

$$\dot{\bar{I}}^{\alpha} = \sum_{\nu}\sum_{j_\nu} \left\{ X^{\alpha}_{\nu,j_\nu} \mathcal{L}_{\vec{X}_{\nu,j_\nu}} D_{\nu,j_\nu} + D_{\nu,j_\nu} [\mathcal{L}_{\vec{X}_{\nu,j_\nu}} X^{\alpha}_{\nu,j_\nu} + X^{\alpha}_{\nu,j_\nu} {\rm div} \vec{X}_{\nu,j_\nu}] \right\}, \tag{12}$$

where $\mathcal{L}_{\vec{X}_{\nu,j_\nu}} = X^{\alpha}_{\nu,j_\nu}(\partial/\partial I^{\alpha})$. The terms in equation (12) amount to the derivative of scalar functions, i.e. D_{ν,j_ν} and X^{α}_{ν,j_ν}, along the integral lines of the vector field $\vec{X}_{\nu,j_\nu}$ together with ${\rm div}\vec{X}_{\nu,j_\nu}$. The latter can also be related to $\mathcal{L}_{\vec{X}_{\nu,j_\nu}}$ in virtue of the identity

$$\mathcal{L}_{\vec{X}_{\nu,j_\nu}}(dV_I) = ({\rm div}\vec{X}_{\nu,j_\nu})dV_I, \tag{13}$$

Journal of Physics: Conference Series **260** (2010) 012001 doi:10.1088/1742-6596/260/1/012001

where $dV_I = \sqrt{|g|}dI$ is the volume form in the space of invariants. The volume element dV_I is approximated by the volume enclosed by the four points of a three-dimensional simplex (convex hull of the simplex) with small-enough edges. On the other hand, the derivative $\mathcal{L}_{\vec{X}_{\nu,j_\nu}}$ applied to a generic function is approximated by symmetric finite differences upon moving upward and backward along the integral lines of $\vec{X}_{\nu,j_\nu}$ with an explicit forth-order Runge-Kutta scheme. In doing this, the properties of the simplex enclosing a convex region significantly simplifies the calculation of the volume dV_I at the new positions. At last, the vectors $\vec{A}_a$ are computed by taking the square root of $\bar{D}^{\alpha\beta}_{\mathrm{rf}}$ regarded as a matrix.

5. Conclusions

We have presented the algorithms and numerical implementation for the evaluation of the orbit-averaged quasilinear operator in the ion cyclotron range of frequencies. Starting from the wave fields calculated by a full-wave solver, such as `TORIC`, and following the guiding-center orbits we have computed the coefficients of the quasilinear diffusion operator in a general axisymmetric configuration. Particular attention has been devoted in calculating the resonance kernel of the wave-particle interaction. This has required to implement a numerical algorithm capable of dealing with all the possible kinds of guiding-center trajectories and relative positions of the resonance points. As a preliminary test, we have estimated the heating rate profile in the limit of Maxwellian distribution function, in order to compare it with the deposition profile calculated by `TORIC`. The agreement in the total heating rate is an important check, in view of the fact that no *ad hoc* normalization of the wave fields have been made. The deposition profile calculated from the quasilinear diffusion coefficient is broader than that determined by `TORIC`: this difference can be explained by the finite orbit width, which are not included in `TORIC` model.

Finally, we have outlined the numerical algorithm necessary to derive the Monte Carlo operators in a general set of invariants starting from the orbit-averaged quasilinear coefficients evaluated in this work.

References

[1] Kaufman A N 1972 *Physics of Fluids* **15** 1063–1069
[2] Bernstein I B and Baxter D C 1981 *Physics of Fluids* **24** 108–126
[3] Eriksson L G and Helander P 1994 *Physics of Plasmas* **1** 308–314
[4] Eriksson L G and Schneider M 2005 *Physics of Plasmas* **12** 072524
[5] Brambilla M 2007 *Nuclear Fusion* **47** 175–180
[6] Boozer A H 1980 *Physics of Fluids* **23** 904–908
[7] Brambilla M 1999 *Plasma Physics and Controlled Fusion* **41** 1–34
[8] Bernstein I B and Molvig K 1983 *Physics of Fluids* **26** 1488–1507
[9] Kasilov S, Pyatak A I and Stepanov K N 1997 *Reviews of Plasma Physics* vol 20 ed Kadomtsev B (New York-London: Consultants Bureau) p 61
[10] Catto P J and Myra J R 1992 *Physics of Fluids B: Plasma Physics* **4** 187–199
[11] Boozer A H and Kuo-Petravic G 1981 *Physics of Fluids* **24** 851–859
[12] Hindmarsh A C 1983 *Scientific Computing* ed Stepleman R S (Amsterdam: North-Holland) p 55

Journal of Physics: Conference Series **260** (2010) 012002 doi:10.1088/1742-6596/260/1/012002

The influence of the edge density fluctuations on electron cyclotron wave beam propagation in tokamaks

N Bertelli[1], A A Balakin[2], E Westerhof[1], O E Garcia[3], A H Nielsen[4] and V Naulin[4]

[1]FOM-Institute for Plasma Physics Rijnuizen, Association EURATOM-FOM, Trilateral Euregio Cluster, Nieuwegein, The Netherlands, www.rijnhuizen.nl
[2] Institute of Applied Physics RAS, Nizhny Novgorod, Russia
[3] Department of Physics and Technology, University of Tromsø, N-9037 Tromsø, Norway
[4] Association EURATOM-Risø National Laboratory, Technical University of Denmark, OPL-128 Risø, PO Box 49, DK-4000 Roskilde, Denmark

E-mail: `N.Bertelli@rijnhuizen.nl`

Abstract. A numerical analysis of the electron cyclotron (EC) wave beam propagation in the presence of edge density fluctuations by means of a quasi-optical code [Balakin A. A. *et al*, *Nucl. Fusion* **48** (2008) 065003] is presented. The effects of the density fluctuations on the wave beam propagation are estimated in a vacuum beam propagation between the edge density layer and the EC resonance absorption layer. Consequences on the EC beam propagation are investigated by using a simplified model in which the density fluctuations are described by a single harmonic oscillation. In addition, quasi-optical calculations are shown by using edge density fluctuations as calculated by two-dimensional interchange turbulence simulations and validated with the experimental data [O. E. Garcia *et al*, *Nucl. Fusion* 47 (2007) 667].

1. Introduction

One of the goals of electron cyclotron current drive (ECCD) in tokamak plasmas is the control of magnetohydrodynamic (MHD) instabilities. The effectiveness of this control depends strongly on the achievable localization of the EC driven current density and therefore it is essential to study the various physics effects determining this localization. In this work, we investigate the effects of density fluctuations associated, in particular, with edge turbulence on the EC wave beam propagation as it may affect the beam width and consequent power deposition and ECCD profile. Such effects are expected to be particularly relevant when the beam propagates over a large distance between the fluctuations and the EC resonance layer. This is the case in ITER, in which the distance between scrape-off layer (SOL) density fluctuations and the EC resonance layer is larger than one meter. An immediate consequence of crossing the turbulent edge region will be a perturbation of the phase front of the beam resulting, in turn, in a perturbation of the wave vector spectrum. As the beam propagates, this can result in a broadening of the beam itself. In the literature, this effect has generally been studied by making use of ray tracing codes based on the geometrical optics approximation [1, 2]. The conventional ray- and beam-tracing codes typically are based on the expansion assuming the different length scales as $\lambda \ll w \ll L$, where λ

is the wavelength, w the wave beam width and L the typical inhomogeneity scale of the dielectric response of the plasma. The presence of the SOL density fluctuations, also known as blobs which can reach the same size of the width of wave beam [3], limits strongly the applicability of geometrical optics. In order to properly account for the effect of the density perturbations, a quasi-optical beam tracing code, as presented in Ref. [4] assuming only $L, w \gg \lambda$ (λ is the wavelength), is required. Here, a systematic numerical analysis of the effects of the edge density fluctuations on the EC beam propagation is presented by using this quasi-optical beam tracing code.

This paper is structured as follows. The main features of the quasi-optical beam tracing code are presented in section 2. Section 3 addresses the relation between the density fluctuations and the perturbations in the phase front of the beam. Quasi-optical calculations are presented in section 4 for an ITER-like beam propagating in vacuum taking into account a single harmonic phase fluctuation, as would result also from a single harmonic density fluctuation in a narrow layer. An analysis of the effective beam intensity is performed in the case of (i) this single harmonic oscillation and (ii) the phase perturbation resulting from a narrow layer of edge density fluctuations as calculated by two-dimensional interchange turbulence simulations [3].

2. Quasi-optical beam tracing code

The quasi-optical beam tracing code solves a parabolic wave equation, which can be seen as a generalization to anisotropic media with spatial dispersion of the parabolic equation for wave propagation in isotropic media of Fock and Leontovich [5]. The equation takes the following form [4]

$$U = u(\tau, \boldsymbol{\xi}) \exp(\mathrm{i}k_0\Psi), \quad \frac{\partial u(\tau, \boldsymbol{\xi})}{\partial \tau} = \mathrm{i}k_0 \hat{H}_0[u(\tau, \boldsymbol{\xi})]. \tag{1}$$

In equation (1), U is the electric field defined in analogy to the geometrical optics assumption, namely, where u is a slowly varying amplitude on the scale of wavelength and Ψ is the fast oscillating phase. The second equation in (1) describes the evolution of the scalar beam amplitude u. In addition, τ is a parameter along the reference ray acting in the role of arc length and $\boldsymbol{\xi} = (\xi_1, \xi_2)$ are coordinates across the beam propagation ($\boldsymbol{\xi} = 0$ on the reference ray). In other words, $\{\tau, \boldsymbol{\xi}\}$ represents an accompanying curvilinear coordinate system of the beam. For more details about this equation, the reader is referred to [4].

This quasi-optical code is capable of treating correctly wave beams of arbitrary shape in the presence of strongly inhomogeneous dispersion and absorption as well as in media with strong spatial dispersion [4]. In this work we make use of this quasi-optical code in order to study the effects of the density fluctuations on the EC wave propagation.

3. Density and phase fluctuations

Crossing the narrow turbulent edge region, also known as scrape-off layer (SOL) [3], results in a perturbation of the phase front of the beam modifying the wave vector spectrum. It is assumed that the density fluctuations are weak, i.e., not strong enough to deflect significantly the geometrical optics ray on the scale of the fluctuations themselves. The slowly varying beam amplitude is now written as $u(\tau, \boldsymbol{\xi}) = \bar{u}(\tau, \boldsymbol{\xi})e^{\mathrm{i}\delta\varphi}$ where $\bar{u}$ is the amplitude in the case without fluctuations. The geometrical optics approximation can then be used to find, through a simple integration along the beam path across the turbulence layer (SOL), the perturbation of the beam phase, $\delta\varphi$, as

$$\delta\varphi(\boldsymbol{\xi}) = k_0 \int_{\mathrm{SOL}} \frac{\delta H(\boldsymbol{\xi}, \tau)}{\partial H/\partial p_\tau} \mathrm{d}\tau. \tag{2}$$

Here, δH is the variation of the Hamiltonian due to the presence of the density fluctuations and p_τ is the normalized longitudinal component of the wave vector. Equation (2) has been

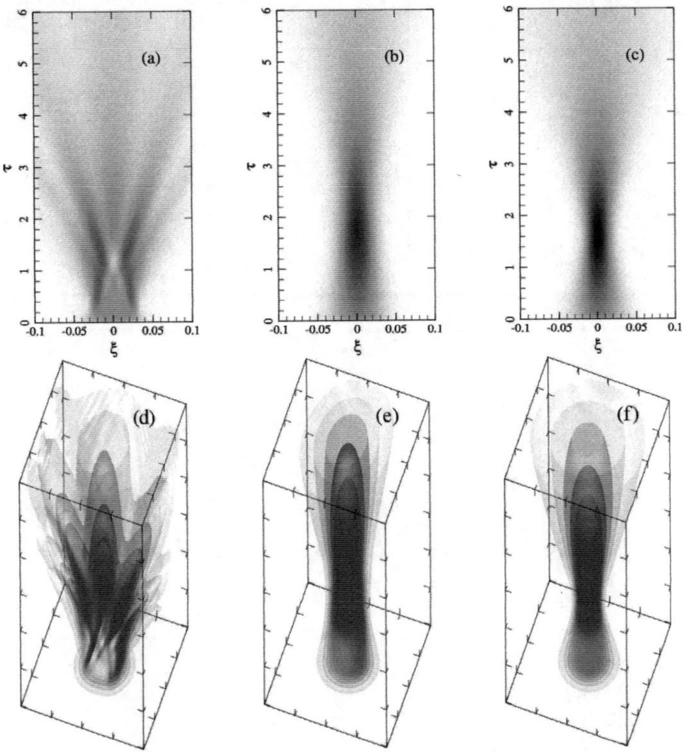

Figure 1. QO beam profiles affected by a harmonic phase perturbation of the type (3) for $\varphi_0 = \pi/2$, $\theta = 0$ and $w\kappa = 3$ (a), $w\kappa = 1$ (b), $w\kappa = 0.3$ (c). Figures (d), (e) and (f) show the 3D beam profiles.

implemented in the quasi-optical code with a Hamiltonian which can take into account the SOL density fluctuations.

4. Numerical results

4.1. Harmonic phase fluctuations

In order to understand the influence of the presence of the phase fluctuations (related to the density fluctuations as described above) on the beam propagation, we start by analyzing a simplified model with phase perturbations given by a single harmonic oscillation,

$$\delta\varphi = \varphi_0 \cos(\kappa\xi + \theta), \tag{3}$$

where φ_0 is the amplitude of the fluctuation, κ^{-1} its scale length, and θ its phase. Such phase perturbations can be seen as resulting from a single harmonic density fluctuation in a narrow layer. Note that κ is perpendicular to the magnetic equilibrium field since along the magnetic field lines the density fluctuations are mostly homogeneous. First we study the case with an amplitude of $\varphi_0 = \pi/2$. Three different scale lengths κ^{-1} are considered: (i) $w\kappa \ll 1$, i.e., large scale as compare to the width of beam waist, w; (ii) $w\kappa = 1$, i.e., the intermediate case; (iii) $w\kappa \gg 1$, i.e., small-scale fluctuations. Four different phases θ are considered: (i) $\theta = 0$, representing a defocusing fluctuation at the beam axis; (ii) $\theta = \pi/2$, representing deflecting fluctuation; (iii) $\theta = \pi$, representing a focusing fluctuation; (iv) random phase θ. For all calculations, ITER beam parameters from the Upper Port Launcher lower steering mirror are

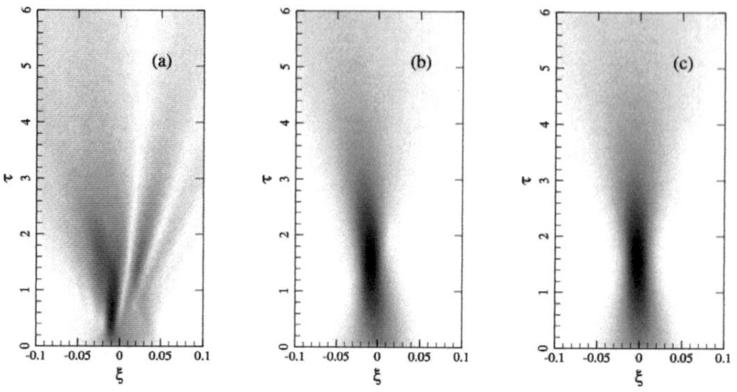

Figure 2. The same as Figure 1 for $\varphi_0 = \pi/2$, $\theta = \pi/2$ and $w\kappa = 3$ (a), $w\kappa = 1$ (b), $w\kappa = 0.3$ (c).

used. In particular, the width of the beam waist is $w = 2.1$ cm and the distance to the beam focus inside the plasma $D_{\text{focus}} = 1.62$ m, which corresponds to the location of the EC resonance layer [6].

4.2. Perturbed beam profiles

The propagation of the beam after being affected by a harmonic phase perturbation of the type (3) and amplitude $\varphi_0 = \pi/2$ is illustrated in Figures 1-4. Figure 1 shows the case of a defocusing phase $\theta = 0$ for small-scale fluctuations (a), an intermediate case (b) and large-scale fluctuations (c). In Figure 1, it appears that (i) for small-scale fluctuations, the beam is fragmented; (ii) in the intermediate case, there is an additional defocusing of the beam together with a forward shift in the location of the waist beam; (iii) for $w\kappa = 0.3$, the influence of fluctuations is practically negligible. In Figure 2, which shows the case for a deflecting phase $\theta = \pi/2$, it appears that, for small-scale fluctuations there is again a fragmented beam but now without symmetry with respect to the beam axis, while intermediate scales show a clear deflection of the beam. The large-scale fluctuation is seen to cause only a very small deflection. Figure 3 shows for the case of a focusing phase $\theta = \pi$, a broadening of the beam for small-scale fluctuations while for $w\kappa = 0.3$ again no effect is seen. On the contrary, for intermediate scales, the beam is clearly focused and

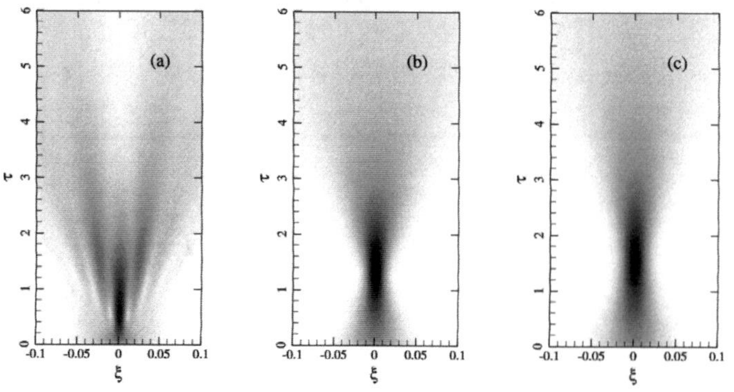

Figure 3. The same as Figure 1 for $\varphi_0 = \pi/2$, $\theta = \pi$ and $w\kappa = 3$ (a), $w\kappa = 1$ (b), $w\kappa = 0.3$ (c).

the location of the waist is shifted backward. Note that this shift is exactly an opposite effect compared to the case of Figure 1(b). Finally, Figure 4 shows an average of the beam profiles over random phase θ. It mainly appears that there is on average (i) a broadening for small-scale fluctuations; (ii) a weak defocusing of the beam for the intermediate case and (iii) practically no effect for large-scale fluctuations.

4.3. Effective beam intensity

Another important aspect to analyze is the beam intensity ($I \propto |\bar{u}|^2$) which is the more relevant number for the power absorption and, in turn, the current drive. Figure 5 shows the effective beam intensity averaged over the phase θ, I_{eff}, normalized to the unperturbed intensity, I_0, as a function of the amplitude, φ_0, of the phase fluctuations. In particular, the normalized effective beam intensity is shown as obtained in the near, mid and far field, namely, for $\tau^* = 0.2$, 0.4 m, $\tau^* = 2$ m and $\tau^* = 6$ m, respectively (τ^* is the distance between the edge of the plasma and the position where the intensity is calculated). Four fluctuation scales are considered: $w\kappa = 10$, 3, 1, and 0.3. One can see that (i) in the near field ($\tau^* = 0.2$, 0.4 m) the decrease of the normalized effective beam intensity is less than in the far field ($\tau^* = 2$, 6 m); (ii) in the far field, a strong decrease of normalized effective beam intensity is found for small-scale fluctuations while small effects appear when the fluctuation scale is larger; (iii) for large-scale fluctuations ($w\kappa = 0.3$) no significant effects can be seen for all cases.

4.4. Scrape-off layer turbulence

Density fluctuations in the TCV scrape-off layer have been investigated by probe measurements and compared with a two-dimensional interchange turbulence simulation. In particular, these numerical simulations are in excellent agreement with the TCV experimental measurements [3, 7]. Here, we make use of these SOL density fluctuation data in order to illustrate the possible consequences on the effective beam intensity. The phase perturbation generated by the passing of the beam through the turbulence SOL is obtained by integration of equation (2) substituting the 2D density perturbation predicted by the turbulence code. The resulting phase perturbation is rescaled in order to study the dependence of the amplitude of the SOL turbulence. The averaged effect of the turbulence is obtained by averaging the perturbed wave beam over its realizations as obtained for each of the predicted 2D turbulence density field at different times as well as at all possible spatial phases. Figure 6 shows the normalized effective beam intensity as a function of the the root mean square amplitude $||\delta\varphi||$ of the fluctuations. The

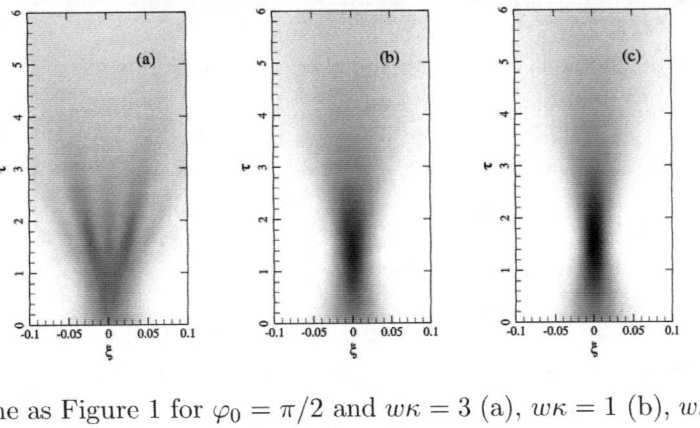

Figure 4. The same as Figure 1 for $\varphi_0 = \pi/2$ and $w\kappa = 3$ (a), $w\kappa = 1$ (b), $w\kappa = 0.3$ (c). Here, QO beam profiles are averaged over random phase θ.

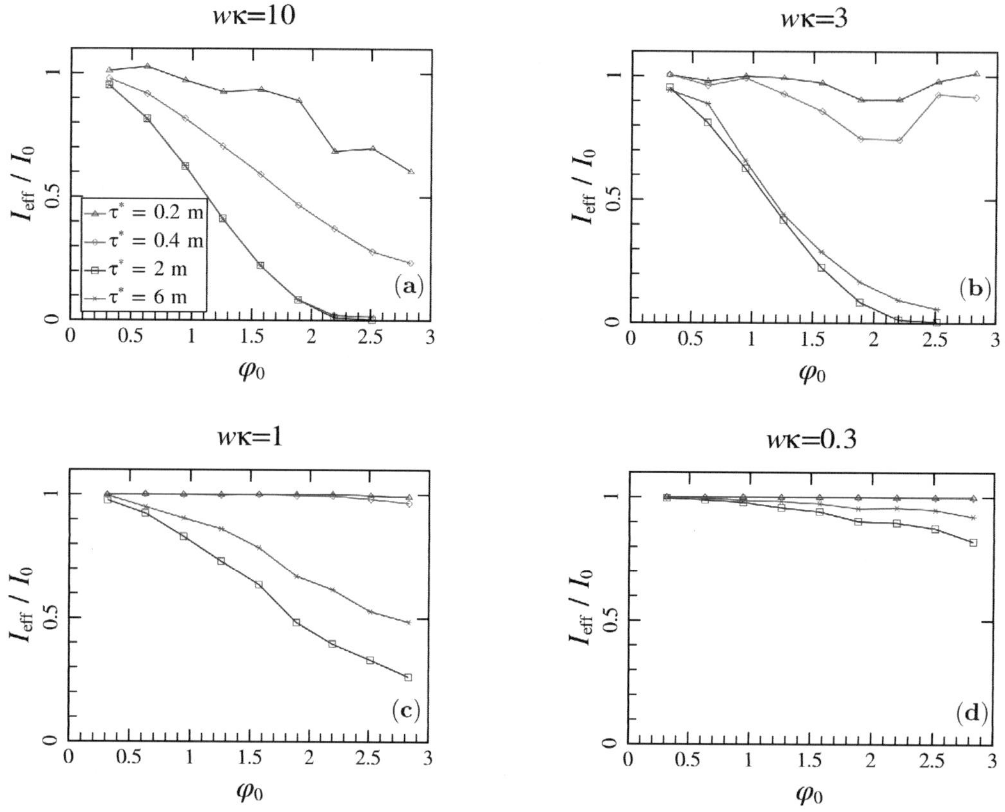

Figure 5. Effective beam intensity, $I_{\rm eff}$, normalized to the unperturbed intensity, I_0, as a function of the amplitude of the harmonic fluctuations for beams averaged over random phase θ and for $w\kappa = 10$ (a), $w\kappa = 3$ (b), $w\kappa = 1$ (c), $w\kappa = 0.3$ (d). The effective intensity is calculated in the near field ($\tau^* = 0.2, 0.4$ m, brown and red curves, respectively), in the mid field ($\tau^* = 2$ m, blue curve) and in the far field ($\tau^* = 6$ m, green curve).

effective intensity is calculated in the near/mid/far field as in Figure 5. One can see a behaviour similar to that shown in Figure 5(a) which corresponds to the case of small-scale fluctuations. For small $||\delta\varphi||$, the effects on the beam intensity are negligible while the effects are larger for increasing $||\delta\varphi||$. In particular, this behaviour is evident for large distances between the edge of the plasma and the position where beam intensity is calculated, namely, $\tau^* = 2$ and 6 m while for $\tau^* = 0.2$ and 0.4 m, the beam intensity decrease is significantly smaller.

5. Summary and Conclusions

The power deposition and driven current density profile can be affected by perturbations of the wave beam due to the edge density fluctuations. Calculations with a quasi-optical code [4] show that, for a simplified model corresponding to a single harmonic fluctuation, this effect can lead to a considerable broadening of the beam and a reduction of the beam intensity, in particular, for fluctuations at scales smaller than or comparable to the beam waist. Furthermore, a TCV-like turbulence case has been investigated confirming that this effect can be relevant particularly when the beam is expected to propagate over a large distance before it reaches the EC resonance. In a future work, ITER simulations with self consistent predictions of ITER edge turbulence will

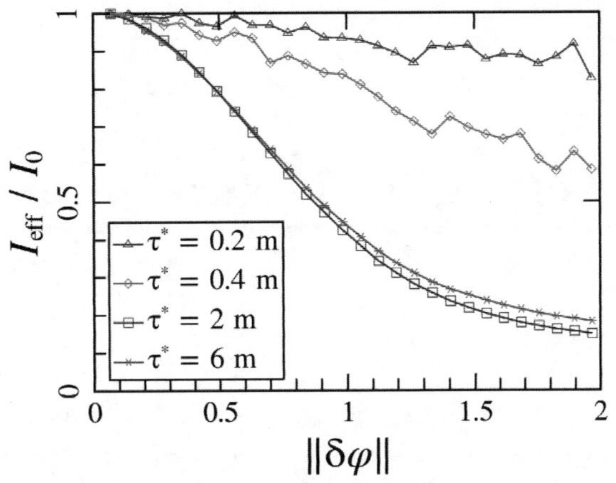

Figure 6. Normalized effective beam intensity as a function of the root mean square of the amplitude $||\delta\varphi||$ of the SOL density fluctuations as obtained from turbulence code [3]. The effective intensity is shown in the near/mid/far field as in Figure 5.

be performed taking into account also the ballooning nature of the SOL density fluctuations, which leads to a reduction of the edge plasma turbulence in front of the location of the ITER ERCH Upper Port Launcher with respect to turbulence in front of the Equatorial Launcher.

Acknowledgments

The work in this paper has been performed in the framework of the NWO-RFBR Centre of Excellence on Fusion Physics and Technology (grant 047.018.002). This work, supported by the European Communities under the contract of Association between EURATOM/FOM, was carried out within the framework of the European Fusion Program. The views and opinions expressed herein do not necessarily reflect those of the European Commission.

References

[1] Tsironis C, Peeters A G, Isliker H, Strintzi D, Chatziantonaki I and Vlahos L 2009 *Phys. Plasmas* **16** 112510
[2] Hansen F R, Lynov J, Michelsen P and Pécseli H L 1998 *Nucl. Fusion* **28** 769
[3] Garcia O E, Horacek J, Pitts R, Nielsen A, Fundamenski W, Naulin V and Juul Rasmussen J 2007 *Nucl. Fusion* **47** 667
[4] Balakin A A, Balakina M A and Westerhof E 2008 *Nucl. Fusion* **48** 065003
[5] Fock V A 1965 *Electromagnetic Diffraction and Propagation Problems* (Oxford: Pergamon)
[6] Ramponi G, Farina D, Henderson M A, Poli E, Saibene G and Zohm H 2007 *Fusion Sci. and Technol.* **52** 193
[7] Garcia O, Horacek J, Pitts R, Nielsen A, Fundamenski W, Naulin V and Juul Rasmussen J 2006 *Plasma Phys. Control. Fusion* **48** L1

Theory of Fusion Plasmas: Joint Varenna–Lausanne International Workshop IOP Publishing
Journal of Physics: Conference Series **260** (2010) 012003 doi:10.1088/1742-6596/260/1/012003

Magnetic chaos healing in the helical reversed-field pinch: indications from the volume-preserving field line tracing code NEMATO

D Bonfiglio[1], M Veranda[1], S Cappello[1], L Chacón[2] and G Spizzo[1]
[1] Consorzio RFX, Euratom-ENEA Association, Padova, Italy
[2] Oak Ridge National Laboratory, Oak Ridge, Tennessee, USA
E-mail: daniele.bonfiglio@igi.cnr.it

Abstract. The emergence of a self-organized reversed-field pinch (RFP) helical regime, first shown by 3D MHD numerical simulations, has been highlighted in the RFX-mod experiment at high current operation (I_P above 1 MA). In fact, a quasi-stationary helical configuration spontaneously appears, characterized by strong internal electron transport barriers. In such regime electron temperature and density become, to a very good approximation, functions of the helical flux coordinate related to the dominant helical magnetic component. In addition, this regime is diagnosed to be associated with the topological transition to a single-helical-axis (SHAx) state, achieved after the expulsion of the separatrix of the dominant mode's magnetic island. The SHAx state is theoretically predicted to be resilient to the magnetic chaos induced by secondary modes. In this paper, we present initial results of the volume-preserving field line tracing code NEMATO [Finn J M and Chacón L 2005 *Phys. Plasmas* **12** 054503] applied to study the magnetic topology resulting from 3D MHD simulations of the RFP. First, a successful 2D verification test of the code is shown, then, initial application to a systematic study of chaos healing in the helical RFP is discussed. The separatrix disappearance is confirmed to play an essential role for chaos healing. The triggering effect of a reversed magnetic shear for the formation of ordered surfaces within magnetic chaos is also diagnosed.

1. Introduction

The reversed-field pinch (RFP) is a toroidal magnetic configuration for plasma confinement. Similarly to what occurs in the tokamak, RFP plasmas are obtained by driving a toroidal current in a plasma embedded in a toroidal magnetic field. However, the plasma current in the RFP is strong enough with respect to the toroidal flux to drive a significant helical shaping of the plasma column. This deformation is related to one or more resistive kink-tearing instabilities which nonlinearly saturate at finite amplitude with the help of the toroidal field reversal at the plasma edge, whence comes the name of the configuration. In the multiple helicity (MH) regime, characterized by the presence of many magnetic fluctuations, plasma confinement is degraded by stochastic transport due to the overlap of core magnetic islands, and only an edge confinement region survives. The turbulent MH state was the first to be studied in RFP devices. During the last years, however, a tendency to develop a regime characterized by a good degree of helical symmetry has been observed in different RFP experiments when the plasma current is increased [1]. The first obtained quasi-single helicity (QSH) regime was characterized by the appearance of a thermal helical structure corresponding to the O-point of the magnetic island associated with the dominant helical mode. Recently, during high current operation (up to 1.7 MA) in the RFX-mod experiment, a quasi-stationary QSH state with a strong internal transport barrier (ITB) enclosing a broad hot core (with electron temperature above 1 keV) spontaneously appeared [2]. In such regime electron temperature, density, and soft X-ray emissivity become, to a very good approximation, functions of the helical flux coordinate related to the dominant mode. This is observed together with the decrease of the secondary modes amplitude, and shows that the high current QSH regime in RFX-mod is close to a single helicity (SH) state.

Many features of the experimental RFP phenomenology have been first shown by theory and numerical modelling. Three-dimensional (3D) visco-resistive magnetohydrodynamic (MHD) numerical simulations of the RFP reproduce quite well the qualitative magnetic behaviour of experimental MH and QSH regimes [3, 4] and shows the existence of pure SH equilibrium solutions [5, 6] which spontaneously appear when the visco-resistive dissipation is increased [7, 4]. As far as transport in concerned, the edge confinement in the MH state has been associated with the presence of a $m = 0$ magnetic islands chain at the field reversal radius [8]. Regarding the QSH regime, chaos healing is theoretically predicted to be induced by the expulsion of the separatrix of the dominant mode's magnetic island [9], corresponding (as described in this paper) to the topological transition from a double-axis (DAx) to a single-helical-axis (SHAx) state. The same transition is diagnosed to occur in the high current QSH regime of RFX-mod [2]. The increased resilience to magnetic chaos predicted for the SHAx state could therefore be related with the good confinement properties of that experimental state, namely the presence of a hot helical core surrounded by an ITB. Similarly to what happens in the tokamak, a null of the magnetic shear could also help to reduce transport at the ITB. In fact, the location of the core transport barrier in RFX-mod is found to be well correlated with the maximum of the non monotonic q profile typical of the SHAx regime [10, 11]. A sheared plasma flow is a possible player for the ITB sustainment [12] against the microturbulence which could arise at the barrier [13].

In this paper, we present an initial application to the study of the conditions for chaos healing in the helical RFP, performed with the volume-preserving field line tracing code NEMATO [14]. The code is used to compute the topology of the magnetic field in a periodic cylinder provided by numerical simulations with the 3D nonlinear visco-resistive MHD code SPECYL [4]. The paper is organized as follows. The features of NEMATO are described in Sec. 2. The verification benchmark of the code against magnetic fields with either helical or poloidal symmetry is reported in Sec. 3. The application of NEMATO to a QSH simulation with separatrix disappearance is discussed in Sec. 4. It is observed that the chaos induced by secondary modes is cured in a short time around the transition to the SHAx state. The role played by the separatrix expulsion is thus confirmed. In addition, the formation of the first conserved surfaces within the chaotic region is shown to occur where the magnetic shear is reversed. As remarked in Sec. 5, the present work will be extended to more general cases of quasi-single helicity RFP states.

2. The field line tracing code NEMATO

The field line tracing code NEMATO is a volume-preserving integrator for solenoidal fields on a (cylindrical or toroidal) 3D grid [14]. It solves the magnetic field line equation $\frac{d\mathbf{x}}{d\tau} = \mathbf{B}(\mathbf{x})$, where $\mathbf{B}$ is the magnetic field given on the 3D grid $\mathbf{x}$ and the "time" τ represents a coordinate along the field line. The two main features of the code are the interpolation scheme, which ensures the solenoidal nature of the field everywhere in the domain, and the volume-preserving integration scheme. Together, they exactly preserve $\nabla \cdot \mathbf{B} = 0$ to numerical round-off along each magnetic field line. As a result, integrable and chaotic field lines are distinguished well one from each other. This makes the code particularly suitable to study weakly chaotic magnetic fields typical of the quasi-single helicity RFP, and represents a novelty with respect to other field-line tracing tools used in past studies [9]. A non volume-preserving version of NEMATO, based on the high-order, adaptive LSODE library [15], is also available for fast preliminary analysis.

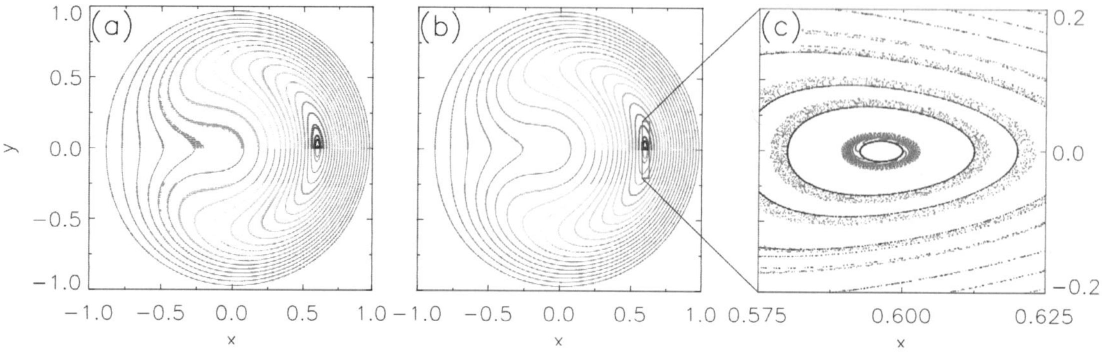

Figure 1. Verification of NEMATO against RFP configurations with helical symmetry provided by the MHD code SPECYL. Poincaré maps on the $z = 0$ plane (colored dots) are compared with the helical flux function χ (black contours levels). Results obtained with a grid resolution of $100 \times 256 \times 256$ and $100 \times 256 \times 2048$ radial $\times$ azimuthal $\times$ axial points are reported in panel (a) and (b), respectively. To make the comparison easier, intersections of the field lines with the Poincaré surface in the $-0.25 \leq y \leq 0$ band are not shown. A blow-up of the most resolved simulation in the region around the helical axis is shown in panel (c), together with the result of the non volume-preserving version of NEMATO with the same resolution (red dots).

3. Verification benchmark against symmetric magnetic fields

If a magnetic field is characterized by a two-dimensional (2D) symmetry, magnetic field lines are integrable in the whole field domain, i.e., they trace out well defined magnetic surfaces. The verification of NEMATO has been performed on the basis of this principle. The code has been run on 2D symmetric magnetic fields provided by the MHD code SPECYL, and it has been checked that magnetic field lines integrated by NEMATO (without any assumption about the magnetic field symmetry) lie on the isosurfaces of the flux function ρ computed from the definition $\mathbf{B} \cdot \nabla\rho = 0$. The verification benchmark has been performed by considering magnetic fields with either helical or poloidal symmetry.

SPECYL solves the MHD equations in a periodic cylinder (rectified torus) with radial coordinate r, azimuthal (poloidal) coordinate θ and axial (toroidal) coordinate z with a period of $2\pi R_0$. The aspect ratio of the rectified torus is set to $R_0 = 4$. In helical symmetry, all quantities depend on the two variables r and $u = m\theta - \frac{n}{R_0}z$, where m and n are the poloidal and toroidal mode numbers, respectively. The helical flux function χ such that $\mathbf{B} \cdot \nabla\chi = 0$ is defined by $\chi(r, u) = mA_z + \frac{n}{R_0}rA_\theta$, where A_z and A_θ are the toroidal and the poloidal components of the vector potential $\mathbf{A}$. For the verification benchmark with NEMATO, we have employed a single helicity RFP equilibrium obtained as the nonlinear saturation state of a resistive kink-tearing instability with $m = 1$, $n = -10$. As reported in the Poincaré maps of Figs. 1(a) and 1(b), results depend on the 3D grid resolution. When a resolution of 100 radial $\times$ 256 azimuthal $\times$ 256 axial grid points is employed, the matching between the field line trajectories and the helical flux surfaces is reasonable, but a certain drift of the field lines away from their initial magnetic surface is observed. This drift is due to the number of mesh points in the axial direction, which is not enough to adequately sample the helical field with its high toroidal mode number $n = -10$. When the grid resolution is increased to $100 \times 256 \times 2048$ points, the drift disappears and the matching of field lines with helical flux surfaces is quite good. The latter grid resolution is hence employed in the remainder of the paper. The volume-preserving scheme employed by NEMATO can be appreciated by comparing it with the non volume-preserving version of the code, as shown in the blow-up of Fig. 1(c). For the same length of the field line integration, NEMATO orbits

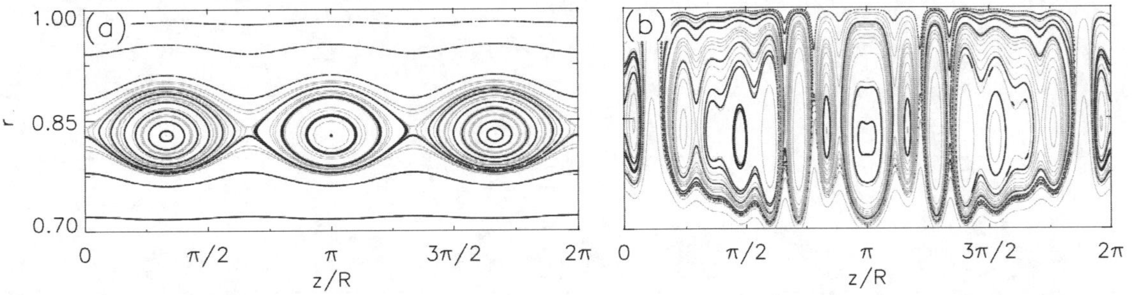

Figure 2. Verification benchmark of NEMATO against RFP configurations with poloidal symmetry provided by the MHD code SPECYL. Poincaré maps on the $\theta = 0$ plane (colored dots) are compared with the toroidal flux function Φ (gray contours levels). The $m = 0$ modes taken into account are (a) the $n = 3$ mode only and (b) all the $1 \leq n \leq 25$ modes involved in the MHD computation.

are perfectly defined, while drifts are produced by the non volume-preserving code, even when the best grid resolution is employed. NEMATO has also been benchmarked against magnetic configurations with poloidal symmetry, obtained from a 3D multiple helicity RFP simulation when $m = 0$ modes only are supplied to the field line tracing code. As reported in Fig. 2, magnetic field lines computed by NEMATO trace out clearly defined magnetic surfaces, like in the case of helical symmetry. These magnetic surfaces match very well with the contour levels of the toroidal flux function $\Phi(r, z) = rA_\theta$ such that $\mathbf{B} \cdot \nabla\Phi = 0$.

4. Study of the chaos healing effect in the helical RFP

As an initial application to the study of chaos healing in the QSH, we consider a 3D MHD simulation with a full spectrum of allowed MHD modes with $0 \leq m \leq 4$ and two helical perturbations of the initial axis-symmetric equilibrium, corresponding to the unstable $m = 1$ modes with $n = -9$ and $n = -10$. The Lundquist and Prandtl numbers are set to $S = 3 \times 10^4$ and $P = 10^3$, respectively. Details of the MHD simulation are reported in Fig. 3. The $n = -9$ mode, the most unstable between the two perturbed modes, is the dominant mode in the second part of the linear growth phase and at the beginning of the nonlinear saturation phase, when the other $m = 1$ modes arise. The nonlinear dynamics makes the $n = -10$ mode to become the dominant one after $t = 2000\ \tau_A$. The exponential decay of the secondary modes brings the system to the final $n = -10$ SH equilibrium configuration. In this study, we focus our attention to the time window where the $n = -9$ mode is the dominant one. The contour levels of the helical flux function χ computed by assuming a $n = -9$ helical symmetry are plotted, for five selected simulation times, in Fig. 3(b). The first two times are characterized by the DAx magnetic topology. In fact, the magnetic field displays two magnetic axes: the unperturbed axis-symmetric one and the one related to the O-point of the $n = -9$ magnetic island. Flux surfaces winding around the two axes are kept apart by the separatrix of the magnetic island, whose width is growing in time. Just before $t = 750\ \tau_A$, a topological transition to the SHAx state occurs: the magnetic separatrix disappears after the coalescence of the axis-symmetric O-point with the island X-point, and the island O-point becomes the only (helical) magnetic axis, as shown for the last three times. For every flux surface which encloses the helical axis we have computed (again by assuming $n = -9$ helical symmetry) the safety factor defined as $q(\chi) = \frac{d\Phi(\chi)}{d\psi(\chi)}$, where $\Phi(\chi)$ and $\psi(\chi)$ are, respectively, the toroidal and poloidal flux across the helical flux surface χ. The safety factor $q(\chi)$ gives the number of toroidal turns field lines

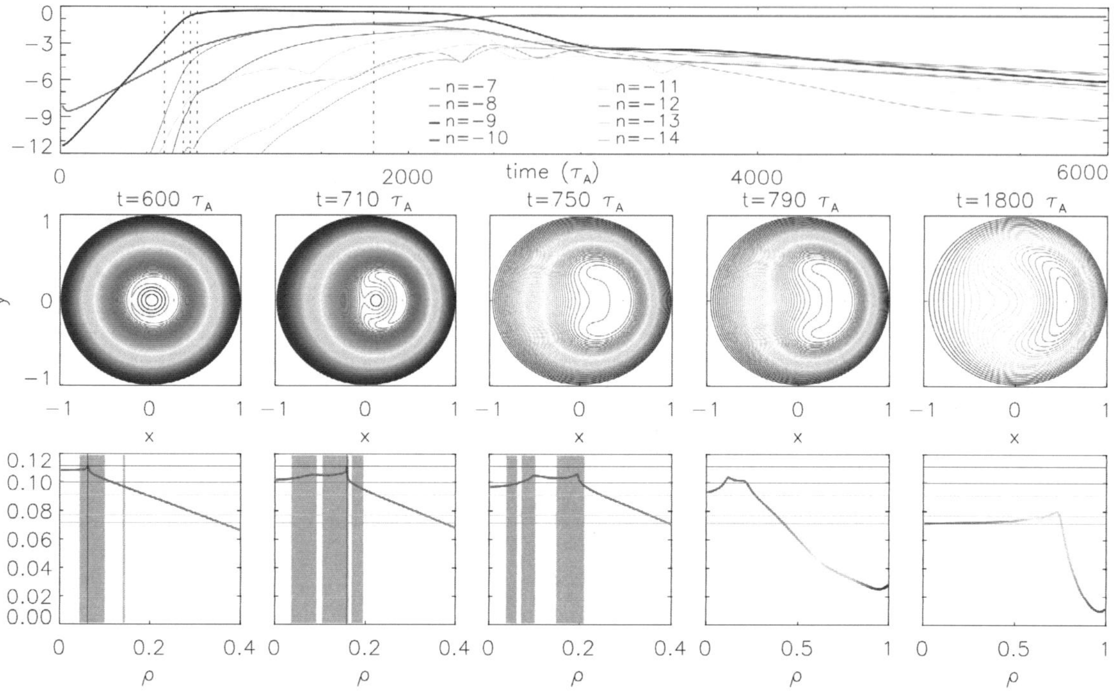

Figure 3. SPECYL simulation employed for the NEMATO study of the chaos healing effect in the RFP. The temporal evolution of the magnetic energy of the main $m = 1$ modes is reported in panel (a). The simulation times chosen for the analysis with NEMATO are marked with dashed vertical lines. For these times, we report (b) the contour levels of the helical flux function χ and (c) the q profile as a function of the effective radius ρ. Both χ and q are computed by assuming $n = -9$ helical symmetry. The color of the flux surfaces in panel (b) refer to their ρ values and reflect the color shifts in the q profiles of panel (c). The ρ value corresponding to the separatrix is marked in panel (c) with a black vertical line, while shaded vertical bands refer to chaotic regions in the Poincaré plots of Fig. 4. The $m = 1$ rational values with n from $n = -9$ to $n = -14$ are shown with horizontal lines with the same color code employed in panel (a).

perform for one poloidal turn *around the helical axis* and is introduced here for the intepretation of the magnetic field topology, as discussed below. The q profile for the selected times is shown in Fig. 3(c) as a function of the effective radius $\rho = (\chi/\chi_b)^{1/2}$, where χ_b is the helical flux at the plasma boundary and $\chi = 0$ at the helical axis. As derived in the framework of the Hamiltonian representation of the magnetic field [16], $q \to 1/9$ at the separatrix. $q(0)$, the safety factor at the helical axis, is close to $1/9$ at $t = 600\ \tau_A$ and decreases in time as the magnetic island width increases. At $t = 710\ \tau_A$, an additional maximum of q appears inside the separatrix. Just after the separatrix expulsion, $t = 750\ \tau_A$, there is still a peak at the "separatrix ghost" but its value is below $1/9$. The hollow between the two q maxima is now more pronounced. In addition, $q(0) < 1/10$, meaning that a second resonance for the $m = 1$, $n = -10$ mode is present. At $t = 790\ \tau_A$, the peak at the separatrix ghost is more depressed. Later in time, at $t = 1800\ \tau_A$, the q profile is much lower and it takes the same non-monotonic shape as reported in Refs. [10, 11] (we note that the maximum of q does not come from the separatrix ghost).

We now discuss the topology of the magnetic field at the five selected times, as computed by NEMATO. The corresponding Poincaré plots on the $z = 0$ and $\theta = 0$ surfaces of section are reported in Fig. 4. At $t = 600\ \tau_A$, the $n = -9$ and $n = -10$ island chains produced by the two perturbed modes are clearly visible along the z direction in the $\theta = 0$ plane. The $n = -10$ island chain is distorted by the $n = -9$ helical perturbation. By overplotting the $n = -10$ conserved surfaces on the $n = -9$ helical flux surfaces of Fig 3(b), it could be shown that they lie on the helical flux surface with $q(\chi) = 1/10$. The separatrices of the two magnetic islands (computed from the helical flux functions with the two related helicities) overlap. According to the Chirikov overlap criterion [17], magnetic chaos develops in the region between the two islands and, in particular, inside the separatrix of the $n = -9$ mode. The magnetic surfaces close to the cylindrical axis remain perfectly conserved, as well as those located in the outer region. Secondary islands due to the beat of the two main islands are also visible inside the chaotic sea. As for the $n = -10$ island, the effective radial positions of the secondary islands with respect to the helical axis correspond to helical flux surfaces with rational q values. As time increases, the chaotic volume inside the separatrix of the $n = -9$ island increases together with the island width. At $t = 710\ \tau_A$, when the strongly deformed $n = -10$ island chain is quite narrow but still visible just outside the $n = -9$ separatrix, a broad chaotic region is observed between the remnant conserved surfaces around the helical axis and those around the (helically displaced) "cylindrical" axis. However, a region with conserved surfaces appears in the middle of the chaotic region. The ρ position of the conserved layer corresponds to the region with negative magnetic shear between the maximum of q inside the separatrix and the separatrix itself, as marked by means of an unshaded band in Fig. 3(c). The extent of the conserved layer increases in time with the deepening of the q hollow. In the meantime, the separatrix expulsion occurs, which is seen by disappearance of the helical axis at $t = 750\ \tau_A$. At this time, the conserved layer is rather large. Again, it turns out to be located in the $\frac{dq}{d\rho} \leq 0$ region of the q profile. Since $q(0)$ is below $1/10$, a $m = 1$, $n = -10$ island chain appears inside the chaotic region at the $q = 1/10$ resonant surface close to the helical axis [18]. Conserved surfaces corresponding to the original $n = -10$ island are not found at this time. At $t = 790\ \tau_A$, when the separatrix is well expulsed, chaos is strongly reduced in the whole domain. It only survives at the separatrix of the outer $n = -10$ island, which reappears with a rather elongated shape at this time. The chaos healing effect is robust and lasts for a long time, even if the amplitude of secondary modes is increasing. At $t = 1800\ \tau_A$, for instance, well defined magnetic surfaces are still observed in the whole domain, either closed around the helical axis or in the form of secondary islands, now with $n = -14$ due to the lower q profile.

5. Summary and future work

In this paper, the verification benchmark of the field line tracing code NEMATO [14] against magnetic configurations with either helical or poloidal symmetry is reported. A first application of NEMATO to the study of the conditions for chaos healing in the helical RFP, of interest for the interpretation of experimental observations in the RFX-mod device [2], is then discussed. In this stage, we avoid dealing with the uncertainties which may affect the magnetic field reconstruction from experimental measurements, and make use of the magnetic field provided by a 3D visco-resistive MHD simulation of the SPECYL code [3]. We consider a simplified QSH case whose dynamics is characterized by the presence of (nearly) only one secondary mode along with the QSH dominant mode. Nevertheless, the observed evolution of the magnetic field topology is quite rich. It is shown that the magnetic chaos produced after the overlap of the magnetic islands associated with the two relevant MHD modes gets healed in a rather short time around the expulsion of the separatrix of the dominant mode's island. The theoretical picture of chaos healing by separatrix disappearance [9] is thus confirmed. In addition, we show that the chaos reduction begins shortly before the separatrix expulsion, when the first conserved surfaces appear within the chaotic region. The position of the conserved layer corresponds to the values of the helical flux function, computed by assuming the helical symmetry of the dominant mode, where the shear of the safety factor q is reversed with respect to the surrounding chaotic region. The

Journal of Physics: Conference Series **260** (2010) 012003 doi:10.1088/1742-6596/260/1/012003

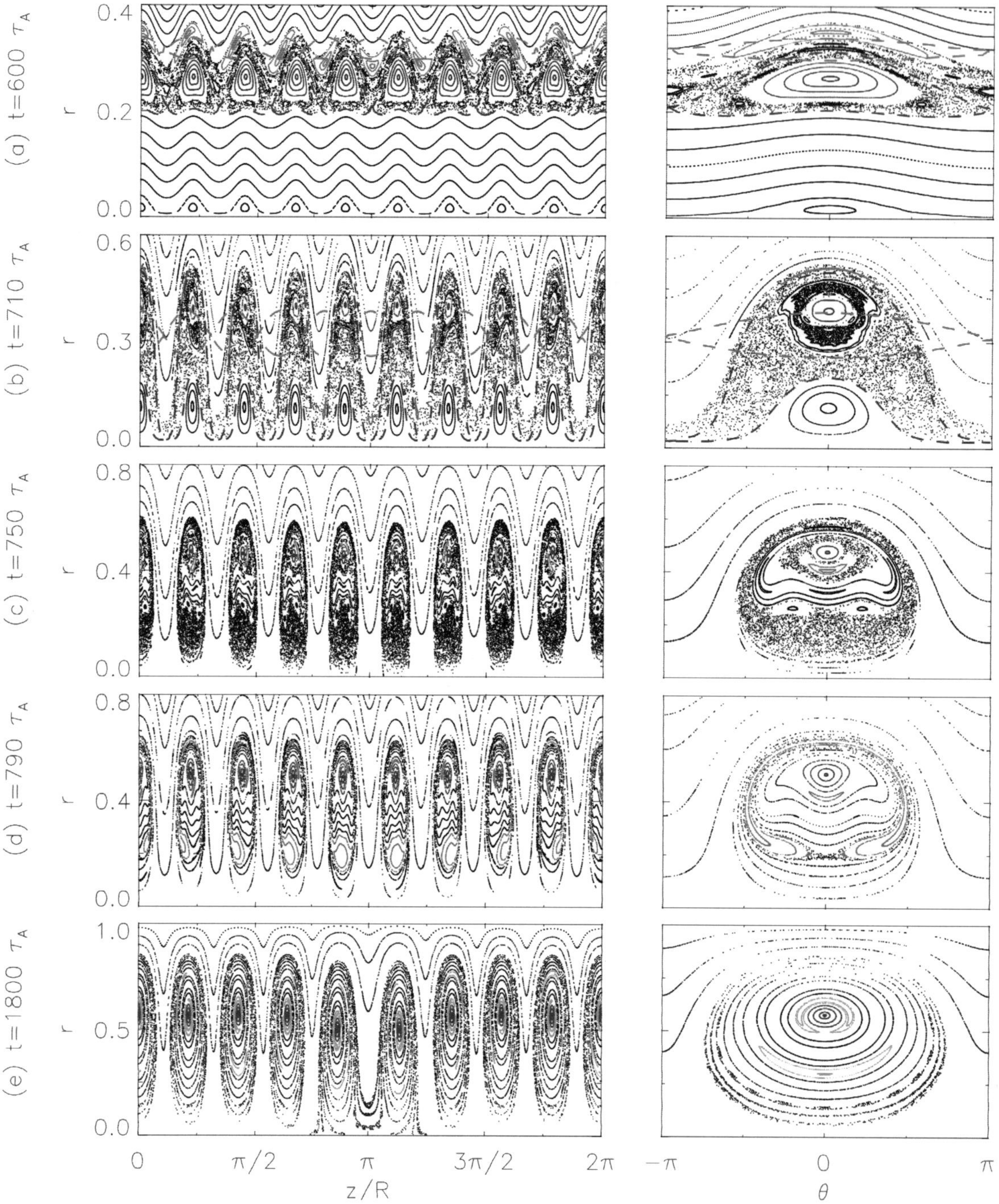

Figure 4. NEMATO simulations for the study of the chaos healing effect in the RFP. Poincaré maps on the $\theta = 0$ plane (first column) and on the $z = 0$ plane (second column) are reported in panels (a-e) for the selected simulation times. Conserved surfaces corresponding to the region close to the $n = -9$ helical axis and to the $n = -10$ magnetic island are shown in blue and red, respectively. The same colors are used in the first two times for dashed curves marking the magnetic separatrices of the two modes, computed from the related helical flux functions.

width of conserved layer increases in time in accordance with the broadening of the reversed shear region. Eventually, after the separatrix disappearance, chaos is completely disappeared from the whole domain, regardless of the shape of the q profile.

Further work is needed to assess the robustness of the described phenomenology in more general QSH conditions. In particular, while the favourable effect of the separatrix expulsion appears to be a robust result, the role played by the magnetic shear, though suggestive of a relation with the experimental observations in the RFX-mod device [10, 11], needs to be interpreted and confirmed. A future detailed study of the conditions for chaos healing in the helical RFP should include, for example, the effect of the following factors: the MHD dynamics (e.g., spontaneous [2] vs. stimulated [19] separatrix expulsion), the shape of the q profile and its shear [20, 21], the number and the amplitude of the secondary modes and of the dominant one, and the geometry (i.e., cylindrical vs. toroidal [22]) of the plasma domain. The use of NEMATO to study the relation between magnetic topology and flow is also envisioned, with reference to both the helical RFP, where a sheared plasma flow could be important for the ITB sustainment [12], and the $m = 0$ island chain located at the reversal surface of the RFP [8]. The latter study turns out to be of interest in relation with the density limit issue in the RFP [23].

Acknowledgments

The authors would like to thank D. F. Escande and F. Sattin for helpful discussions in the course of this study. This work was supported by the European Communities under the Contract of Association between EURATOM/ENEA.

References

[1] Martin P *et al.* 2003 *Nucl. Fusion* **43** 1855
[2] Lorenzini R *et al.* 2009 *Nature Phys.* **5** 570
[3] Cappello S and Biskamp D 1996 *Nuclear Fusion* **36** 571
[4] Cappello S 2004 *Plasma Phys. Control. Fusion* **46** B313
[5] Cappello S and Paccagnella R 1992 *Phys. Fluids B* **4** 611
[6] Finn J M, Nebel R and Bathke C 1992 *Phys. Fluids B* **4** 1262
[7] Cappello S and Escande D F 2000 *Phys. Rev. Lett.* **85** 3838
[8] Spizzo G *et al.* 2006 *Phys. Rev. Lett.* **96** 025001
[9] Escande D F, Paccagnella R, Cappello S, Marchetto C and D'Angelo F 2000 *Phys. Rev. Lett.* **85** 3169
[10] Puiatti M E *et al.* 2009 *Plasma Phys. Control. Fusion* **51** 124031
[11] Gobbin M *et al.* 2010 Vanishing magnetic shear and electron transport barriers in the RFX-mod reversed field pinch *submitted to Phys. Rev. Lett.*
[12] Bonomo F *et al.* 2010 *Proc. 37th EPS Conf. on Plasma Physics, 21-25 June 2010, Dublin* p O2.101
[13] Sattin F *et al.* 2010 Microturbulence studies in RFX-mod *This conference*
[14] Finn J M and Chacón L 2005 *Phys. Plasmas* **12** 054503
[15] Hindmarsh A C 1983 ODEPACK, a systematized collection of ODE solvers *Scientific Computing* ed Stepleman R S (Amsterdam: North-Holland) pp 55–64
[16] Momo B *et al.* 2010 *Proc. 37th EPS Conf. on Plasma Physics, 21-25 June 2010, Dublin* p P4.147
[17] Chirikov B 1979 *Phys. Rep.* **52** 263
[18] Note that, at $t = 750\ \tau_A$, the $n = -10$ islands can only be counted 9 times in the $\theta = 0$ Poincaré surface of Fig. 4(c), instead of the expected 10. This is because the $n = -10$ island (in red) is in fact precessing around the $n = -9$ helical axis (in blue) with a precessing number of 1. The expected winding number of the $n = -10$ island (10) results by adding its strobing number with respect to the Poincaré surface (9) plus the precessing number around the $n = -9$ helical axis (1).
[19] Lorenzini R *et al.* 2008 *Phys. Rev. Lett.* **101** 025005
[20] del-Castillo-Negrete D, Greene J M and Morrison P J 1996 *Physica D* **91** 1
[21] Nasi L and Firpo M C 2009 *Plasma Phys. Control. Fusion* **51** 045006
[22] Sovinec C R *et al.* 2003 *Phys. Plasmas* **10** 1727
[23] Spizzo G *et al.* 2009 *Plasma Phys. Control. Fusion* **52**

Journal of Physics: Conference Series **260** (2010) 012004 doi:10.1088/1742-6596/260/1/012004

Pfirsch-Schlüter impurity transport in stellarator edge plasmas with large radial gradients

S Braun and P Helander

Max Planck Institute for Plasma Physics, Wendelsteinstr. 1, 17491 Greifswald, Germany

E-mail: `stefanie.braun@ipp.mpg.de`

Abstract. Large radial gradients of the bulk ion temperature and density in edge plasmas cause heavy impurity ions to redistribute themselves *within* flux surfaces, thereby making the impurity density vary poloidally (and, in a stellarator, toroidally). This redistribution reduces the radial impurity flux. In this paper, the effects of stellarator geometry on this mechanism are investigated, which lead to qualitatively different results compared with axisymmetric systems, in which the plasma can rotate freely. We derive analytic expressions for the radial impurity flux and the impurity density. The resulting differential equations are solved analytically in some limiting cases and numerically for W7-X and W7-AS.

1. Introduction

Toroidal edge plasmas are often characterised by large radial gradients of the bulk ion density and temperature. In fusion relevant experiments, these gradients can easily become larger than allowed for by conventional neoclassical theory based on a small-gyroradius expansion, $\delta_a \equiv \rho_{a_\theta}/L_\perp \ll 1$, of the kinetic equation, where ρ_{a_θ} is the poloidal gyroradius and $L_\perp$ the radial scale length of the density and temperature gradients. This problem may be of particular interest in the context of the H-mode. Although it is fundamentally difficult to construct a rigorous neoclassical transport theory if δ_a is not a small parameter, it is possible to extend conventional theory to include cases of larger gradients. In this new ordering, δ_i is still treated as small, though not as the *smallest* expansion parameter any more, as is implicitly assumed in conventional theory. Products of δ_i with large parameters can easily be of order unity. In the present work, we take $\Delta_i \equiv \delta_i \bar{\nu}_{ii} Z^2 = \mathcal{O}(1)$, where $\bar{\nu}_{ii} \equiv L_\parallel/\lambda_{ii}$ is the collisionality, which is large in the Pfirsch-Schlüter regime, λ_{ii} being the ion mean free path and $L_\parallel$ the connection length, and Z is the impurity charge. The consequence is that the lowest order impurity distribution function is only locally Maxwellian, contrary to the conventional case where all lowest order quantities are flux functions. In practice, as a consequence of the large gradients, the impurities start developing density variations *within* the flux surface, which, depending on the plasma conditions, can lead to an accumulation of impurities at certain points on the flux surface. In tokamaks, this redistribution has previously been predicted to lead to impurity accumulation of the order of up to 30% of the total density at the high-field side of the torus [1, 2], a result which is in qualitative agreement with experimental results of present machines such as Alcator C-mod or MAST. However, there has been no work at all on how this effect is affected by stellarator geometry, and the calculations presented in this paper are aiming at calculating this effect. As a stellarator plasma, contrary to the tokamak case, cannot rotate freely [3] but instead has a radial electric field which adjusts itself so as to make the radial transport ambipolar, the impurity redistribution is different in a stellarator due to effects from the radial electric field which cancel in axisymmetric systems. The remainder of this paper is organised as follows. In section 2, the kinetic theory is described and a differential equation for the impurity density variation is derived. This equation is solved, first in different analytical limits in section 3, and then fully with the help of numerical tools in section 4. The results are discussed and summarised in section 5.

2. Kinetic equations

We consider a stellarator plasma in the collisional regime consisting of hydrogenic bulk ions and a single species of heavy, highly charged impurities with charge $Z \gg 1$. The effect of electron collisions on the other species is small and therefore neglected, and the impurities are assumed to be not too numerous, so that the assumption $Z^2 n_z/n_i \ll 1$ holds (trace impurities). The drift kinetic equation for each species is

$$C_a(f_a) = v_\parallel \nabla_\parallel f_a + \mathbf{v}_d \cdot \nabla f_a - \frac{e_a}{m_a}\nabla_\parallel \phi \frac{\partial f_a}{\partial v_\parallel}, \tag{1}$$

where ϕ is the electric potential, e_a the charge of particle species a and $\mathbf{v}_d$ is given by

$$\mathbf{v}_d = \frac{\mathbf{E}\times\mathbf{B}}{B^2} + \frac{v_\perp^2}{2\Omega_a}\frac{\mathbf{B}\times\nabla B}{B^2} + \frac{v_\parallel^2}{\Omega_a}\frac{\mathbf{B}\times\kappa}{B},$$

where $\mathbf{E}$ and $\mathbf{B}$ denote the electric and the magnetic field, respectively, Ω_a is the gyrofrequency and $\kappa \equiv (\mathbf{b}\cdot\nabla)\mathbf{b}$ is the magnetic curvature. $\mathbf{b}$ is the unit vector along the magnetic field, $v_\parallel = \mathbf{b}\cdot\mathbf{v}$ and $\nabla_\parallel = \mathbf{b}\cdot\nabla$. These equations are solved employing the ordering $\delta_i \ll 1, \Delta_i = \mathcal{O}(1)$. For the bulk ions, this ordering corresponds to the conventional ordering, leading to a Maxwellian distribution in lowest order, and the first order equation

$$C_i(f_{i1}) = v_\parallel \nabla_\parallel f_{i1} + \mathbf{v}_d\cdot\nabla f_{i0} + \frac{e_i}{T_i} v_\parallel \nabla_\parallel \phi f_{i0}$$

has the solution [4]

$$f_{i1} = -\frac{2}{5}\frac{m_i}{p_i T_i} q_{i_\parallel} v_\parallel f_{i0}\left(L_1^{(3/2)}(x_i^2) - \frac{4}{15}L_2^{(3/2)}(x_i^2)\right) + \left(\frac{p_{i1}}{p_{i0}} + \frac{m_i}{T_i}v_\parallel V_{i_\parallel} + \left(x_i^2 - \frac{5}{2}\right)\right) f_{i0},$$

where $L_j^{(k)}$ denote the Sonine polynomials (generalised Laguerre polynomials), $x \equiv v/v_{th}$, where v_{th} denotes the thermal velocity. Since we consider trace impurities, ion-impurity collisions are negligible compared with ion self collisions, so that the solution corresponds to the pure-plasma solution. The parallel flow velocity $V_{i_\parallel}$ and the parallel heat flow $q_{i_\parallel}$ can be determined using particle conservation and the approximate conservation of energy in collisions between particles with disparate masses. It is convenient to use the equilibrium current $\mathbf{j}_0$ and the equilibrium pressure p_0 to introduce a geometric quantity u, defined by

$$\mathbf{h} \equiv \frac{\mathbf{j}_0}{p_0'} = \frac{1}{p_0'}\left(\frac{\mathbf{b}\times\nabla p_0}{B} + j_{0_\parallel}\mathbf{b}\right) \equiv \frac{\mathbf{b}\times\nabla\psi}{B} + u\mathbf{B},$$

where a prime denotes derivation with respect to the radial spatial coordinate ψ. Since the divergence of the equilibrium current must vanish, $\nabla\cdot\mathbf{h} = 0$, u must satisfy

$$\nabla_\parallel u = \frac{1}{p_0'}\nabla_\parallel\left(\frac{j_{0_\parallel}}{B}\right) = \frac{2}{B^2}(\mathbf{b}\times\nabla\psi)\cdot\nabla\ln B. \tag{2}$$

In this notataion, particle and energy conservation yield [5]

$$B\nabla_\parallel\left(\frac{n_a V_{a\parallel}}{B}-\frac{p_a}{e_a}A_{a_1}u\right) = 0, \qquad B\nabla_\parallel\left(\frac{q_{a\parallel}}{BT_a}-\frac{5}{2}\frac{p_a}{e_a}uA_{a_2}\right)=0,$$

and thus for the parallel ion flow velocity and heat flux

$$V_{i_\parallel} = \frac{T_i}{e_i}A_{i_1}uB+K_i(\psi)B,$$
$$\frac{q_{i_\parallel}}{T_i} = \frac{5}{2}\frac{p_i}{e_i}A_{i_2}\left(uB-B\frac{\langle uB^2\rangle}{\langle B^2\rangle}\right),$$

where the thermodynamic forces are defined as

$$A_{a_1} \equiv \frac{d\ln p_a}{d\psi}+\frac{e_a}{T_a}\frac{d\phi}{d\psi}, \qquad A_{a_2}\equiv\frac{d\ln T_a}{d\psi}.$$

Since, for highly charged impurities, the ion and impurity temperatures are narly equilibrated, $A_{i_2}\approx A_{z_2}$. The integration constant in the equation for the heat flux was determined from the requirement that $\langle Bq_{i_\parallel}\rangle=0$, where angular brackets denote an average over the flux surface, and, in the trace impurity limit, $K_i(\psi)$ can, be determined from the next-order equations [4]

$$K_i=-\frac{T_i n_i}{e_i}\left((A_{i_1}+1.82A_{i_2})\frac{\left\langle u(\nabla_\parallel B)^2+\frac{1}{4}\nabla_\parallel u\nabla_\parallel B^2\right\rangle}{\langle(\nabla_\parallel B)^2\rangle}-0.05A_{i_2}\frac{\langle u(\nabla_\parallel B)^2\rangle}{\langle(\nabla_\parallel B)^2\rangle}-1.77A_{i_2}\frac{\langle uB^2\rangle}{\langle B^2\rangle}\right).$$

For the impurities, the lowest-order equation in the new ordering is

$$C_z(f_z^{(0)}) = 0, \tag{3}$$

where C_z is the full linearised impurity collision operator. The solution to this equation is the *perturbed* Maxwellian

$$f_z^{(0)} = \left(\frac{p_z^{(0)}}{p_{z0}}+\frac{m_z}{T_z}v_\parallel V_{z\parallel}^{(1)}+\left(x_z^2-\frac{5}{2}\right)\frac{T_z^{(1)}}{T_{z0}}\right)f_{z0}, \tag{4}$$

leaving the lowest order density $n_z^{(0)}$, temperature $T_z^{(0)}$ and parallel flow velocity $V_{z\parallel}^{(0)}$ free to vary on the flux surface as they are functions of the poloidal and toroidal angles, θ and φ, respectively, as well as ψ. Quantities ordered according to the new scheme are indicated by superscript indices whereas subscript indices symbolise the expansion in the conventional ordering $\delta_i \ll 1, \Delta_i \ll 1$. Quantities with a subscript 0 are always flux functions. Eq. (3) implies equal temperatures and flow velocities of all involved particle species, thereby making the lowest order impurity temperature a flux function, and the lowest order flow vanishes since this is true for the bulk ions . Therefore, the first quantity to vary on the flux surface when the gradients are large is the impurity density n_z. The variation of the impurity temperature and flow velocity is one order smaller in δ_i than the density variation.

Taking the particle moment of the impurity version of (1), one can exploit particle conservation to find

$$\nabla\cdot\left(n_z^{(0)}\left(V_{z\parallel}^{(1)}\mathbf{b}+\mathbf{v}_d\right)\right) = 0$$
$$\Rightarrow B\nabla_\parallel\left(\frac{n_z^{(0)}V_{z\parallel}^{(1)}}{B}\right) = -\frac{d\phi}{d\psi}\left(\frac{\mathbf{b}\times\nabla\psi}{B}\cdot\nabla n_z^{(0)}-Bn_z^{(0)}\nabla_\parallel u\right). \tag{5}$$

Both the impurity density and flow velocity are unknown, so another equation is needed, which can be obtained from the $v_\parallel$-moment of (1). It shows that the friction force is related to the impurity density variation via

$$\int m_z v_\parallel C_{zi}(f_z^{(1)}) = R_{zi_\parallel} = \int m_z v_\parallel^2 \nabla_\parallel f_z^{(0)} = \nabla_\parallel p_z^{(0)}$$
$$\Rightarrow \nabla_\parallel n_z^{(0)} = -\frac{1}{T_i} R_{iz_\parallel}. \tag{6}$$

Introducing the notation

$$\tilde{v}(\psi) \equiv \frac{4}{3\sqrt{\pi}} \frac{m_i n_i}{e_i} \left\langle \frac{\hat{\nu}_{iz}}{n_z} \right\rangle, \qquad \hat{\nu}_{iz} \equiv \frac{n_z Z^2 e^4 \ln\Lambda}{4\pi\epsilon_0^2 m_i^2 v_{th,i}^3}$$
$$K_i^*(\psi) \equiv \frac{K_i(\psi)}{n_i} + \frac{d\ln T_i}{d\psi} \frac{\langle uB^2 \rangle}{\langle B^2 \rangle}$$
$$\gamma \equiv u + (A_{i_1} - A_{i_2})^{-1} K_i^*(\psi)$$

and using a Lorentz operator plus a term guaranteeing momentum conservation to model ion-impurity collisions, one obtains from (5) and (6) the two equations

$$\nabla_\parallel n_z = \tilde{v}\left((A_{i_1} - A_{i_2})\gamma B n_z - n_z V_{z_\parallel}\right) \tag{7}$$

and

$$B\nabla_\parallel \left(\frac{n_z V_{z_\parallel}}{B}\right) = -\frac{d\phi}{d\psi}\left(\frac{\mathbf{b}\times\nabla\psi}{B} \cdot \nabla n_z - B n_z \nabla_\parallel \gamma\right). \tag{8}$$

In order to know the impurity density variation, eqs. (7) and (8) have to be solved. However, it is not possible to do so analytically for stellarator geometry. Therefore, different limits are solved analytically in the next section, and numerical results for the full equation will be given in section 4.

3. Analytical limits

It is possible to gain some useful information about how the impurity redistribution is affected by the various terms by considering different limits, i.e. strong or weak radial gradients or radial electric field, which will be done in the subsequent subsections.

3.1. Weak radial gradients

In the case of weak radial gradients, i.e. $\Delta_i \ll 1$, eq. (8) yields

$$B\nabla_\parallel \left(\frac{n_z V_{z_\parallel}}{B}\right) \ll \frac{n_z V_{z_\parallel}}{L_\parallel},$$

which leads to

$$V_{z_\parallel} \approx \frac{B}{n_z} K_z(\psi).$$

The constant $K_z(\psi)$ can be determined from the constraint that $\langle B\nabla_\parallel n_z \rangle$ must vanish, and becomes $K_z(\psi) = (A_{i_1} - A_{i_2})\langle \gamma B^2 n_z \rangle / \langle B^2 \rangle$. The right-hand side of the remaining equation for n_z,

$$\nabla_\parallel n_z = \tilde{\nu}(A_{i_1} - A_{i_2})\left(\gamma n_z B - B\frac{\langle \gamma B^2 n_z \rangle}{\langle B^2 \rangle}\right), \tag{9}$$

Journal of Physics: Conference Series **260** (2010) 012004 doi:10.1088/1742-6596/260/1/012004

is small as well, and thus

$$\nabla_{\parallel} n_z \ll \frac{n_z}{L_{\parallel}}$$

with the solution $n_z \approx n_z(\psi)$. Thus, the conventional limit of weak gradients, leading to the densities being constant on flux surfaces, is correctly reproduced.

3.2. Large radial gradients, vanishing radial electric field

An interesting and analytically tractable limit is the case of small radial electric field, $e_i/T_i d\phi/d\psi \ll d\ln n_i/d\psi$, in which the terms containing the perpendicular derivative in eq. (8) become negligible and the partial differential equations reduce to ordinary ones. Although this case would not occur in a realistic experimental equilibrium where density gradient and radial electric potential tend to cancel each other (ion root operation) and are thus of comparable magnitude, it is nonetheless enlightening to study this case as it has the appealing property of mathematically resembling the tokamak case and giving some fundamental insight into the different mechanisms involved in the redistribution process. In this limit, the right-hand side of (8) is approximately equal to zero, and thus the impurity flow velocity can be calculated in terms of the impurity density, leading to eq. (9) as in the opposite limit. The solution to this equation depends fundamentally on the properties of the function γ, especially on whether γ has any zeroes. If this is not the case, and if the gradients are steep in the sense that $\Delta_i \gg 1$, the left-hand side of (9) is negligible due to the large multiplier $A_{i_1} - A_{i_2}$ on the right-hand side. The solution then becomes rather simple,

$$n_z \approx \left\langle \frac{1}{\gamma} \right\rangle^{-1} \frac{1}{\gamma}.$$

Employing the value for γ in the axisymmetric limit,

$$\gamma|_{axisym} = -\frac{I(\psi)}{B^2(1 + 2.77d\ln T_i/d\psi)}$$

where $I(\psi)$ is the toroidal current, reproduces the tokamak result found in [1] correctly.

However, if γ does have zeros, the problem is fundamentally different. As there are points at which the right-hand side of (9) vanishes, it is not possible to neglect the parallel gradient in these regions, in which a boundary layer forms. The calculation for this case, based on the method of integrating factors and an expansion around the points where γ vanishes, is given in the following. The result is that, in the limit of very large gradients, the impurities become strongly localised around these zeros, an effect which can be understood physically by noticing that the friction force, which is the drive for the redistribution, is proportional to γ and thus vanishes at these points. Due to periodicity, there must exist points from where all impurity ions are carried away, but also points at which they accumulate. In the limit $\Delta_i \to \infty$, the impurity density becomes a delta distribution at the accumulation points.

This situation can in principle also occur in a tokamak, but parametric dependencies of $\gamma|_{axisym}$ show that this is only possible if the radial temperature and density gradients have opposite signs, which is usually not the case.

4. Numerical solution of the full problem

To study realistic scenarios, in which the pressure gradient and the radial electric field are of comparable magnitude and approximately cancel each other, one has to resort to numerical means. This has been done via a Fourier decomposition with respect to the poloidal and toroidal angle using the MConf library, which has been developed for transport analyses in stellarators [6]. The Fourier series was truncated after 54 components in both angles, and periodicity was used for the toroidal direction.

Fig. 1 shows the results for W7-AS for different values of $\tilde{v}(A_{i_1} - A_{i_2})a^2$. The plasma parameters were chosen to match experimental values in W7-AS during H-mode operation (ion root operation). For a bulk ion density of $5 \cdot 10^{19} - 1 \cdot 10^{20}$, a bulk ion temperature of $100eV$ and a radial density scale length of $2cm$ in the pedestal, realistic values for $\tilde{v}(A_{i_1} - A_{i_2})a^2$ lie between 0.01, corresponding to weak gradients, and 10 for very large gradients and high impurity charge. As an upper limit and to show what happens qualitatively in the limit of very large gradients and very high impurity charge, values of up to 100 have been included in the calculations. The flux surface was chosen to be $s = 0.9$, where s is the magnetic flux normalised to the flux at the separatrix.

As is visible from the figure, for weak radial gradients the impurity density is nearly constant on the flux surface, and thus the limit of conventional theory is correctly recovered. A slight in-out asymmetry exists, with the impurities accumulating at the inboard side of the torus, but the overall density variation is moderate with only 5% variation. When the gradients get larger, the impurities accumulate at a certain toroidal angle. In the limit of very large gradients, the distribution function approaches a delta distribution at these points. When looking at the pattern of the function γ for these plasma parameters, one finds that γ crosses zero, and thus points of vanishing friction force exist, which explains the very strong accumulation of the impurities. In fig. 3, the corresponding radial impurity flux is plotted as a function of the largeness of the gradients. One can show that, in leading order, the particle flux across the field is given by [5]

$$\langle \mathbf{\Gamma}_z \cdot \nabla \psi \rangle = \frac{1}{e_z} \left\langle uBR_{z\|} \right\rangle ,$$

and this is still true even when the impurity density is no longer a flux function. In comparison with a conventional calculation, where the flux increases linearly with increasing gradients, it is reduced by the redistribution. However, even for extreme values of $\tilde{v}(A_{i_1} - A_{i_2})a^2$, the flux is still a monotonically increasing function of the gradients.

Qualitatively different results are obtained for W7-X. Since the order of magnitude of the radial gradients in this device can be expected to be comparable to that found in W7-AS, the same range of simulation parameters should be appropriate. In this parameter regime, one finds that γ does not have any zeros. The impurity accumulation is thus not as strong as in W7-AS, leading to up to 15% density variation as shown in fig. 2, again for different radial gradients. The pattern of the redistribution has two regions of accumulation separated by elongated regions with reduced density. The corresponding impurity flux is shown in fig. 3. Interestingly, although the density variation is much smaller than in W7-AS, the effect on the radial transport is much more pronounced, leading to a considerable reduction of the impurity flux and thus making it a non-monotonic function of the radial gradients.

Running the code for tokamak devices leads to the results found previously of impurity density peaking on the high-field side of the torus [1], with an overall variation between 10–30%, depending mainly on the aspect ratio of the device.

From both simulations and the analytic formulas, it is obvious that the drive of the redistribution is parallel friction between the bulk ions and the impurities, as in a tokamak. However, the way in which the impurities are redistributed is qualitatively influenced by, on the one hand, the radial ion temperature gradient, and, on the other hand, the radial electric field, which does not play a role in axisymmetric configurations. The effect of the radial temperature gradient comes from its entering the constant $K_i^*(\psi)$. Thereby, it influences the magnitude and possibly even the sign of the friction force, which has a large impact on the redistribution if

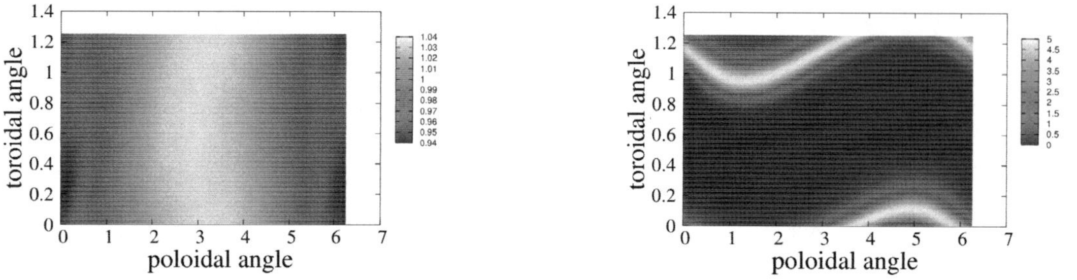

Figure 1. Impurity density in W7-AS. Left: $\tilde{v}(A_{i_1} - A_{i_2})a^2 = 0.02$, right:$\tilde{v}(A_{i_1} - A_{i_2})a^2 = 78.0$

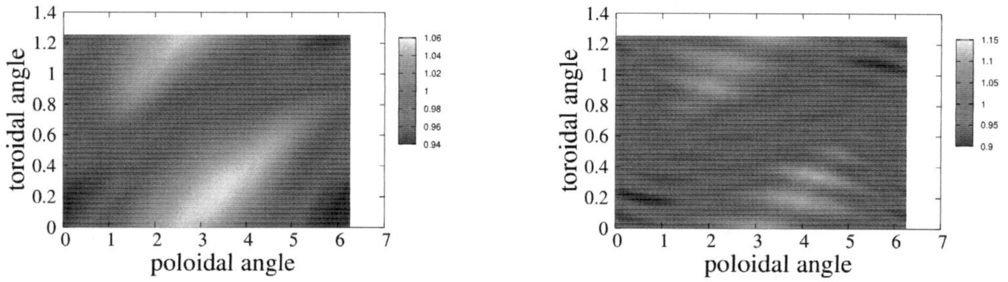

Figure 2. Impurity density in W7-X. Left: $\tilde{v}(A_{i_1} - A_{i_2})a^2 = 0.02$, right:$\tilde{v}(A_{i_1} - A_{i_2})a^2 = 43.5$

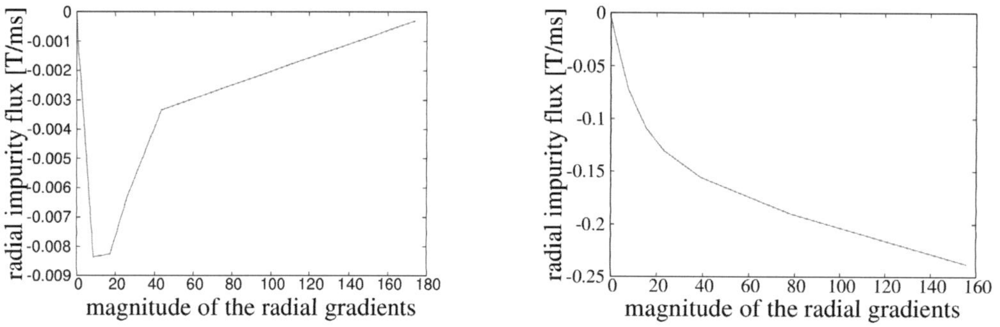

Figure 3. Radial impurity flux vs. radial gradients, left: W7-X, right: W7-AS

γ is modified from being entirely positive (negative) to a state where both signs occur. The radial electric field influences the redistribution in two different ways. Since it also enters the constant $K_i^*(\psi)$, it can directly affect the friction force in this way. Additionally, it drives an $\mathbf{E} \times \mathbf{B}$ rotation, which varies on the flux surface and thereby changes the pattern of the redistribution. Therefore, the dependence of the impurity density on the radial electric field is rather complicated and analytically not tractable.

However, the numerical results suggest that, even when a radial electric field is included, the impurity redistribution and consequently also the radial particle transport are strongly affected by whether or not γ, which is proportional to the friction force, has zeros. If this in not the case, the density variations are moderate. The exact redistribution pattern depends on the magnetic geometry of the device. Regarding the radial impurity transport, as long as the gradients are weak, the relation between the gradients and the radial impurity flux is approximately linear. When the gradients get larger, the impurities are redistributed in such a way that they experience less friction. Therefore, the radial flux is reduced and becomes very small for large gradients. Mathematically, this is manifested by the fact that the left-hand side of (7), which is proportional to the friction force and thus to the particle flux, becomes negligible when the gradients are large enough.

If γ has zeros, the friction force vanishes at certain points and drives the impurity distribution function towards a delta distribution at "half" the zeros, namely, at those toward which the friction force is directed. If the radial electric field is strong, this process can be counteracted by the $\mathbf{E} \times \mathbf{B}$ rotation, which carries the impurities away from these points. However, in the simulation for W7-AS, the ambipolar electric field seems to be unable to prevent strong accumulation. Although the redistribution slightly reduces the friction force, since the impurity flux now increases less rapidly than linear with increasing magnitude of the radial gradients, it is still a monotonic function. Unlike the case when γ does not have any zeros, there exist regions where the left-hand side of (7) is not negligible (namely, where γ vanishes). Thus, the radial flux cannot become as small as when γ does not have any zeros. Again, whether or not this is the case depends on the magnetic geometry.

5. Conclusions & Summary

The presence of large gradients in the plasma edge region has a significant effect on heavy impurity ions. Their density develops variations within the flux surface of the order of up to $10-100\%$, depending on the device. The effect is qualitatively different from that previously found in tokamaks since the radial electric field plays a major role in the redistribution process, on the one hand by driving a non-constant (on the flux surface) $\mathbf{E}\times\mathbf{B}$ flow, which may carry particles away from regions of accumulation, and on the other hand by influencing the magnitude and especially the sign of the friction force, which is the main driving mechanism of the redistribution. The exact pattern of the impurity density on the flux surface depends sensitively on the exact geometry of the device. A crucial aspect for the redistribution process is whether there exist points at which the friction force vanishes. If this is the case, the redistribution can become much more strongly pronounced than otherwise. The radial impurity transport is greatly affected by the redistribution. Whereas it becomes very small when the gradients are large if there are no such points, it might still be a monotonically increasing function of the radial gradients when such points do exist, as in W7-AS, although the flux is still reduced in comparison with a situation without impurity redistribution. Apart from the effects due to the radial electric field, the main difference between tokamaks and stellarators with respect to impurity redistribution is the occurrence of strong localisation and the corresponding consequences for the transport for realistic values of the plasma parameters in stellarators. In tokamaks, this is in principle also possible, but only if the radial density and temperature gradients have opposite signs. The numerical solution of the full impurity density differential equation reproduces the analytical limits found correctly

References

[1] Fülöp T and Helander P 2001 *Phys. Plasmas* **8** 3305
[2] Helander P 1998 *Phys. Plasmas* **5** 3999
[3] Helander P and Simakov A N 2008 *Phys. Rev. Lett.* **101** 145003
[4] Simakov A N and Helander P 2009 *Phys. Plasmas* **16** 042503
[5] Braun S and Helander P 2010 *Phys. Plasmas* **17** 072514
[6] Turkin Y and Maaßberg H and Beidler C and Marushchenko N B 2006 *Fusion Science Techn.* **50** 387

Journal of Physics: Conference Series **260** (2010) 012005 doi:10.1088/1742-6596/260/1/012005

One-dimensional particle models for heat transfer analysis

H Bufferand[a], **G Ciraolo**[a], **Ph Ghendrih**[b], **P Tamain**[b], **F Bagnoli**[c], **S Lepri**[d] **and R Livi**[e]

[a] M2P2, UMR 6181 - CNRS Aix-Marseille Universite - Ecole Centrale, Marseille, Technopole de Château Gombert, 38, Rue F. Joliot-Curie, 13451 Marseille, France.
[b] CEA, IRFM, 13108 St Paul-Lez-Durance, France.
[c] Dipartimento di Energetica, Università di Firenze, Via S. Marta 3, I-50139 Firenze, Italy.
[d] Instituto dei Sistemi Complessi, Consiglio Nazionale delle Ricerche, via Madonna del Piano 10, I-50019 Sesto Fiorentino, Italy.
[e] Dipartimento di Fisica, Università di Firenze, via G. Sansone 1 I-50019, Sesto Fiorentino, Italy.

E-mail: hugo.bufferand@gmail.fr

Abstract. For a better understanding of Spitzer-Härm closure restrictions and for estimating the relevancy of this expression when collisionnality decreases, an effort is done in developing simple models that aim at catching the physics of the transition from conductive to free-streaming heat flux. In that perspective, one-dimensional particle models are developed to study heat transfer properties in the direction parallel to the magnetic field in tokamaks. These models are based on particles that carry energy at a specific velocity and that can interact with each other or with heat sources. By adjusting the particle dynamics and particle interaction properties, it is possible to generate a broad range of models of growing complexity. The simplest models can be solved analytically and are used to link particle behavior to general macroscopic heat transfer properties. In particular, some configurations recover Fourier's law and make possible to investigate the dependance of thermal conductivity on temperature. Besides, some configurations where local balance is lost require defining non local expression for heat flux. These different classes of models could then be linked to different plasma configurations and used to study transition from collisional to non-collisional plasma.

1. Introduction

Heat transport in the presence of a temperature gradient imposed by heat reservoirs is a classical problem in the physics of plasmas, that has been tackled over decades by different approaches. In order to improve heat transfer modeling in weakly collisional plasma, the development of so called "toy models" is of a great interest. These models have the goal of catching the specificities of different regimes of heat transfer in the simplest way. In this paper, we focus on heat transfer in the direction parallel to the magnetic field in tokamak plasma. These plasmas are known to be weakly collisional, however, the computation of the heat flux in these plasma often relies on Spitzer-Härm expression [1] of the heat flux that is based on a strong collisionality assumption. In fact, this closure equation is used in the fluid approximation where local balance should be verified. In order to account for the departure from local balance at high temperature, the Spitzer-Härm formulation is sometime adjusted. Many examples of these flux limiter corrections adjusting the flux expression to work at low collisionalities can be found in the literature [2; 3]. In this manuscript we aim at revisiting this problem by considering very simple models that epitomize the minimal ingredients for reproducing some general properties that are known to characterize the diffusive as well as the ballistic regimes typical of heat transfer in plasmas.

Far from a complex kinetic treatment, our models rely on particles that move along a closed magnetic field line and that exchange energy with each other. They can also interact with heat sources. In our approach, the properties of the tokamak magnetic field line are simplified and we consider that the magnetic field is constant on the field line. By comparison with a tokamak, one can consider that all the particles are passing and trapped particles are not considered. The models implemented in this work can be sorted in two categories. The first ones that will be presented in section 2 consider particles traveling with a constant velocity. The second ones illustrated in section 3 assume that particle velocities are linked to their respective energy following the relation $v = A\sqrt{E}$. We further show that these two models can be related one to another by a suitable relation.

This particle models could be studied in parallel with the kinetic treatment performed two decades ago to study transition from strongly collisional to collisionless regime [4; 5].

2. Models with constant particle velocity

The domain considered in the following model is a periodic line of length L where a heat source of energy E_{s1} is located at x_{s_1} and another one of energy E_{s2} is located at x_{s_2}. On this line, $2 \times N$ particles are uniformly spread on the mesh $\{x_i\}_{i \in [\![1,N]\!]}$, with $x_i = iL/N$. Hence, on each node of the grid x_i are located two particles. One is travelling forward with velocity v^+, one is travelling backward with velocity v^- with $|v^+| = |v^-| = v$. E_i^+ and E_i^- denote the forward (+) and backward (−) moving particle volumic energies, respectively. The energy $\mathcal{E}$ of a particle is given by the relation $\mathcal{E} = \mathcal{V}E = S\delta x E$, where $\mathcal{V} = S\delta x$ is the volume of a particle, S is the arbitrary surface of the particle pependicular to the direction considered so far and δx is the characteristic length of a particle equal to the mesh length $\delta x = L/N$.

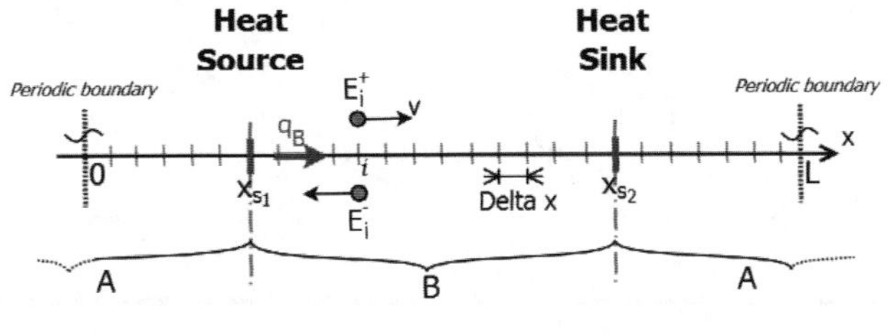

Figure 1. Sketch of the domain

The evolution law of the system relies on two steps: a *dynamic step* ($t \to t + 1/2$) when the particles jump from one node to the closest neighboring node according to the sign of their velocity, and an *interaction step* ($t + 1/2 \to t + 1$) during which the + and − particles that are located on the same node exchange part of their energy. The evolution is given by Equations 2.1 where the $t+1/2$ step is denoted with a prime and the $t+1$ step is denoted with two primes.

$$\textbf{Dynamic step} \quad \begin{cases} E_i^{+\prime} &= E_{i-1}^+ \\ E_i^{-\prime} &= E_{i+1}^- \end{cases} \qquad \textbf{Interaction step} \quad \begin{cases} E_i^{+\prime\prime} &= E_i^{+\prime} + \frac{E_i^{-\prime} - E_i^{+\prime}}{\alpha} \\ E_i^{-\prime\prime} &= E_i^{-\prime} + \frac{E_i^{+\prime} - E_i^{-\prime}}{\alpha} \end{cases} \tag{2.1}$$

α is a number that quantify the exchanged amount of energy between the particles. If $\alpha = 2$, the energy of the two particles after the interaction is the same. In this case, one can say that the interaction is total. Moreover, when the particles pass through the nodes $x_{i_{s1}}$ and $x_{i_{s2}}$ where the heat sources are located, they take the energy of the source. In stationary conditions, we have for all i, $E_i^{\pm\prime\prime} = E_i^{\pm}$. Hence, using Equations (2.1), the solution can be obtained by iterating equations (2.2) with the "boundary conditions" $E_{i_{s1}}^{\pm} = E_{s1}$ and $E_{i_{s2}}^{\pm} = E_{s2}$.

$$\begin{cases} E_i^+ &= E_{i-1}^+ + \Delta E_i \\ E_i^- &= E_{i+1}^- - \Delta E_i \end{cases} \quad \text{with} \quad \Delta E_i = \frac{E_{i+1}^- - E_{i-1}^+}{\alpha} \tag{2.2}$$

From recursive serie analysis, one can prove that the solution takes the form $E_i^+ = ai + b$ and $E_i^- = ci + d$. Reporting these expressions in Equations 2.2 and considering the boundary conditions, one obtains the value of a, b, c and d:

$$\begin{cases} E_i^+ &= \frac{E_{s1}-E_{s2}}{i_{s1}-i_{s2}+(2-\alpha)} \cdot i + \frac{E_{s2}\cdot i_{s1} - E_{s1}\cdot i_{s2} + E_{s1}\cdot(2-\alpha)}{i_{s1}-i_{s2}+(2-\alpha)} \\ E_i^- &= \frac{E_{s1}-E_{s2}}{i_{s1}-i_{s2}+(2-\alpha)} \cdot i + \frac{E_{s2}\cdot i_{s1} - E_{s1}\cdot i_{s2} + E_{s2}\cdot(2-\alpha)}{i_{s1}-i_{s2}+(2-\alpha)} \end{cases} \quad \text{for } i \in [\![i_{s1}, i_{s2}]\!] \tag{2.3}$$

It is also possible to compute the energy exchanged with the source at each time step. For instance, for source s_1, this energy increment is given by the relation

$$\delta\mathcal{E} = \mathcal{V} \cdot \left(2E_{s1} - E_{i_{s1}}^{+\prime} - E_{i_{s1}}^{-\prime}\right) = \mathcal{V}\left(2E_{s1} - E_{i_{s1}-1}^{+} - E_{i_{s1}+1}^{-}\right) \tag{2.4}$$

Also, since $E_{s1} = E_{i_{s1}}^- = E_{i_{s1}}^+$, the expression of the $\delta\mathcal{E}$ can be expressed as

$$\delta\mathcal{E} = \mathcal{V} \cdot \left\{ (E_{i_{s1}}^- - E_{i_{s1}-1}^+) + (E_{i_{s1}}^+ - E_{i_{s1}+1}^-) \right\} \tag{2.5}$$

One notices from Equation 2.2 that the quantity $\xi_i = E_{i-1}^+ - E_i^-$ verifies $\xi_{i+1} = \xi_i = \xi$. This quantity is conserved wherever the particles interact with each other and do not interact with the sources. These subdomains are denoted with the letter A and B on Figure 1. Consequently, the energy exchanged with the source s_1 can also be written as $\delta E = \mathcal{V}(-\xi_A + \xi_B)$. If we divide this energy by the time step $\delta t = \delta x/v$ and by the surface S, one can determine the heat flux q by

$$q = \frac{\delta\mathcal{E}}{S \cdot \delta t} = v(-\xi_A + \xi_B) = q_A + q_B \tag{2.6}$$

Using Equations 2.3 gives the following expression for q_B

$$q_B = v \cdot \xi_B = v(E_{i_{s1}}^+ - E_{i_{s1}+1}^-) = v \cdot \frac{(E_{s1} - E_{s2})(1-\alpha)}{i_{s1} - i_{s2} + (2-\alpha)} \tag{2.7}$$

Besides, one can calculate the energy gradients,

$$\nabla E^{\pm} = \frac{E_{i+1}^{\pm} - E_i^{\pm}}{x_{i+1} - xi} = \frac{1}{\delta x} \cdot \frac{E_{s1} - E_{s2}}{i_{s1} - i_{s2} + (2-\alpha)} \tag{2.8}$$

Reporting equation 2.8 into Equation 2.7 gives

$$q_B = -(\alpha - 1) \cdot \delta x \cdot v \cdot \nabla E^{\pm} \tag{2.9}$$

As noticed previously, the case where $\alpha = 2$ is of particular interest since it implies a thermalization between the + and − particles. This condition guarantees local equilibrium, i.e. $E_i^+ = E_i^-$. This balance leads to defining the temperature as $T \sim E^{\pm}$. The heat flux is then expressed by the standard Fourier law for heat conduction : $q = -v\delta x \nabla T = -\kappa \nabla T$. Moreover, this expression is compatible with the standard derivation of heat conductivity [6]:

$$\kappa = k_B n \lambda v_{th} \tag{2.10}$$

where λ is the mean free path, v_{th} is the thermal speed, k_B is Boltzmann constant and n is the plasma density. In the case where $\alpha = 2$, the equivalent mean free path is δx since thermalization occurs at the lattice scale. v plays the role of the thermal speed.

For $\alpha \neq 2$, $E_i^+ \neq E_i^-$ and it is not possible to define local balance at the mesh scale. However, if we define $\bar{E} = (E^+ + E^-)/2$, we have a Fourier's like expression for the heat flux that is $q = -v(\alpha - 1)\delta x\ \nabla\bar{E}$. In this case, we have $\kappa = v(\alpha - 1)\delta x$. To be coherent with the usual derivation of heat capacity, one should find that the equivalent mean free path in this case is $\lambda = (\alpha - 1)\delta x$. In fact, the interaction rule guarantees an exponential decay to equilibrium after a characteristic number of interactions $n = \alpha - 1$ and thus a characteristic decay length compatible with the above expression for the mean free path.

For α going to infinity, the $+$ and $-$ particles almost do not interact with each other. A standard collisionless regime is then expected. Indeed, using Equations 2.3, one can find directly that $\nabla E^{\pm} \to 0$. Hence, in the domain B of Figure 1 for instance, we have $E_{s1} = E^+ \neq E^- = E_{s2}$ and there is no local equilibrium. As expected, the mean free path tends to infinity and the particle behavior is ballistic. Moreover, making $\alpha \to \infty$ in Equation 2.7 gives $q = v(E_{s1} - E_{s2})$ which is compatible with the ballistic expression of heat flux.

On Figure 2 are plotted results for $\alpha = 2$ and $\alpha = 30$. In the first case, one can notice that $E^+ = E^- = \bar{E}$ that defines local equilibrium.

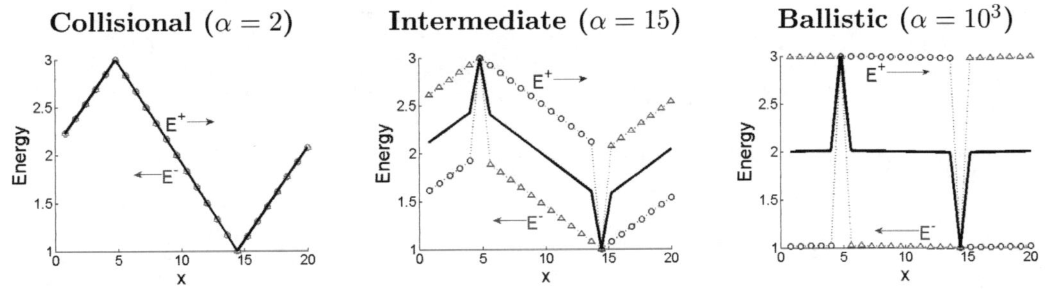

Figure 2. Simulations performed with constant velocity for different values of α. E^+ (circles), E^- (triangles) and $\bar{E}$ (solid thick line). The sources are located at $x = 5$ and $x = 15$. Their respective energies are 3 and 1. The length of the domain is 20. The number of particles is 2×50

To go further, since the quantity $\lambda = (\alpha - 1)\delta x$ plays the role of mean free path, it can be interesting to introduce also a dependance of α on the energy. This is done by considering the following law for α:

$$\alpha_i = 2 \cdot \left(\frac{\max(E_i^+, E_i^-)}{E_{\text{ref}}}\right)^r \tag{2.11}$$

In this expression, E_{ref} is chosen small enough to ensure that the energy ratio is always larger than one such that $\alpha > 2$. r is a positive real number. The idea is to consider the case where $\alpha \gg 1$ yielding the approximation $\lambda = \delta x(\alpha - 1) \approx 2\delta x \left(\max(E_i^+, E_i^-)/E_{\text{ref}}\right)^r \propto {E_i}^r$. For plasma physics, the case where $r = 2$ is of particular interest since Coulombian interactions between plasma particles give a mean free path proportional to the energy square. The assumption $\alpha \gg 1$ does not imply that collisionless cases are the only ones considered. Indeed, as long as $\lambda \ll L \Leftrightarrow (\alpha - 1) \ll N$ where L is the domain legnth anf N the number of interactions undergone by a particle as it travels through the domain, the model can be supposed to be collisional.

We are interested in finding analytical solutions for this system. We will solve it in the domain

B on Figure 1 where $E^+ > E^-$. The system to solve is then

$$\left\{ \begin{array}{lcl} E_i^+ & = & E_{i-1}^+ + \frac{E_{i+1}^- - E_{i-1}^+}{2\cdot(E_{i-1}^+/E_{\text{ref}})^r} = E_{i-1}^+ + \varepsilon\cdot(E_{i+1}^- - E_{i-1}^+) \\ E_i^- & = & E_{i+1}^- - \frac{E_{i+1}^- - E_{i-1}^+}{2\cdot(E_{i-1}^+/E_{\text{ref}})^r} = E_{i+1}^- - \varepsilon\cdot(E_{i+1}^- - E_{i-1}^+) \end{array} \right. \tag{2.12}$$

where $\varepsilon = 1/\alpha \ll 1$. One must notice that the numerator of the exchange term can be developped as

$$\begin{aligned} E_{i+1}^- - E_{i-1}^+ &= E_{i+1}^- - E_i^+ + (E_i^+ - E_{i-1}^+) \\ &= E_{i+1}^- - E_i^+ + \varepsilon\cdot(E_{i+1}^- - E_{i-1}^+) \\ &\approx E_{i+1}^- - E_i^+ = \xi = \text{constant} \end{aligned}$$

Hence, the problem is reduced to solving recursive Equation 2.13 for E_i^+:

$$E_i^+ = E_{i-1}^+ + \frac{B}{(E_{i-1}^+)^r} \quad \text{with} \quad B = \xi\cdot E_{\text{ref}}^r/2 \tag{2.13}$$

An approximate solution is

$$E_i^+ = \left(E_{s1}^{1+r} + (i - i_{s1})\cdot B\cdot(1+r)\right)^{\frac{1}{1+r}} \tag{2.14}$$

The unknown B is determined using $\xi = E_i^- - E_{i-1}^+ = 2B/E_{\text{ref}}^r$ and $E_{i_{s2}}^- = E_{s2}$. Reporting in Equation 2.14 for $i = i_{s2} - 1$, one find

$$2B/E_{\text{ref}}^r = E_{s2} - \left(E_{s1}^{1+r} + (i_{s2} - 1 - i_{s1})\cdot B\cdot(1+r)\right)^{\frac{1}{1+r}} \tag{2.15}$$

Solving equation 2.15 gives B, ξ and finally the heat flux q. The simulation results are compared with the analytical solution on Figure 3. One expects to find a good agreement between the simulation results and the analytical solution in the cases where $\alpha \gg 1$. For the first case, we have chosen $E_{\text{ref}} = 0.5$, $E \in [2,\ 5]$ and $r = 2$ hence $\alpha \in [16,\ 100] \gg 1$. The hottest particles will thus require about 100 interactions to thermalize. The number of particles used in the simulation is $N = 2 \times 3000$. Consequently, each particle undergoes 3000 interactions when it covers the domain. Since $3000 \gg 100$, this case can then be considered as a collisional case, that is $\lambda \ll L$. In fact, on Figure 3,a, one can notice that for all i, $E_i^+ \approx E_i^-$. In the second case, we choose the same parameters except that we limit the number of particles to 300. Thus, the relation $\lambda \sim L$ applies and the case is not collisional. One can notice the departure between E^+ and E^- on Figure 3,b

3. Models with non-constant velocity

Unlike the previous models where the velocity of particles is constant, a dependancy of velocity on the energy is added such that $v^\pm = A\sqrt{E^\pm}$. The previous energy can now be seen as the particle's kinetic energy. In the following simulations, we consider the same domain as described on Figure 1. Particles are randomly placed in the domain with a given initial energy. One half is moving forward and one half is moving backward. The algorithm detects when two particles meet or when they meet the sources. In that case, particles exchange part of their energy with the same rules as those described in the previous section. The particle's velocity modules are recalculated and their velocity sign is conserved. Since the sign of velocity is unchanged, particles exchange energy with the two source alternatively. The U-turn is not allowed since making the

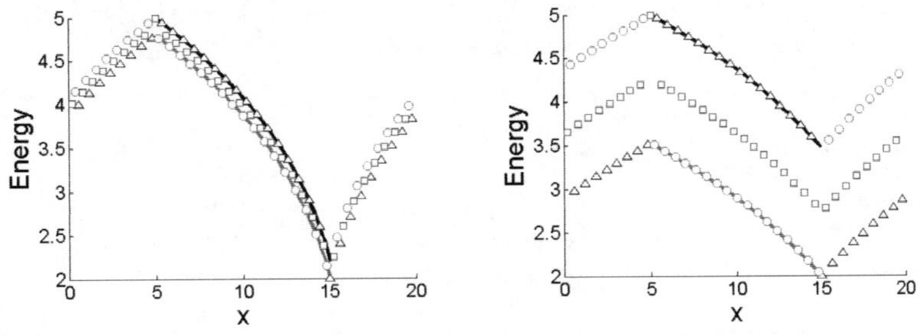

(a) (b)

Figure 3. Simulation results with $\alpha \propto \max(E^+, E^-)^2$. $E_{s_1} = 5$ and $E_{s_2} = 2$. (a): Energy profiles obtained when the simulation is performed with 2×3000 particles and guarantee that for all i, $1/\alpha \ll 1$ and $\lambda \ll L$. (b): The simulation is performed with 2×300 particles for $\lambda \sim L$. Simulations results are plotted with symbols: E^+ (triangles), E^- (circles) and $\bar{E}$ (squares). Analytical estimations corresponding to Equation (2.14) are plotted with lines.

two traveling directions equally probable would lead to defining an automatic local equilibrium. As a consequence, some fluctuations apart, the rate of interaction with the two sources is the same. This interaction rate is also characteristic of the rate with which particles interact with each other. Analytical solutions are more difficult to obtain for this kind of model since there is no well defined lattice for solving the recursive equations. However, in the collisional case where $E^+ \approx E^-$, an approximate solution can be found by solving Equations (2.1,2.12) on the lattices $\{x_i^+\}$ and $\{x_i^-\}$, see Figure 4, where

$$\begin{cases} |x_{i+1}^+ - x_i^+| &= |\delta t \cdot v_i^+| = \delta t \cdot A\sqrt{E_i^+} \\ |x_i^- - x_{i-1}^-| &= |\delta t \cdot v_i^-| = \delta t \cdot A\sqrt{E_i^-} \end{cases} \tag{3.1}$$

Since $E_i^+ \approx E_i^-$, one has $\forall i$, $\Delta x_i^+ \approx \Delta x_{i-1}^-$ and one can define the lattice $x_i \sim x_i^+ \sim x_i^-$, see Figure 4. To determine the values of x_i for all i, one has to solve $\Delta x_i = A\delta t\sqrt{E_i}$ constrained by $\sum_N \Delta x_i = L$ where the values of E_i are given by the solutions found in the previous section. One has to remember that this approach is only valid for the collisional cases.

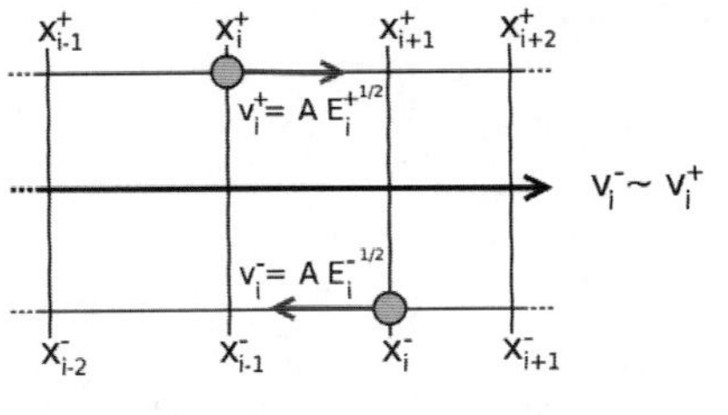

Figure 4. Lattice definitions in the collisional approximation for non-constant velocity

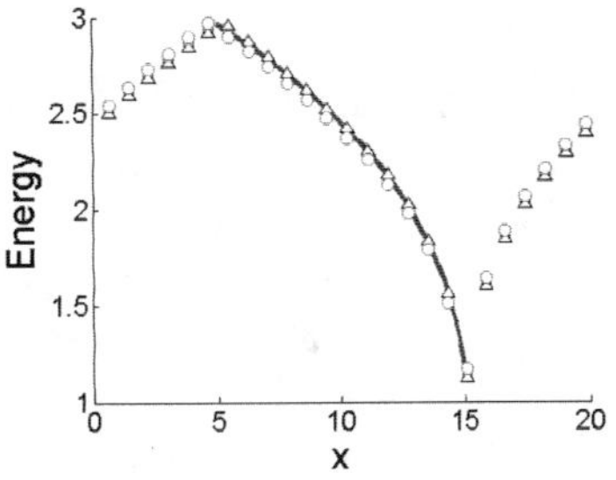

Figure 5. Energy profiles in the non-constant velocity case for $r = 2$. Simulation results: E^+ (triangles), E^- (circles) and $\bar{E}$ (squares). Line: analytical estimation. Number of particles: 2×500.

Journal of Physics: Conference Series **260** (2010) 012005 doi:10.1088/1742-6596/260/1/012005

In particular, one finds $\{x_i\}$ in the case where the mean free path depends on the energy. From Equation 2.14, the energy is given by the relation

$$E_i = (a + b \cdot i)^{1/(1+r)} \tag{3.2}$$

where $a = E_{s_1}^{1+r} - Bi_{s_1}(1+r)$ and $b = B(1+r)$. In the following, we just focus on the domain B defined on Figure 1. In order to determine the value of x_i, we use the approximation $\Delta x_i/\Delta i \approx \mathrm{d}x/\mathrm{d}i$. Hence,

$$\begin{aligned} \mathrm{d}x &= A\delta t (a + b \cdot i)^{\frac{1}{2(1+r)}} \mathrm{d}i \\ x_i - x_{s1} &= A\delta t \int_{i_{s1}}^{i} (a + b \cdot i)^{\frac{1}{2(1+r)}} \mathrm{d}i \\ x_{s2} - x_{s1} &= A\delta t \int_{i_{s1}}^{i_{s2}} (a + b \cdot i)^{\frac{1}{2(1+r)}} \mathrm{d}i \end{aligned} \tag{3.3}$$

The solution of these equations is

$$\begin{cases} x_i = c + d(a + b \cdot i)^{\frac{3+2r}{2(1+r)}} \\ i = \frac{1}{b}\left(\frac{x_i - c}{d}\right)^{\frac{2(1+r)}{3+2r}} - \frac{a}{b} \end{cases} \tag{3.4}$$

c and d can be determined using $x_{i_{s1}} = x_{s_1}$ and $x_{i_{s2}} = x_{s_2}$. Reporting the expression of i in Equation 3.2 gives

$$E_i = \left(\frac{x_i - c}{d}\right)^{\frac{2}{3+2r}} \tag{3.5}$$

Since in the collisional case, one has $E(x_{s1}) = E_{s1}$ and $E(x_{s2}) = E_{s2}$, the constant can be determined and the energy is finally given by Equation 3.6.

$$E = \left(E_{s1}^{1/\beta} + \frac{E_{s2}^{1/\beta} - E_{s1}^{1/\beta}}{x_{s2} - x_{s1}}(x - x_{s1})\right)^{\beta} \quad \text{with} \quad \beta = \frac{2}{3+2r} \tag{3.6}$$

In this collisional case where a lattice has been determined, the heat flux can be calculated as in the previous section, see Equation 2.7. The velocity is now a function of the temperature:

$$\begin{aligned} q &= \frac{\mathcal{V}\cdot(E_i^+ - E_{i+1}^-)}{S\cdot\delta t} = \ell \cdot A\sqrt{E_i}\frac{E_i^+ - E_i^-}{x_{i+1} - x_i} \\ &= -A \cdot \ell \cdot (\alpha - 1)\sqrt{\bar{E}}\nabla\bar{E} \end{aligned} \tag{3.7}$$

On Figure 5, a comparison between simulation results in the collisional case for $r = 2$ and the above analytical estimation is plotted. The + and − particles energy are shown. One notices that these two variables are almost equal which justifies the above approximations.

We also study the dependance of the heat flux on the energy in the case $r = 2$. To do so, the number of particles is kept constant equal to 500. The energy of the source is chosen as $E_{s_1} = 1.05 \times E_0$ and $E_{s_2} = 0.95 \times E_0$ where E_0 is a parameter that describes the energy of the system. We choose $E_{\text{ref}} = 0.8$ and simulations are performed for values of E_0 between 2 and 200. A particle undergoes approximately $N_{\text{inter}} = 500$ interactions as it travels from one source to the other. Since α represents the number of interactions necessary to thermalize, the regime is expected to be collisional for $\alpha = 2(E_0/E_{\text{ref}})^2 < 500$. To make it clearer, we define the collisionality ν^* as $\nu^* = N_{\text{inter}}/\alpha$. We are in a collisional regime if $\nu^* > 1$ that is $E_0 < 13$. For $E_0 > 13$, the regime is ballistic. The distance between the source is denoted $\mathfrak{L}$. For each case, one computes the average gradient of $\bar{E}$. On Figure 6,a is plotted the equivalent heat condictivity $\kappa = q/\bar{E}$ as a function of E_0. On Figure 6,b is plotted $q\mathfrak{L}/(E_{s_1} - E_{s_2})$.

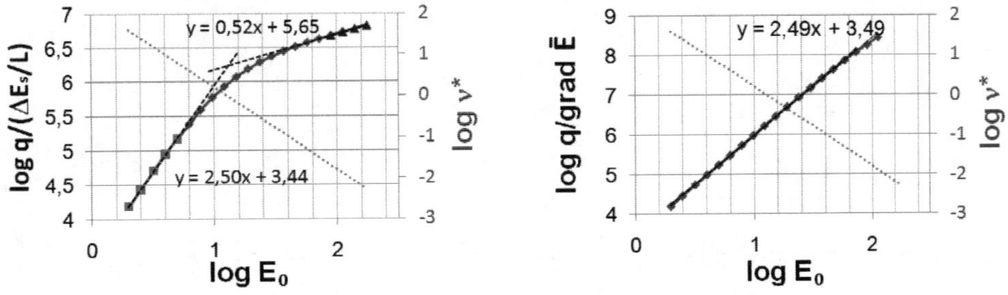

(a) (b)

Figure 6. (a). $q/\{(E_{s_1} - E_{s_2})/L\}$ as a function of the energy of the system E_0. Simulation results (symbols) exhibiting the transition from conductive to ballistic regime. Power law fittings are represented with lines. (b). Effective heat conductivity $\kappa = q/\nabla\bar{E}$ as a function of the energy of the system E_0, simulation results (symbols), power law fitting (line) exhibiting $\kappa \propto E_0^{5/2}$. The collisionality is plotted with dot-line.

On Figure 6,a we notice that for low energies we have $q \propto E_0^{5/2}(E_{s2} - E_{s1})/\mathfrak{L}$. Besides, for low energies, we have $\nabla T \approx (E_{s_2} - E_{s_1})/\mathfrak{L}$ thus $q \propto E_0^{5/2}\nabla T$. This characteristic of the collisional Spitzer-Härm regime. For high energy, we have $q \propto \sqrt{E_0}(E_{s_2} - E_{s_2})/\mathfrak{L} \sim v_{th}(E_{s_2} - E_{s_2})$. This is characteristic of the ballistic regime. The transtion occurs for $\nu^* = 1$, as expected. On Figure 6,b we notice that with the definition of $\bar{E} = (E^+ + E^-)/2$, we recover the Spitzer-Härm like law for the heat flux, even in the collisionless case ($\nu^* < 1$).

4. Conclusions

The analysis of simple models where energy transfer takes place exhibits the well-known Spitzer-Härm diffusive behavior in the collisional regime. Also, the transition from collisional to ballistic behavior was observed as the equivalent mean free path of the particles increased. This transition can be compared with the one observed with a kinetic treatment limited to considering 4 moments of the distribution function [4; 5]. In the present "toy-model", no distinction is made between thermal and suprathermal particles. A perspective of this work could be to introduce this ingredient to investigate flux-limiter formulations that are used when thermal particles are collisional and suprathermal particles are ballistic.

Acknowledgements
This work supported by the ANR project ESPOIR and by the European Communities under the contract of Association between EURATOM and CEA, was carried out within the framework of the European Fusion Development Agreement. The views and opinions expressed herein do not necessarily reflect those of the European Commission.

References
[1] Spitzer L and Härm R 1953 *Phys. Rev.* **89-5**
[2] Luciani J F and Mora P 1983 *Phys. Rev. Lett.* **51** 1664–1667
[3] Fundamenski W 2005 *Plasma Phys. Control. Fusion* **47** R163–R208
[4] Bond D J 1981 *J. Phys. D: Appl. Phys.* **14**
[5] Shirazian M H and Steinhauer L C 1981 *Phys. Fluids* **24**
[6] Stangeby P C 2000 *The Plasma Boundary of Magnetic Fusion Devices* (IoP)

Journal of Physics: Conference Series **260** (2010) 012006 doi:10.1088/1742-6596/260/1/012006

On the effects of the equilibrium model in gyrokinetic simulations: from s-α to diverted MHD equilibrium

A Burckel[1], O Sauter[1], C Angioni[2], J Candy[3], E Fable[2] and X Lapillonne[1]

[1] *Ecole Polytechnique Fédérale de Lausanne (EPFL), Centre de Recherches en Physique des Plasmas, Association EURATOM- Confédération Suisse, CH-1015 Lausanne, Switzerland*
[2] *Max-Planck-Institut für Plasmaphysik, IPP-EURATOM Association, D-85748 Garching bei München, Germany*
[3] *General Atomics, P.O. Box 85608, San Diego, California 92186, USA*

E-mail: `olivier.sauter@epfl.ch`

Abstract. In order to better identify the role of the magnetic topology on ITG and TEM instabilities, different MHD equilibria with increasing complexity are calculated using the CHEASE code [1]. We start from the geometry of the s-α cyclone benchmark case [2], consider the corresponding circular numerical equilibrium, and then successively add a non zero value of α consistent with the kinetic profiles, an elongation of 1.68, a triangularity of 0.15, and finally an up-down asymmetry corresponding to a single-null diverted geometry. This gives the opportunity to study separately the effect of each main characteristics of the equilibrium on microinstabilities in core plasmas. Linear local electrostatic gyrokinetic simulations of these different numerical equilibria and of their corresponding analytical descriptions (Miller-type representations [3]) are performed using the codes GS2 [4, 5] and GYRO[6]. It is observed that each modification of the equilibrium has an influence on the results of gyrokinetic simulations. The effect of the α parameter can compensate the stabilizing effect of an increase in the elongation. A comparison between the up-down symmetric shaped equilibrium and its corresponding diverted configuration show a non negligible effect on the growth rate of ITG and TEM turbulence. The comparison between the local Miller model and using a full equilibrium shows that it is mainly the indirect change of elongation in the plasma core which influences the results. The global aim is to provide well defined benchmark cases including real geometry and kinetic electrons physics, since this is not analyzed by the cyclone case. In addition, the goal is to define a procedure for testing of local simulations inspired by experimental constraints and results.

1. Introduction

The magnetic flux surface shape defines the geometry of the plasma considered in a gyrokinetic simulation. It is generally defined either using an analytical model or solving numerically the Grad-Shafranov equation to obtain the MHD equilibrium. There are still many gyrokinetic simulations which are based on the $s - \alpha$ model for the equilibrium and used to compare with experimental results. In addition, these simulations are usually local flux tube simulations in order to save CPU time. However experiments do use the stabilizing effects of plasma shaping. In addition, the comparison of local transport properties between different plasma regions inherently include a difference in plasma shaping. Therefore gyrokinetic simulations need to include the effective magnetic topology [7, 8], either through the Miller model [3] or coupled to a full 2D MHD equilibrium representation. Therefore, before validation of experimental results can occur, code benchmarking including these features is necessary. The aim of this work is to provide a benchmark framework which builds upon the cyclone case [2], but which would include the main new physics assumption which are usually used in present simulations, namely kinetic electrons, toroidal geometry, consistent α and collisionality. In addition, electromagnetic effects should be tested, but this can be performed at a later stage. On top of this, we would like this framework to allow for the comparison between local and global simulations. In doing so, we shall see that it provides a framework for comparison with experiments as well.

There have been many valuable published results of comparison between gyrokinetic simulations and experimental results or of high level code benchmarking. We cannot review them here and our purpose is to add to these activities with a different point of view, which provides new physics insights while performing code benchmarking. The most recent benchmarking activity has used the cyclone base case and adiabatic electrons, for core transport benchmark [9] using nonlinear codes. In this paper we first test linear simulations, since one cannot expect quantitative analyses of nonlinear simulations if the linear results do not agree already. A first step is to go beyond the $s - \alpha$ model, both in terms of magnetic topology and of finite value of α. It was thought that using an $s - \alpha$ model would ease the benchmarking exercise, removing the difficulties of checking the implementation of the equilibrium interface in different codes. It was shown recently that it was actually not true and even the $s - \alpha$ model introduced discrepancies and errors [10]. These can lead to differences sufficient to bring doubt about one code's simulations. This was recently resolved with respect to the $\rho*$ scaling [11], but it shows that benchmarking including the full MHD equilibrium is a timely exercise. These recent results [11] also show that comparing global and local results are very useful.

It is well known that the effect of kinetic electrons on ITG is significant. In addition, a recent paper using a quasi-linear model has shown that the transition between ITG and TEM is very important to determine the zero particle flux stationary state and therefore for particle transport [12]. This is confirmed by nonlinear results [13, 14]. Therefore it is important to benchmark as well the treatment of kinetic electrons included in the codes, both the effect on the ITG and the transition point to the TEM. Since finite α [15, 16] and collisionality have significant effects on TEM, they should be included as well. In this paper, we shall discuss the inclusion of finite α, however the inclusion of collisionality effects will be studied at a later stage.

2. The five MHD equilibria and the s-α model

There are many choices that can be made to define various equilibria from the cyclone test case, which is local, circular and based on the s-α model with α=0. Since the cyclone case is based on DIII-D experimental results [17], we found interesting to aim towards an equilibrium equivalent to the actual experiment. The latter has $q_{95} = 3.8$, $\kappa = 1.68$, $I_p = 1.3$MA, $B_0 = 2.09$, $\beta_N = 1.85$, $Z_{\text{eff}} = 2.3$ and we reconstructed the pressure profile from Fig. 5 of [17] as well, with $p_{\text{edge}} = 0$. For the plasma boundary, we have taken a typical single-null (SN) from DIII-D and the above geometrical characteristics as seen in Fig. 1(e). To match the top triangularity, we have chosen $\delta = 0.15$. These choices allow to define five equilibria with increasing shaping effects from circular-$\alpha = 0$ (Fig. 1a), circular-$\alpha \neq 0$ (Fig. 1b), elongated (Fig. 1c), elongated+triangularity (Fig. 1d) and up-down asymmetric (Fig. 1e).

The q profile still needs to be determined. Since the central q value is usually close to one, experimentally, we imposed $q_0 = 0.95$. One could also impose the $q = 1$ radius, however matching q_0, q_{95} and the value of $q = 1.4$ and $s = 0.78$ at $r/a = 0.5$, as in the cyclone case, was sufficently constraining. This has been performed by tuning the current density profile given to the CHEASE equilibrium code [1]. The resulting flux surfaces are shown in Fig. 1, where $a = 0.36$ and $R_0 = 1.68$ was also used. The corresponding eqdsk files are available as output of the code. Note that both the q profile and the magnetic shear profiles do match very well for the

Journal of Physics: Conference Series **260** (2010) 012006 doi:10.1088/1742-6596/260/1/012006

five equilibria. One could have chosen to keep I_p fixed. However experimental comparison across machines are usually performed at similar q_0 and q_{95} values and thus at similar q profile. Note also that we want to compare the ITG and TEM growth rates with local physics parameters as close to the cyclone base case as possible, in order to compare them within the same context. It should be noted that usually, when the plasma elongation is increased in particular, the q profile broadens and is essentially flat and near one over a large volume. However in the suite of equilibria computed here, we have kept the q profiles self-similar.

Table 1 gives the main parameters of these equilibria, with κ and δ being the local values of elongation and triangularity and not the edge values. The shift is defined by dR_0/dr where R_0 is the flux surface geometrical center.

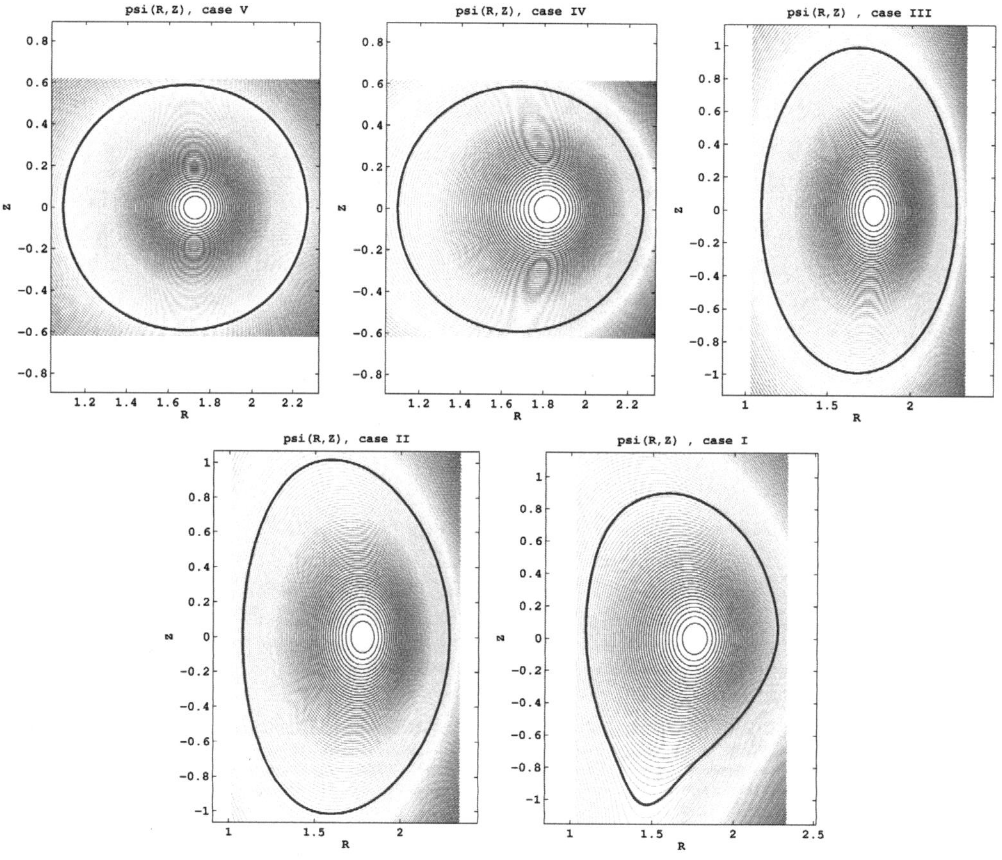

Figure 1. The five equilibria: (V) circular $\alpha = 0$; (IV) circular $\alpha \neq 0$; (III) elongated $\alpha \neq 0$; (II) $\kappa + \delta$ and $\alpha \neq 0$; (I) single-null down.

3. Some specifications about the simulations

Other parameters, non related directly to the magnetic topology are taken from cyclone case (based on DIII-D discharge 81499): $n_e = n_i$, $T_e = T_i$, $\epsilon_0 = \frac{r0}{R_0} = 0.18$, $r/a = 0.5$, R_0=1.68m, a=0.6m , R/L_n=2.22 ,R/L_T=6.91. These are the local values at $r/a = 0.5$, the full profiles are shown below. In this paper, only the local Eulerian gyrokinetic simulations have been performed with GS2 and GYRO.

number	shape	q	q_{95}	s	α	shift	κ	δ	I_p [MA]	β_N
V	circular	1.42	3.72	0.84	0.0	-0.06	1.01	0.00	0.64	0.0
IV	circular	1.45	3.66	0.76	**0.47**	-0.23	1.04	0.00	0.78	3.60
III	elongaged	1.38	3.70	0.74	0.49	-0.16	**1.47**	0.00	1.34	2.16
II	shaped	1.37	3.87	0.77	0.49	-0.16	1.43	**0.03**	1.33	2.18
I	**diverted**	1.37	3.92	0.80	0.48	-0.15	1.30	0.08	1.31	2.17
Exp.[17]	diverted	1.4	3.8	0.78	-	-	-	-	1.3	1.85

Table 1. Caracteristics of the 5 equilibria created to pass from a concentric circular geometry to a geometry close to an experimental discharge. The value of κ at the last closed flux surface is 1.68 and the value of δ is 0.15 .

In GS2, the Miller flux surface shape is given by (in normalized units):

$$R(r,\theta) = R_0(r) + r\,cos[\theta + \delta(r) sin(\theta)] \quad (1)$$
$$Z(r,\theta) = \kappa(r)\, r\, sin(\theta) \quad (2)$$

The Miller-type flux surface shape of GYRO is given by:

$$R(r,\theta) = R_0(r) + r\,cos(\theta + arcsin(\delta(r)) sin(\theta)) \quad (3)$$
$$Z(r,\theta) = Z_0(r) + \kappa(r)\, r\, sin(\theta + \zeta(r) sin(2\theta)) \quad (4)$$

One can notice that $\delta_{GS2} = sin^{-1}(\delta_{GYRO})$. In both codes, $\kappa(r)$ and $\delta(r)$ are calculated using limited developments around r the flux surface considered. The codes also need in input $\kappa' = d\kappa/d(r/a)$ and $\delta' = d\delta/d(r/a)$. The values considered for the five cases are in table 2.

number	κ'	δ'
V	-0.015	0.002
IV	-0.050	0.007
III	0.205	-0.019
II	0.217	0.078
I	0.200	0.209

Table 2. Additional Miller parameters

The pressure profile used for the equilibrium reconstruction was based on the profiles presented in [17]. Using CHEASE input parameters, we can obtain from total pressure the n_e and T_e profiles, assuming $T_e = T_i$, $n_e = n_i$, $L_{ne}/L_{Te} = 3$, $T_{e0keV}/n_{e0keV} = 0.69$ and $p_e/p = 0.5$. In this way, we make sure the kinetic profiles are consistent with the total pressure used for the equilibrium. We obtain the profiles shown in Fig. 2. We also show the profiles of R/L_{ne}, R/L_{ne} and $\rho* = \rho_s/a$, with $c_s = \sqrt{T_e/m_i}$ and $\rho_s = c_s/\Omega_i$, assuming a deuterium plasma. The values near $r/a = 0.5$ are very close to the cyclone base case and thus can be used with global codes for realistic benchmarking. Note that with global codes, not only the profile gradients vary over the plasma radius, but also the $\rho*$ value. However, we have constructed the equilibrium such that the pressure profile is consistent with the local scalelengths at $r/a = 0.5$ used for the local calculations (within 10%). In addition, the q and magnetic shear profiles are very close to one another throughout the scan in equilibria as shown in Fig. 3. Therefore this set of equilibria and profiles is very well suited for global code benchmarking as well.

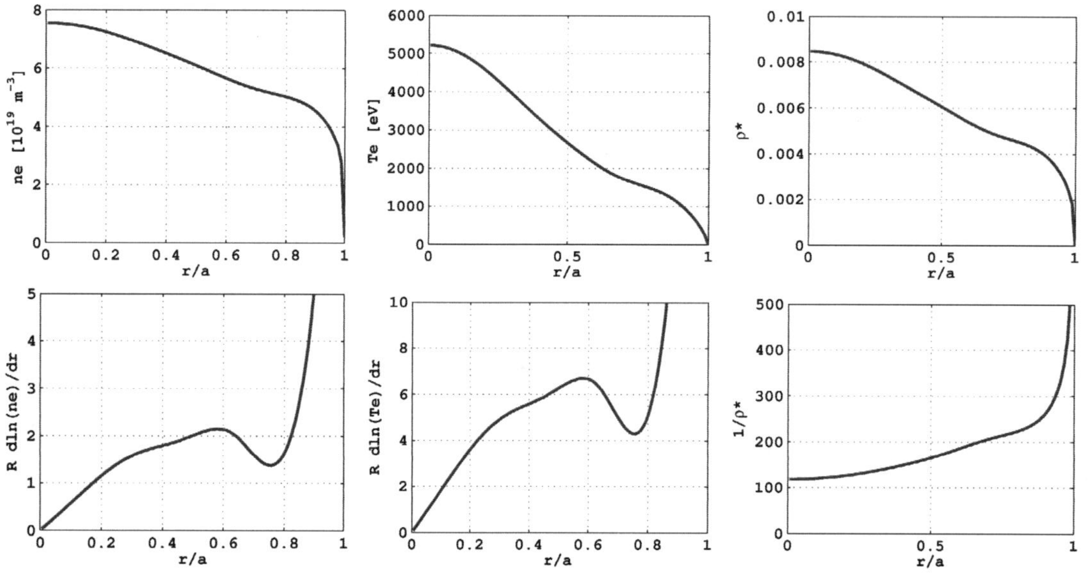

Figure 2. Profiles extracted from the pressure profile used for the equilibrium calculation with prescribed η and p_e/p with specific CHEASE input parameters.

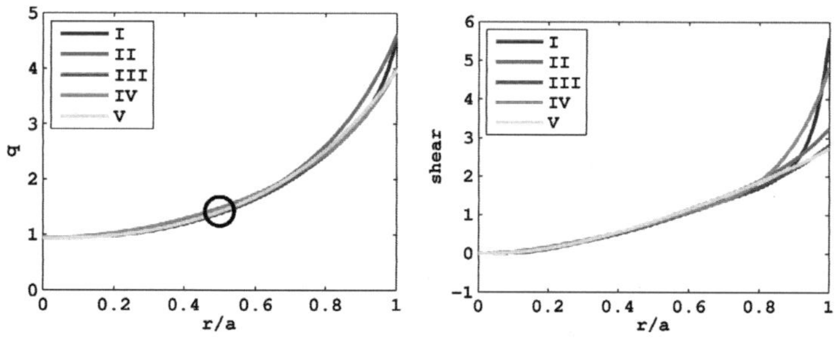

Figure 3. Safety factor q and magnetic shear s for the various equilibria.

4. Comparison of the three equilibrium models

There are basically three equilibrium models considered in this study: (1) the $s-\alpha$ model, (2) the Miller model, and (3) the full global MHD equilibrium. We compare these on the cyclone case, that is with $\alpha = 0$ and case V shown in Fig. 1. This is a similar comparison as performed in [10]. We find the same result, namely that the modes frequencies and growth rates obtained with the $s-\alpha$ model as implemented in GS2 and GYRO, does not match the ones obtained with a circular shaped configuration. The difference is explained in [10] and is due to an over-simplification of a metric term. On the other hand, both the Miller model and the full numerical equilibrium give the same results. In addition, for all cases, GS2 and GYRO give very similar results for both the ITG and the TEM branches. Note that the comparison performed here includes kinetic electrons, contrary to [2, 10]. It shows that the difference between the $s-\alpha$ model and the real geometry is not only with respect to the value of γ, but also the value of $k_y \rho_i$ at the transition between the ITG and TEM.

In Fig. 5, we compare these three equilibrium models with case IV as well, which has a finite value of $\alpha = 0.5$. We find the same effects as with case V, $\alpha = 0$, Fig. 4. The value $\alpha = 0.5$ is such as to be consistent with the profiles corresponding to the input physics parameters $R/L_{n,T}$.

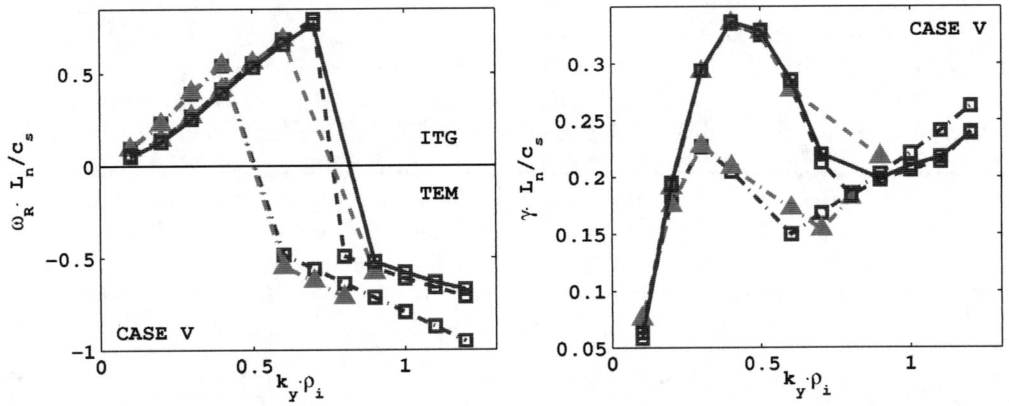

Figure 4. Real frequency (left) and growth rate (right) as a function of the poloidal wave vector k_y for a linear simulation of case V (Table 1). Solid line: numerical equilibrium, dashed line: Miller geometry and dashed-dotted line: $s - \alpha$ model. Red triangles: GYRO results, blue squares: GS2 results. The frequency follows GS2 convention, ion direction for $\omega > 0$.

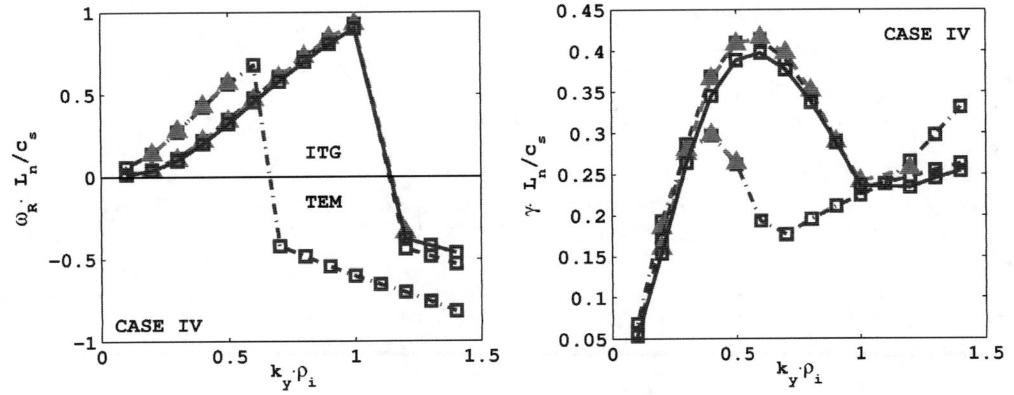

Figure 5. Same as in Fig. 4 but for the equilibrium case IV.

5. Miller and full equilibrium comparison at similar κ

In this Section we compare the Miller and numerical equilibrium models for the cases III, II and I, that is at fixed elongation and including finite δ, keeping up-down symmetry (case II), and including a full single-null down (SND) geometry with the same upper triangularity (case I). This is mainly to compare the codes with different equilibrium interface, within small modifications of the equilibrium, in order to have a better measure of the level of code and equilibrium model agreement.

Fig. 6 shows the modes frequencies and growth rates versus $k_y\rho_i$. Comparing solid lines (GS2-numerical equilibrium) with dashed lines (GS2-Miller), we see that there is essentially no difference for the up-down symmetric cases and a small difference (about 5%) for the SND case. The GYRO results, with Miller, match very well the GS2 results as well, even for the SND case although the Miller representation is slightly different. We did not have time to compare with GYRO and the full MHD equilibrium and this will be performed at a later stage.

With respect to the influence of small shaping effects on the modes, we see that adding an edge $\delta = 0.15$ does not affect the modes much at mid-radius, as expected from Ref. [18]. On the other hand, the SND geometry has higher maximum growth rates. This is because the effective elongation decreases faster from the edge and is therefore smaller at $r/a = 0.5$ as seen in table 1. Since elongation is stabilising, this yields larger growth rates for the SND case. Note that with case I, there is a small difference between the Miller and numerical equilibrium, with GS2. This difference is larger than for cases II and III and is therefore significant. It is due to the non up-down symmetric shape of the flux surface, which can be encapsulated only with the numerical equilibrium, and the larger triangularity at the bottom. However the effect is small and Fig. 6 shows that the main parameter which needs to be matched is the flux surface elongation.

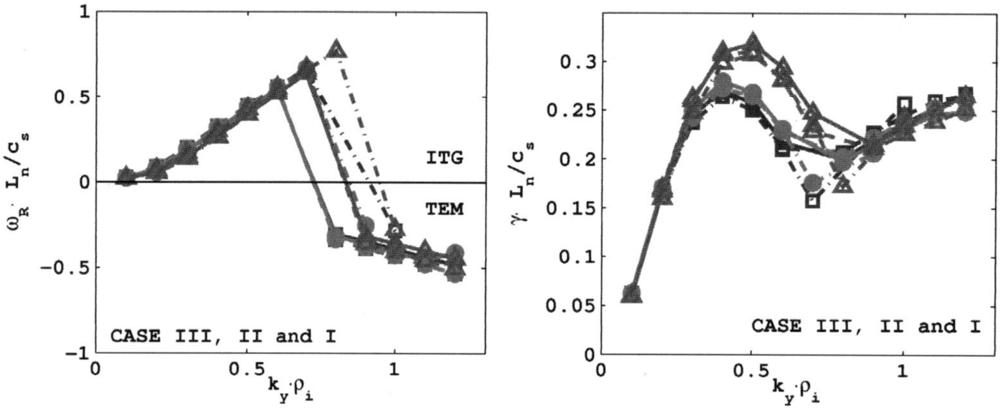

Figure 6. Real frequency (left) and growth rate (right) spectra for cases III (blue squares), II (red circles) and I (green triangles). Solid line: GS2 results with numerical equilibrium, dashed line: GS2 results with Miller geometry and dash-dotted line: GYRO results with Miller geometry.

6. From circular to single-null equilibria and the effects on ITG and TEM

We can now look at the general effects of shaping on the ITG and TEM modes when the global MHD equilibrium is modified from circular-non-shifted (V), to circular shifted (IV), to elongated-shifted (III), to elongated-shifted with triangularity (II) and finally to SND-shifted (I), while keeping the local value of ϵ, q and s constant as well as the global values of q_0 and q_{95}. This suite of equilibria mimics the various approximations performed when comparing with an actual experimental case (SND) and an "$s - \alpha$" equivalent case. Note that we do not compare with the $s - \alpha$ model since it is incomplete, as discussed above.

Fig. 7 shows the modes frequencies and growth rates versus $k_y\rho_i$ obtained with GS2 and the five numerical equilibria shown in Fig. 1. Comparing just the "experimental" case I and the simplest case V, circular-α=0, one could have concluded that there is little effects of shaping and no reason to use a realistic equilibrium. Indeed, both curves are nearby for both the frequency and the growth rate. However this is a coincidence and can be explained through the various steps. First, including a finite α value consistent with the imposed profiles, the modes are destabilized. This is also somewhat coincidental since the effect of α can be either stabilizing or destabilizing dependending on the magnetic shear as will be discussed below.

Adding elongation is strongly stabilizing and therefore the high growth rates of case IV are brought down by case III to the lowest values of the various cases (Fig. 7) and below the starting circular-α=0 case. The addition of positive triangularity, case II, only slightly destabilizes the modes, as discussed before. On the other hand, the final step, using a SND equilibrium which reduces the effective plasma elongation inside the plasma, is destabilizing and brings the growth rates back up to near the levels of the circular-$\alpha = 0$ case.

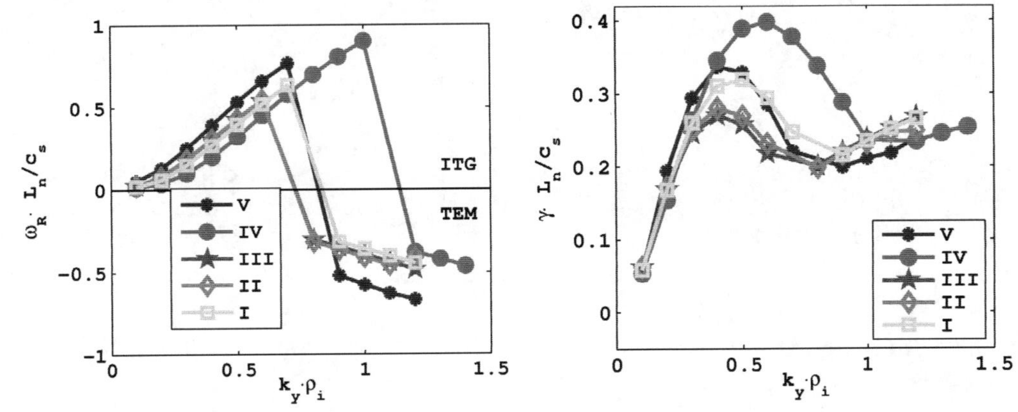

Figure 7. Real frequency (left) and growth rate (right) spectra for cases V, IV, III, II and I. Results obtained with GS2 and numerical equilibria.

The destabilization with α can be surprising since it is usually said that it is stabilizing. Actually, from the Ballooning representation one sees that it is $(s - \alpha)$ which is important. In addition, it is well known that the growth rate has a maximum near $(s - \alpha) = 0.5$. Therefore, depending on the starting parameters, increasing α can be either stabilizing or destabilizing, except if very large values of α are added, in which case it will be stabilizing. To have a better idea of the situation with this set of parameters, namely the cyclone case with kinetic electrons, we have performed a scan of shear and α values and determined the value of γ of the most unstable mode at $k_y\rho_i = 0.4$. The contour plot for the cyclone case with kinetic electrons and $s - \alpha$ model, for simplicity, is shown in Fig. 8(a). One clearly sees that not only the maximum value, but the whole contour lines align essentially along $(s - \alpha)$. The black line indicates that actually $s = 0.28 + 0.65\alpha$ reflects better the location of $\max(\gamma)$ at $k_y\rho_i = 0.4$. Assuming adiabatic electrons, we obtain almost the same slope for the ridge, namely $s = 0.37 + 0.63\alpha$. In Fig. 8(b), we show the case with $q = 2.84$ instead of $q = 1.42$. The fit changes slightly to $s = 0.315 + 0.70\alpha$, but the contours remain self-similar.

In Fig. 9 we then plot both γ and the mode frequency versus $(s - 0.65\alpha - 0.28)$, for various values of s, for the cyclone case with kinetic electrons. We see that all the curves align indeed very well and the offset of 0.28 is such as to bring the maximum γ's at $x = 0$. In this way one sees that the mode frequency is not at a minimum at this point. Of course, $(s - 0.5\alpha)$ would look almost as good, similarly to $(s - \alpha)$, which explains why either have been mentioned in the literature. The value of 0.65 is not universal and depends on the plasma parameters, as seen from the examples above. Note that the cyclone case has $s = 0.78$ and $\alpha = 0$, leading to $s - 0.65\alpha - 0.28 = 0.5 > 0$. This is why adding $\alpha = 0.5$, yielding $s - 0.65\alpha - 0.28 = 0.175$, one gets closer to the maximum and α is destabilizing. Figs. 8 and 9 show that the benchmark case

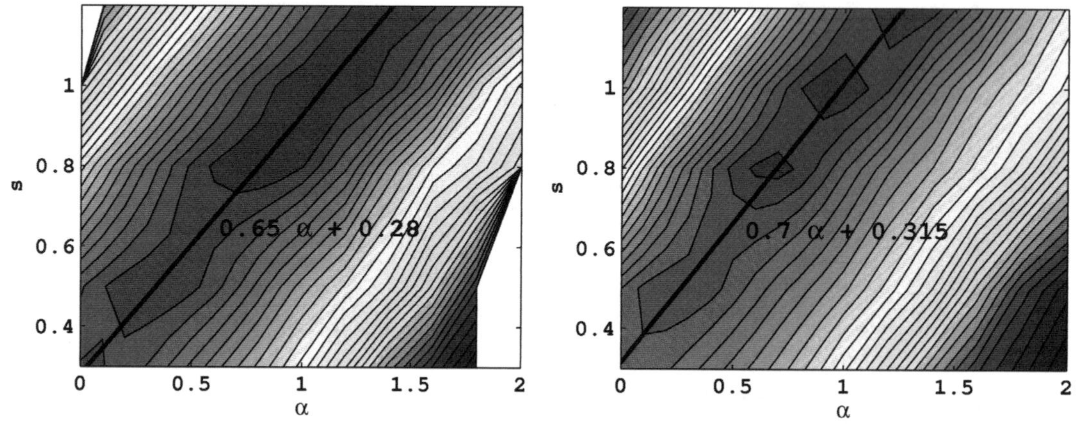

Figure 8. Contours of γ of the most unstable mode at $k_y\rho_i = 0.4$ for different values of s and α, using GS2 and the $s-\alpha$ model. (cyclone case with kinetic electrons and finite α. (b) same but with $q = 2.84$ instead of 1.42.

is close to the ridge and therefore small differences in the implementation of s or α in the codes might not be seen. Therefore, the comparison of a full scan is useful.

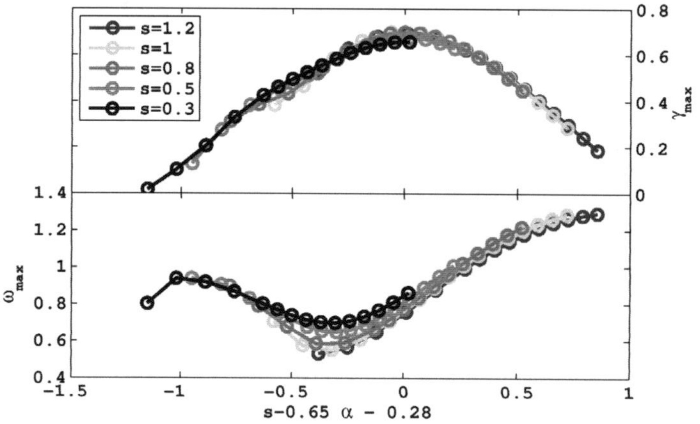

Figure 9. Growth rate (top) and real frequency (bottom) of the cases shown in Fig. 8(a).

7. Conclusion

We have extended the cyclone benchmark case [2] to include kinetic electrons, self-consistent finite α values and realistic magnetic geometry. This has lead to the construction of five equilibria in order to be able to compare with circular shape simulations (case V) up to full SND geometry (case I). The intermediate cases allow to check the equilibrium interface and to understand the effects of adding elongation, triangularity and up-down asymmetry. The calculation of the five equilibria is not unique. We have constrained the pressure profile to be consistent with the values of $R/L_{n,T}$ and with the original experimental data [17]. We have also imposed $q_0 = 0.95$

and the value of q_{95}, since this allows also to be closer to the comparison with experimental results.

We have analyzed the linear electrostatic results using the GS2 and GYRO codes, including three equilibrium models in the codes: $s-\alpha$, Miller [3], and MHD equilibrium (calculated with CHEASE in our case [1]). We have confirmed that the $s-\alpha$ model, as implemented in the codes, cannot reproduce the results with the circular non-shifted equilibrium (case V), as found in [10]. This difference remains important with kinetic electrons and finite α. We have also shown that both codes give similar results in all cases, and that the difference in the Miller model does not lead to significant differences with these parameters.

Both the ITG and TEM modes have been studied and their growth rates and mode frequencies vary with each equilibrium. Due to the original parameters of the cyclone case, adding a self-consistent α value to the circular equilibrium is destabilizing (case IV), while adding elongation is stabilizing (case III). The addition of triangularity on top of elongation (case II) has little effect and one would need a much larger triangularity to see a significant effect [18]. Finally, using a SND geometry (case I) with the same elongation and top triangularity as case II, is destabilizing. This is due to the reduction of the effective elongation of the core flux surfaces with respect to the up-down symmetric case. This could explain why double-null geometries tend to have better properties. In any case, it shows that the variations through these five equilibria are very instructive both for code benchmark purposes and to understand the effects of various assumptions when gyrokinetic simulations are applied to compare with experimental results.

These results and code comparisons are preliminary and for linear simulations only. The next step will be to compare nonlinear results. We have also only used local simulations, flux tube geometry at $r/a = 0.5$. However the framework of this set of equilibria and physics parameters are such as to allow the comparison with global simulations as well. The relevant files can be found in [19].

Acknowledgments

The authors thank M. Kotschenreuther and W. Dorland for having made available the GS2 code, and J Candy "and team" for the GYRO code. The simulation have been run on the Pleiades2 cluster at EPFL Lausanne, on the HPC-FF cluster at Juelich and on the vip machine at IPP-Garching. This work is partly supported by the Swiss National Science Foundation.

References

[1] H. Lütjens, A. Bondeson and O. Sauter, Comp. Phys. Comm. **97** (1996) 219.
[2] A. M. Dimits *et al*, Phys. Plasmas **7** (2000) 969.
[3] R. L. Miller *et al*, Phys. Plasmas **5** (1998) 973.
[4] M. Kotschenreuther, G. Rewoldt, and W. M. Tang, Comput. Phys. Commun. **88** (1995) 128.
[5] W. Dorland, F. Jenko, M. Kotschenreuther, and B. N. Rogers, Phys. Rev. Lett. **85** (2000) 5579.
[6] J. Candy, R.E. Waltz , J. Comp. Phys. **186** (2003) 545.
[7] E. A. Belli, G.W. Hammett, and W. Dorland, Phys. Plasmas **15** (2008) 092303.
[8] P. Angelino *et al*, Phys. Rev. Lett. **102** (2009) 195002.
[9] G. Falchetto *et al*, Plasma Phys. Contr. Fus. **50** (2008) 124015.
[10] X. Lapillonne *et al*, Phys. Plasmas **16** (2009) 1.
[11] B. F. McMillan *et al*, Phys. Rev. Lett., accepted (2010).
[12] E. Fable, C. Angioni and O. Sauter, Plasma Phys. Control. Fusion **52** (2010) 015007.
[13] T. Görler, Ph.D. thesis University of Ulm, Germany, 2009.
[14] X. Lapillonne *et al*, in Proc. of the 2010 Varenna-Lausanne conf., Journal Conf. Proc. Series 2010.
[15] C. Bourdelle *et al*, Nuclear Fusion **45** (2005) 110.
[16] A. Bottino *et al*, Plasma Phys. Control. Fusion **48** (2006) 215.
[17] C. M. Greenfield *et al*, Nuclear Fusion **37** (1997) 1215.
[18] A. Marinoni *et al*, Plasma Phys. Control. Fusion **51** (2009) 055016.
[19] http://crpp.epfl.ch/~sauter/benchmark

Journal of Physics: Conference Series **260** (2010) 012007 doi:10.1088/1742-6596/260/1/012007

Analytical and numerical studies of the cold electromagnetic LH wave equation in the mode conversion regime

A Cardinali, V Fusco

Associazione Euratom-ENEA sulla Fusione, C.P. 65 - I-00044 - Frascati, Rome, Italy

E-mail: valeria.fusco@enea.it

Abstract. The purpose of this study is to investigate the effect of the presence of the mode conversion ($n_z<n_{z\ critical}$), on the evolution of the electric field of the lower hybrid (LH) propagating wave. The propagation of the LH wave in a magnetized plasma, for both mode polarization (slow and fast), is described by a fourth order ordinary differential equation, which here has been analyzed analytically and numerically. A complete WKB solution of the 4th order equation has also been obtained showing the failure of the method near the cut-offs and the mode conversion point.

1. Introduction.

The study and the analysis of the lower hybrid waves (LH) are a broad and interesting field of research. The LH waves [1], in fact, exhibit the highest current drive efficiency at low plasma temperature thus allowing the control of the safety factor (q) profile: this is a desirable feature in ITER where it is important to maintain the burning plasma performance on a long time scale. Moreover the LH waves are suitable for scenarios based on internal transport barrier [2]; their control on q can produce a safety factor profile with negative magnetic shear thus rising an internal barrier to the electron energy transport.

The difficulties in coupling the LH waves in severe edge condition, the concern of an array structure near the plasma and the impossibility for the LH of penetrating the core of a Q=10 ITER plasma, excluded the LH from the ITER first phase heating system. Anyway, the advantages mentioned above encourage the effort devoted to the study of the LH waves in many tokamak experiments like JET [1], Tore Supra [3], JT-60U [4], FTU [5].

Of course, the efficient transfer of electromagnetic power to the plasma is an important goal. First, the power is transferred from the radiofrequency generator to the plasma; this can be achieved if the generator impedance is made equal to the conjugate of the plasma impedance. Second, the electromagnetic wave, propagating in the plasma, has to transfer its energy to electrons or ions. In real plasmas, the non-uniformity of magnetic field and the plasma densities give rise to critical layers where dissipation may occur with a consequent loose of efficiency in energy transfer. In the region of this critical layer, the wave may be reflected (cut-off), absorbed (resonance) and converted into a companion mode (mode conversion).

LH waves, generally, have two propagating modes depending on the polarization: the slow and the fast modes characterized by an E_z/E_x and E_y/E_x wave field polarization respectively. Usually the LH antenna (waveguide grill) is able to launch the slow wave mode, with a power spectrum (power vs parallel wave-number, n_z), which depends on the structural elements of the antenna. The parallel wave

Journal of Physics: Conference Series **260** (2010) 012007 doi:10.1088/1742-6596/260/1/012007

number changes when the wave is propagating into the plasma, and if the condition $n_z \leq n_{z\text{-crit}}$ is encountered, part of the slow-wave power can be transformed into the fast wave (mode conversion). Diffraction effects, localized near the mode conversion layer, can play a crucial role in spreading the power to higher components of the parallel wave-number spectrum, and consequently induce a more effective and peripheral wave absorption. An analytical and numerical approach to the problem is presented in the following section. The full electromagnetic wave equation (a fourth order ordinary differential equation for the electric field) has been analytically and numerically solved in order to verify this conjecture. A solution based on the WKB expansion [6] has also been obtained in order to show the failure of this method near the cut-offs and the mode conversion.

To simplify the study, it will be assumed that the wave propagates in a plane-stratified plasma and the wavelength λ of the radiation is much smaller than the characteristic length L of the non-homogeneities (λ<<L).

2. Wave equation for an LH wave

It is possible to obtain the wave equation valid for the LH waves from the Maxwell-Vlasov system of equations

$$\nabla \times \nabla \times \underline{E}(\underline{r},\omega) - \frac{\omega^2}{c^2}\underline{\underline{\varepsilon}}(\underline{r},\omega)\cdot \underline{E}(\underline{r},\omega) = 0 \tag{1}$$

where $\underline{\underline{\varepsilon}}(\underline{r},\omega) = \underline{\underline{I}} + j\frac{4\pi}{\omega}\underline{\underline{\sigma}}(\underline{r},\omega)$ is the cold dielectric tensor,

$$\underline{\underline{\varepsilon}} = \begin{pmatrix} \varepsilon_{xx}(x) & \varepsilon_{xy}(x) & 0 \\ -\varepsilon_{xy}(x) & \varepsilon_{xx}(x) & 0 \\ 0 & 0 & \varepsilon_{zz}(x) \end{pmatrix} = \begin{pmatrix} S(x) & -jD(x) & 0 \\ jD(x) & S(x) & 0 \\ 0 & 0 & P(x) \end{pmatrix}$$

$\underline{\underline{I}}$ the unit tensor, $\underline{\underline{\sigma}}$ the conductibility and E is the oscillating electric field of the wave. Note that in Stix's notations the elements of the cold plasma dielectric tensor are:

$$S = 1 - \sum_s \frac{\omega_{ps_0}^2}{\omega^2 - \omega_{cs_0}^2 g(\vec{r})} f(\vec{r})$$

$$D = \sum_s \frac{\omega_{cs}}{\omega} \frac{\omega_{ps}^2}{\omega^2 - \omega_{cs_0}^2 g(\vec{r})}$$

$$P = 1 - \sum_s \frac{\omega_{ps_0}^2}{\omega^2} f(\vec{r})$$

and f and g are suitable space functions of density and external magnetic field, and ω_{ps}, ω_{cs} are the plasma and cyclotron frequencies for the species s.

Several assumptions have been made to obtain the above equation from the Maxwell-Vlasov system. First of all it has been considered the field amplitude sufficiently small so that $W_{wave} \equiv |E|^2 << W_{thermal} \equiv \kappa T$; this implies that the kinetic equation can be linearized. The second assumption concerns the possibility of assuming the local thermodynamic equilibrium, while the third assumption relies on the drift orbit approximation (weak space non-homogeneity). Moreover the steady state (Fourier analysis in time) and the cold plasma limit have been assumed. It's worth noting, although of numerous simplifying assumptions, the propagation of the electromagnetic field in a non-homogenous medium is not a straightforward problem [7], [8] because the wave equation has non-constant coefficients.

Fourier transforming from space coordinate (y,z) domain to the wave vector (k_y,k_z) domain and assuming $k_y = 0$ (grill-antenna radiation pattern) for slab geometry, the electric field becomes $\underline{E}(\underline{r},\omega) = \underline{E}(x,\omega)e^{j(k_y y + k_z z)}$ and the wave equation (1), in matrix form, reads:

$$\begin{pmatrix} k_z^2 & 0 & jk_z \dfrac{d}{dx} \\ 0 & k_z^2 - \dfrac{d^2}{dx^2} & 0 \\ jk_z \dfrac{d}{dx} & 0 & -\dfrac{d^2}{dx^2} \end{pmatrix} \begin{pmatrix} E_x \\ E_y \\ E_z \end{pmatrix} - \frac{\omega^2}{c^2} \begin{pmatrix} \varepsilon_{xx} & \varepsilon_{xy} & 0 \\ -\varepsilon_{xy} & \varepsilon_{xx} & 0 \\ 0 & 0 & \varepsilon_{zz} \end{pmatrix} \begin{pmatrix} E_x \\ E_y \\ E_z \end{pmatrix} = 0 \quad \textbf{(2)}$$

If B lies along the z-axis and in the hypothesis of slightly inhomogeneous plasma, equation (2) can be expanded to obtain:

$$\begin{cases} \dfrac{d^2E_y}{dx^2} + \dfrac{k_z D}{(n_z^2 - S)} \dfrac{dE_z}{dx} - \dfrac{\omega^2}{c^2} \dfrac{(n_z^2 - S)^2 - D^2}{(n_z^2 - S)} E_y = 0 \\ \dfrac{d^2E_z}{dx^2} + \dfrac{k_z D}{S} \dfrac{dE_y}{dx} - \dfrac{\omega^2}{c^2} \dfrac{(n_z^2 - S)P}{S} E_z = 0 \end{cases} \quad (3)$$

where $n_z = k_z c/\omega$ is the refractive index and $E_x = (-jDE_y - jk_z \frac{c^2}{\omega^2} dE_z/dx)/(n_z^2 - S)$. Equation (3) will be written using normalized quantities as follow:

$$\begin{aligned} &\frac{d^2y(\bar{x})}{d\bar{x}^2} + \delta_0^{-1}\alpha(\bar{x})\frac{dz(\bar{x})}{d\bar{x}} + \delta_0^{-2}\beta(\bar{x})y(\bar{x}) = 0 \\ &\frac{d^2z(\bar{x})}{d\bar{x}^2} + \delta_0^{-1}\alpha'(\bar{x})\frac{dy(\bar{x})}{d\bar{x}} + \delta_0^{-2}\beta'(\bar{x})z(\bar{x}) = 0 \end{aligned} \quad (4)$$

where

$$\begin{aligned} &\bar{x} = \frac{x}{a} && \delta_0^{-1} = \frac{\omega a}{c} \\ &y(\bar{x}) = \frac{E_y(\bar{x})}{\sqrt{E_x^2 + E_y^2 + E_z^2}} && z(\bar{x}) = \frac{E_z(\bar{x})}{\sqrt{E_x^2 + E_y^2 + E_z^2}} \\ &\alpha(\bar{x}) = \frac{n_z D}{n_z^2 - S} && \alpha'(\bar{x}) = \frac{n_z D}{S} \\ &\beta(\bar{x}) = -\frac{(n_z^2 - S)^2 - D^2}{n_z^2 - S} && \beta'(\bar{x}) = -\frac{(n_z^2 - S)P}{S} \end{aligned} \quad (5)$$

a is the plasma radius.

3. Dispersion relation and accessibility condition

The coefficients of equations (4) are not constant because the plasma and cyclotron frequency are functions of the magnetic field and the plasma density. These quantities can be chosen to change along x as follow, (figure 1):

Journal of Physics: Conference Series **260** (2010) 012007 doi:10.1088/1742-6596/260/1/012007

$$B_z(\bar{x}) = B_0(1 + \frac{a}{R}\bar{x})^{-1}$$

$$n(\bar{x}) = N\left(1 - (M\bar{x})^2\right)^{\alpha}$$

where B_0 is the magnetic field in the center of the plasma, R is major radius and N and M are constants whose values determine the plasma density in the center of the plasma and the plasma density variation along x.

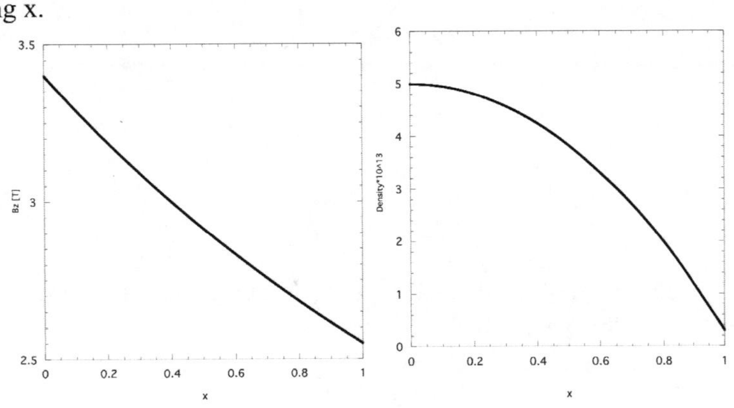

Figure 1. Magnetic field (right hand side) and density along the slab coordinate.

For this reason the perpendicular wave number k_x (or the refractive index n_x) changes when the wave crosses the inhomogeneous plasma, while the parallel k_z, remains constant and settled by the antenna periodicity. In realistic cases the magnetic field and the metric of the chosen geometry depend, generally, by two space variables (e.g. radius and poloidal angle when assuming a pseudo-toroidal geometry). This means that the parallel wave number is varying too; obviously in this case the space Fourier analysis in z direction is not yet allowed, and a partial differential system of equations must be considered instead of equation (4). In order to investigate the mode conversion phenomenon, and due to the weak variation of the parallel wave number in a 2D realistic geometry [6], we can assume the invariance of the parallel wave number. The dispersion relation and the associated accessibility condition describes how k_x changes; Fourier transforming equations (4), knowing that $d/d\bar{x} \rightarrow j\delta_0^{-1} n_x$ and $d^2/d\bar{x}^2 \rightarrow -\delta_0^{-2} n_x^2$ one obtains:

$$\begin{cases} (-n_x{}^2 + \beta)y(\bar{x}) + j\alpha n_x z(\bar{x}) = 0 \\ j\alpha' n_x y(\bar{x}) + (-n_x{}^2 + \beta')z(\bar{x}) = 0 \end{cases}$$

The previous system can be solved if the determinant of the coefficients is equal to zero, that is:

$$n_x{}^4 - (\beta + \beta' - \alpha\alpha')n_x{}^2 + \beta\beta' = 0$$

Solving the fourth order equation:

$$n_{x\pm}{}^2 = \frac{\beta + \beta' - \alpha\alpha' \pm \sqrt{(\beta + \beta' - \alpha\alpha')^2 - 4\beta\beta'}}{2} \tag{6}$$

which are the perpendicular refractive indexes (or wave number) for the slow wave (plus sign) and the fast wave (minus sign).

As one can't have damped oscillations, $n_{x\pm}^2$ has to be purely propagative ($n_{x\pm}^2 > 0$) or purely evanescent ($n_{x\pm}^2 < 0$) and this imposes a condition on the determinant of equation (6) establishing the accessibility condition as follow:

$$(\beta + \beta' - \alpha\alpha')^2 - 4\beta\beta' \geq 0 \tag{7}$$

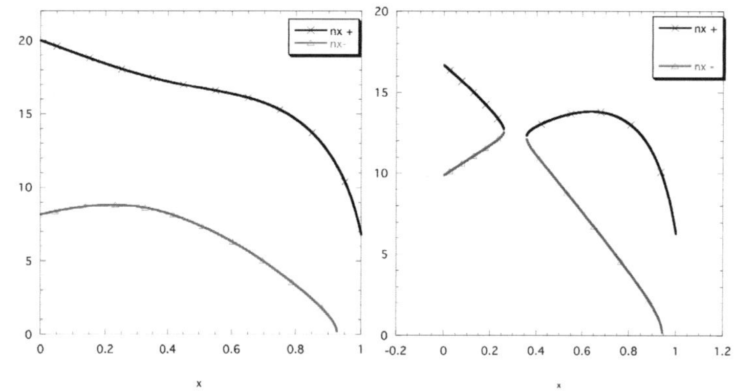

Figure 2. Slow (black-cross) and fast (red-triangle) refractive indexes along the slab coordinate, when the accessibility condition is fulfilled (left hand side) and when it is not (right hand side).

The accessibility condition, in fact, is the condition for avoiding the coalescence of the roots of the dispersion relation. This is depicted on the right hand side of figure 2, where the mode conversion happens around the slab coordinate x=0.25 and x=0.35.

If one retains the equal sign and explicits the β, β', α, α' using equation (5), the n_z index can be expressed as a function of x just solving the following fourth degree algebraic equation:

$$(S-P)^2(n_z^2-S)^2+2(S+P)D^2(n_z^2-S)+(D^4+4PSD^2)=0$$

This equation can be solved analytically for n_z, by considering that $S << P$ (this condition is valid everywhere into the plasma, except the cut-off point $P \approx 0$ at very low density), so we have

$$n_{z-crit}=\sqrt{S}+\frac{D}{\sqrt{P}}$$

This condition fixes a relation between n_z (fixed by the antenna periodicity) and the space variable x because S, D and P depend on the space variable x through the density and the magnetic field. The region where the accessibility condition $\left(n_z > n_{z-crit}\right)$ is fulfilled is represented in the green shadow of figure 3.

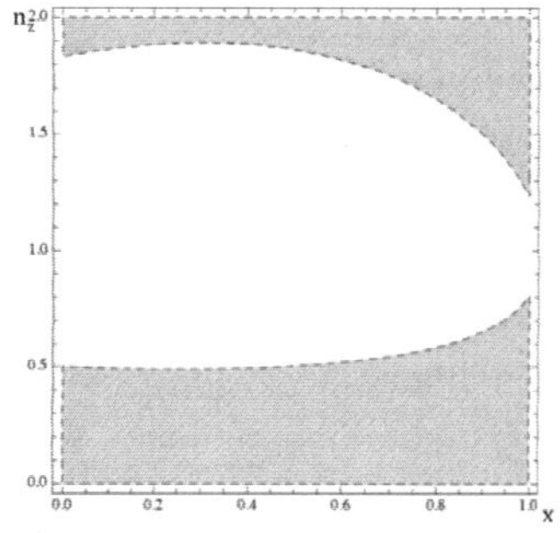

Figure 3. Accessibility condition represented by the shadow.

The same equation can be solved in terms of the space variable x, instead of n_z: in this case, it is possible to have an analytical expression that shows the critical layer x_{crit} (where the roots of the dispersion relation have a coalescence) as function of n_z. To this purpose, let's consider the following approximated expressions for S, D and P:

$$S \approx 1; \; P \approx -\frac{\omega_{pe}^2(x)}{\omega^2}; \; D \approx \frac{\omega_{pe}^2(x)}{\omega\Omega_{ce}(x)} = \frac{\omega}{|\Omega_{ce}|}\frac{\omega_{pe}^2(x)}{\omega^2} = -dP(x)$$

where d has been considered as constant. The solution is given in terms of P (which is an explicit function of x)

$$\left(\sqrt{|P|}\right)_{crit} = \frac{(n_z - 1)}{|d|} = (n_z - 1)\frac{|\Omega_{ce}|}{\omega}$$

Inserting in the above equations n_z, and the plasma parameters (central density and magnetic field) used to obtain figure 2 (right hand side), we find $x_{crit} \approx 0.25$ and $x_{crit} \approx 0.35$, which corresponds to the exact values found by solving numerically equation (7).

4. Solution of the LH wave equation

The second order coupled ordinary differential equation system can be combined to obtain the following fourth order differential equations:

$$\begin{aligned} &\frac{dy(\bar{x})^4}{d\bar{x}^4} + \delta_0^{-2}(\beta + \beta' - \alpha\alpha')\frac{dy(\bar{x})^2}{d\bar{x}^2} + \delta_0^{-4}\beta\beta' y(\bar{x}) = 0 \\ &\frac{dz(\bar{x})^4}{d\bar{x}^4} + \delta_0^{-2}(\beta + \beta' - \alpha\alpha')\frac{dz(\bar{x})^2}{d\bar{x}^2} + \delta_0^{-4}\beta\beta' z(\bar{x}) = 0 \end{aligned} \tag{8}$$

Just to have an idea of the behavior of the solution, it is supposed that the coefficients of equation (8) are constant (indeed they are weakly varying with space). The above differential equation can be solved analytically, obtaining:

$$y(\bar{x}) = c_1 e^{j\delta_0^{-1} n_{x+}\bar{x}} + c_2 e^{-j\delta_0^{-1} n_{x+}\bar{x}} + c_3 e^{j\delta_0^{-1} n_{x-}\bar{x}} + c_4 e^{-j\delta_0^{-1} n_{x-}\bar{x}} \tag{9}$$

where c_1, c_2, c_3, c_4 are constants to be determined with the initial conditions. Of course the same solution holds for the slow wave. It is obvious that the nature of the solution depends on the reciprocal value of the coefficients of the above equation; thus if the accessibility condition is not fulfilled the solution is evanescent, on the contrary if the accessibility condition is fulfilled the solution is oscillating ($n_{x\pm}^2 > 0$) or evanescent ($n_{x\pm}^2 < 0$).

The solution (9), in the oscillating case, is plotted in figure 4 for the slow and the fast wave:

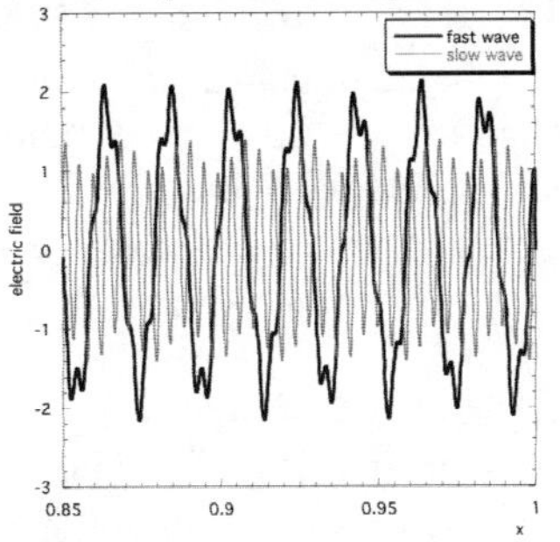

Figure 4. Transverse (fast wave) and longitudinal (slow) normalized electric field.

In particular the relevant case:

$$\delta_0^{-4}\left(\beta+\beta'-\alpha\alpha'\right)^2 = 4\delta_0^{-4}\beta\beta'$$

has been considered. This relation means, as seen before, that the roots of the dispersion relation coalesce and the LH slow wave converts into the LH fast wave. In this case the above equation assumes the form:

$$\frac{d^4y(\bar{x})}{d\bar{x}^4} + 2\delta_0^{-2}\sqrt{\beta\beta'}\frac{d^2y(\bar{x})}{d\bar{x}^2} + \delta_0^{-4}\beta\beta' y(\bar{x}) = 0$$

and the solution can be written as

$$y(\bar{x}) = (c_1+c_2\bar{x})\cos(\delta_0^{-1}n_x\bar{x}) + (c_3+c_4\bar{x})\sin(\delta_0^{-1}n_x\bar{x}) \quad \text{if } n_x^2 = \sqrt{\beta\beta'}$$

$$y(\bar{x}) = (c_1+c_2\bar{x})e^{\delta_0^{-1}n_x\bar{x}} + (c_3+c_4\bar{x})e^{-\delta_0^{-1}n_x\bar{x}} \quad \text{if } n_x^2 = -\sqrt{\beta\beta'} \qquad (10)$$

Finally for slightly inhomogeneous plasma, the length of the non-homogeneities is greater than the wavelength of the radiation involved (λ<<L). Thus a WKB approximation (Wentzel, Kramers, Brillouin) can be applied to the system (8). Making the following ansatz:

$$y(\bar{x}) = y_0(\bar{x})e^{j\delta_0^{-1}S_1(\bar{x})}$$

$$z(\bar{x}) = z_0(\bar{x})e^{j\delta_0^{-1}S_1(\bar{x})}$$

Substituting in equation (8) and retaining only the leading terms in the δ_0^{-1} expansion, one obtains:

$$y(\bar{x}) \approx w_1\left(n_{x+}^2-\frac{\gamma}{2}\right)^{-1/2} n_{x+}^{-1/2}\exp\left(j\delta_0^{-1}\int n_{x+}dx\right) + w_2\left(n_{x+}^2-\frac{\gamma}{2}\right)^{-1/2} n_{x+}^{-1/2}\exp\left(-j\delta_0^{-1}\int n_{x+}dx\right)$$

$$+w_3\left(n_{x-}^2-\frac{\gamma}{2}\right)^{-1/2} n_{x-}^{-1/2}\exp\left(j\delta_0^{-1}\int n_{x-}dx\right) + w_4\left(n_{x-}^2-\frac{\gamma}{2}\right)^{-1/2} n_{x-}^{-1/2}\exp\left(-j\delta_0^{-1}\int n_{x-}dx\right)$$

where w_1, w_2, w_3, w_4 are constants to be determined with the initial conditions and $\gamma = \beta+\beta'-\alpha\alpha'$. The same solutions hold for z.

It's worth noting the WKB approximation fails when the accessibility condition is not fulfilled (n_z=$n_{z\ crit}$ means that $(\beta+\beta'-\alpha\alpha')^2 - 4\beta\beta' = 0$) and in the cut offs ($n_x \to 0$) where the condition of a slightly varying medium is not more satisfied ($\lambda << \left(dn_e/n_e dx\right)^{-1} \approx L$).

5. Numerical solution of the system

The ordinary differential equation system, described by equation (8), has also been solved by means of a numerical scheme based on the finite difference method. In this case the complication arises when considering the coupling terms $\alpha(\bar{x})$ and $\alpha'(\bar{x})$; the resulting matrix to be inverted loses its tridiagonal peculiarity and becomes a banded matrix of seven diagonals. A routine of the IMSL mathematical library has been used to invert this matrix; this routine solves a system of linear algebraic equations, which have a real banded coefficient matrix. First it uses a routine to compute an "LU" factorization of the coefficient matrix and to estimate the condition number of the matrix: large condition number means the matrix is poorly invertible and the accuracy of solution of a linear system may be bad.

The results are shown in figure 5. The electric field for both polarization E_y and E_z is plotted vs the slab variable x when both wave are propagative, n_z=2 (figure 5-left hand side), and when the mode conversion is present in the plasma, n_z=1.7 (figure 5 right hand side).

It is worth noting that the numerical solution, when the accessibility condition is satisfied (propagative slow and fast modes), is very similar to the analytical solution obtained for constant coefficients (equation (9)).

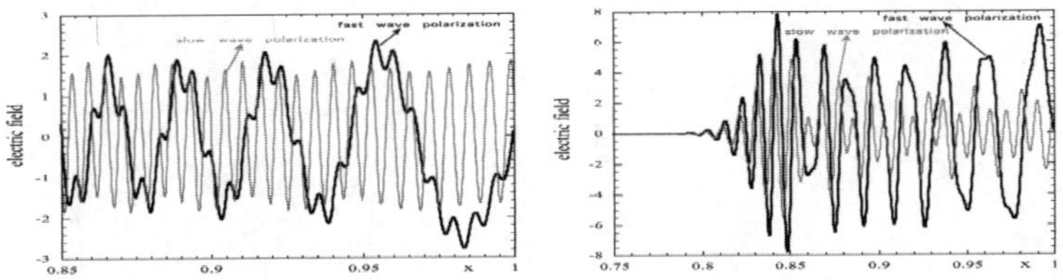

Figure 5. Transverse and longitudinal normalized electric field for (left hand side) propagative modes and (right hand side) when a confluence between fast and slow modes is present.

6. Conclusions

An analysis of the fourth order ordinary differential equation, describing the propagation of the LH wave for both polarizations (slow and fast), has been performed when a confluence point is located inside the plasma (mode conversion layer).

The equation has been studied analytically and numerically and the wave oscillating electric field has been determined before and after the mode conversion.

A complete analytical expression for the electric field has also been obtained when treating the wave equations by means of the WKB approximation, showing the failure of the solutions near the cut-offs and mode conversion points. This problem has been examined thoroughly in order to establish the behavior of the propagating slow wave in presence of the mode conversion, the role played by diffraction in the off-axis absorption of the wave and the consequent loose of efficiency in energy transfer.

7. References

[1] Tuccillo A A *et al.* 2005 Plasma Phys. Controlled Fusion **47**, B363
[2] Mailloux J *et al.* 2001 *Proceedings of the 28th EPS Conference on Controlled Fusion and Plasma Physics*, Funchal, Madeira, Portugal, 2001, edited by R. M. Pick and C. Bastian EPS, Vol 25A, p. 793.
[3] Peysson Y *et al.* 2001 Nucl. Fusion **41**, 1703
[4] Ide S *et al.* 1996 Proceed. Of the 16th Int. Conf. on Fusion Energy, Montreal IAEA-CN-64/E-3
[5] Cesario R *et al.,* Nat. Commun., **1,** 1:55 doi: 10.1038/ncomms1052 (2010).
[6] Cardinali A *et al.* 2007 Phys. Plasmas **14** 112506.
[7] Stix T H 1992 Waves in Plasma AIP
[8] Brambilla M 1998 Kinetic Theory of Plasma Waves, Oxford Science Publication.

Journal of Physics: Conference Series **260** (2010) 012008 doi:10.1088/1742-6596/260/1/012008

Parametric instability and lower hybrid current drive at plasma densities required for thermonuclear reactors

R Cesario[1], L Amicucci[2], D De Arcangelis[2], M. Ferrari[2], F Napoli[3], E Pullara[2]

[1]Associazione Euratom-ENEA sulla Fusione, C.P. 65 - I-00044 - Frascati, Rome, Italy
[2]Università di Roma Sapienza, Dipartimento Ingegneria Elettronica, Via Eudossiana 18, 00184 Rome, Italy
[3]Università di Roma Tre, Facoltà di Ingegneria Dip. Elettronica Applicata, Via della Vasca Navale 84, 00146 Rome, Italy

E-mail: roberto.cesario@enea.it

Abstract. The lower hybrid current drive (LHCD), produced by externally launched lower hybrid (LH) wave, potentially provides the most suitable tool for driving current at large radii of fusion relevant plasmas, as required for the advanced scenarios of operation foreseen in the International Tokamak Experiment Reactor (ITER). In this regard, a critical problem exists due to the circumstance that the LH waves do not penetrate into the plasma cores neither of experiments aimed at heating the plasma ions nor of those aimed at driving current in presence of plasma densities approaching those required for ITER. We present here results of the analysis of parametric instability (PI) of LH waves showing that high growth rates occur for ion-sound quasi-mode-driven PIs, considering the plasma parameters of the plasma periphery of the FTU tokamak (Frascati Tokamak Upgrade). The non-linear PI phenomenon can broaden the launched antenna spectra in frequency and parallel wavenumber ($n_{//}$), and, consequently, produce via quasi-linear effect a deposition at the very periphery of high-density plasmas. As result, higher electron temperatures at the plasma edge have the effect of diminishing the PI growth rate and, potentially, the broadening of the launched spectrum, thus enabling the penetration of the coupled LH power in the core of high density reactor grade plasmas.

1. Introduction.

Although the bootstrap current [1], self-generated by particle Coulomb scattering is a non-inductive current and would help producing steady-state radial profiles of plasma current density off-axis, it would not have the flexibility required for plasma profile control in the International Thermonuclear Experiment Reactor (ITER), now in construction phase [2]. Non-inductive current tools based on power externally launched in the plasma, would thus be necessary, able to operate at relatively high plasma densities that occur, in particular, in the layers located at the edge were the first interaction with the launched power occurs. ITER requires at the periphery of the plasma [3]: $n_{e\_0.8}{\approx}0.7{\div}1{\cdot}10^{20}\,\mathrm{m}^{-3}$ at normalised minor radius $r/a{\approx}0.8$, where $r{=}a$ is the plasma minor radius at the LCMS (last closed magnetic field surface); similar values are also required at the magnetic axis (i.e.: $n_{e0}{\approx}n_{e\_0.8}$). Lower hybrid waves, coupled to plasmas by means of phased waveguide array antennas [4], produce LHCD [5–7] as confirmed in many experiments in tokamak plasmas [6–9]. However, these experiments operated at plasma densities that are markedly lower than those required for ITER, so that for plasma densities at the periphery (at $r/a{\approx}0.8$) also lower of about a factor two than in ITER, the LH wave does not penetrate into the plasma bulk or the driven current decreases much more strongly than expected due to increasing density (i.e., than the trend $1/n_{e\_av}$, where n_{e_av} is the line-averaged plasma density) [10–12]. In particular in JET, when a certain value of the plasma density at the edge was exceeded (of a factor two lower than in ITER), the coupled LH power resulted to be fully deposited at the plasma edge, and signatures of wave interaction with the main plasma disappeared, possibly due to the occurrence of a strong broadening of the launched LH spectrum due to parametric instability (PI) [12]. Recent experiments have been performed on FTU following theoretical predictions that this spectral broadening produced by PI may be reduced via higher edge electron temperatures. The latter enable LH power penetration in the core of high density plasmas [13–14] and this was confirmed by the LHCD observed at high plasma densities, even higher than in ITER [15].

The same behaviour of coupled LH power not penetrating in the core occurred also in the early LH experiments aimed at the plasma core ion heating. It is worth noting that linear plasma wave theory expects the LH waves to propagate in the high density plasma, since this condition allows exceeding the cut-off for LH waves: $\omega_{pe} \geq \omega_0$ (where ω_0 is the LH operating frequency). At high density, the condition of LH accessibility [4] in the core should also be satisfied in order to avoid mode-conversion into fast waves: this condition is produced by operating with sufficiently high values of confinement magnetic field and parallel wavenumber ($n_{//}$) of the launched LH waves. The aforementioned on-off behaviour of the coupled LH power in producing effects in the core when increasing plasma density looks very similar to that exhibited by the early LH experiments aimed at heating the plasma ions of the core. These experiments operated at densities much higher than in LHCD scenarios in order to locate the layer $\omega_{LH} \approx \omega_0$ in the plasma core (where ω_{LH} is the cold lower hybrid resonant frequency). The negative results of the ion heating experiments have not been sufficiently understood yet. Therefore, the dependence on density of the properties of LH waves of propagation and damping in tokamak plasmas appears to be a critical problem.

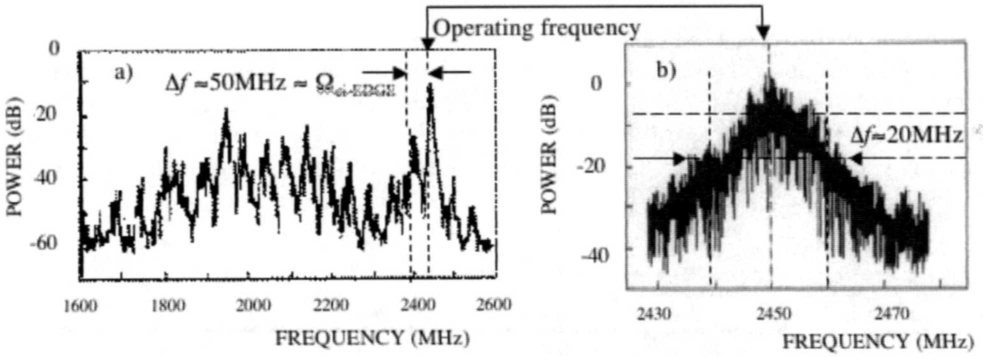

Fig. 1 (from Ref. 16) Typical frequency spectra of the RF probes obtained during the experiment aimed at heating the plasma bulk ions on the FT tokamak. Fig 1a: broadband spectrum exhibiting the *ion-cyclotron* sidebands (span filter bandwidth: 20 MHz). Fig 1b: pump broadening (span filter bandwidth: 5 MHz).

The solution should be properly indicated by the circumstance that, in these experiments, signatures of LH wave interaction occurred only at the very plasma edge, which is the first region in which the launched waves travel. These signatures were contained in the frequency spectrum of signals measured by RF (radio-frequency) probes located outside the vacuum vessel in experiments aimed at the LH ion heating [16] and LHCD as well [13–15]. These spectra showed an enormous broadening of the operating line frequency (orders of magnitude higher than the line width of the RF power sources), accompanied by several *ion-cyclotron* sidebands, i.e., sidebands shifted by harmonics of the ion-cyclotron frequency at the plasma edge, which exhibit a typical non monotonic envelop. An example of such spectra observed during the LH experiments on FT (Frascati Tokamak) aimed at heating ions is shown in the Fig. 1.

The broadband frequency spectrum, Fig. 1a, shows the *ion-cyclotron* sidebands (with maximum of the sidebands located at the tenth harmonic). The spectrum around the operating line frequency is shown in Fig. 1 b. Such spectra were interpreted in terms of sidebands of PIs in which the launched LH power acts as the pump, and low frequency modes in the ion-sound and ion-cyclotron frequency ranges drive the instabilities. In summary, the RF probe spectra are the signature of cascades of ion-sound into ion-cyclotron quasimode-driven-PIs, as the level and the shape envelop of the ion-cyclotron sidebands could be produced only provided that, at the plasma edge, an LH pump with an $n_{//}$ spectrum broader than that launched by the antenna is available. Such pump broadening, possibly produced by ion-sound quasimode-driven PIs, is responsible of the behaviour of the cyclotron sidebands and of the deposition at the very periphery. The LH experiments aimed at producing ion heating, by operating at high plasma densities, identified also a regime in which fast ions were observed at the plasma periphery. This phenomenon might also be explained as an effect of the strong spectral broadening produced by PI at the plasma edge, which makes possible the local quasilinear LH power damping on plasma ions.

The aforementioned results indicate that experiments utilising externally launched LH waves for different aims (i.e, ion heating or current drive at high plasma densities) exhibit the same behaviour consisting in the circumstance that the coupled LH power does not penetrate in the core and a strong PI-produced spectral broadening occurs [12,15]. It is therefore reasonable to believe that the same non-linear mechanism that was identified for interpreting the RF probe spectra in the LH experiments aimed at the ion heating, i.e., PI driven by ion-sound quasimodes [16], should be properly considered also for the LH current drive regime. In the present paper, we present the analysis of the PI considering the parameters of the experiment of FTU, which for the first time demonstrated LHCD at densities also higher than these required for ITER, by means of operations aimed at diminishing the spectral broadening produced by ion-sound quasimode-driven PI [15].

2. Parametric instability analysis

The parametric instability of the launched LH wave is a non-linear phenomenon characterising LHCD regimes of tokamak plasmas that can broaden the launched spectra in frequency and phase velocities in direction parallel to the toroidal magnetic field, provided that the growth rate should be sufficiently strong to compensate the convective loss, which are intrinsic to PI [15]. Frequencies and growth rates of the coupled modes are identified by solving the parametric dispersion relation in the limit of a homogeneous plasma slab [17,18]:

$$\varepsilon(\omega,\mathbf{k})-\frac{\mu_1(\omega_1,\mathbf{k}_1,\mathbf{k}_0,E_0)}{\varepsilon(\omega_1,\mathbf{k}_1)}-\frac{\mu_2(\omega_2,\mathbf{k}_2,\mathbf{k}_0,E_0)}{\varepsilon(\omega_2,\mathbf{k}_2)}=0 \qquad 1$$

In Eq. (1), the quantities without suffix refer to the low-frequency mode driving the instability: ε is the dielectric function, ω, $\boldsymbol{k}$ are, respectively, the complex frequency and wavevector of this low frequency perturbation. The suffix i = 0,1,2, refers to the pump, the lower and the upper sidebands, respectively. $\mu_{1,2}$ are the coupling coefficients referring to the lower and upper sidebands, respectively, defined in Eq. (2) [18]:

$$\mu_{1,2}=\frac{\chi_e(\omega)-\varepsilon(\omega)}{\chi_e(\omega)}\frac{\omega_{pi}^2}{\omega_0^2}\frac{\omega_{pi}^2}{4k^2c_s^2}(1+\frac{\omega}{k_z v_{the}}Z)^2\sin^2\delta_{1,2}\frac{u^2}{c_s^2} \qquad 2$$

In Eq. 2, χ_e is the electron susceptibility, c_s is the sound speed, Z is the plasma function which retains, in the performed calculations, a sum over many (≈100) ion-cyclotron frequency harmonic terms that is necessary for having sufficiently accurate solutions of growth rate [17]. $\delta_{1,2}$ are the angles between the perpendicular wavevector of the lower and upper sidebands and the perpendicular wavevector of the pump, and $u=ek_0\Phi_0/m_e\omega_{ce}$. The parametric instability dispersion relation, Eq. (1), holds in the limit of *adiabatic electrons*, i.e., for $\omega \le k_{//}v_{the}$, where $k_{//}$ is the component of the wavevector parallel to the confinement magnetic field and v_{the} is the thermal velocity of the plasma electrons. The occurrence of the condition of adiabatic electrons limit has been generally verified to occur for standard tokamak parameters, namely in the range of ion-sound frequencies ($\omega << \omega_{ci}$). As general results, the highest growth rates, $\gamma \equiv \mathrm{Im}(\omega) \ge \mathrm{Re}(\omega)$, are found for this PI channel at the lowest values of the plasma densities (n_e) and electron temperatures (T_e). However, the plasma density should be high enough to exceed the cut-off of LH waves.

Eq. 1 has been numerically solved for the complex frequency ω of the low frequency driving mode, considering the following typical parameters of the LHCD experiment relevant for the plasma periphery of FTU: operating frequency f_0= 8.0 GHz, toroidal magnetic field (considered at a normalised minor radius r/a=0.8) B_T= 4.54 T, lower sideband angle between directions of $\mathbf{k}_{\perp 1}$, and $\mathbf{k}_{0\perp}$: δ_1= 16 degs, ion temperature T_i= 10 eV, deuterium plasma, coupled LH power P_{LH}= 0.35 MW, parallel wavenumber of the lower sideband: $n_{//}$ =7. Keeping fixed the aforementioned parameters, the solutions of Eq. 1 have been obtained considering different values of plasma density and electron temperature that have been measured in the layer of the plasma periphery located between the antenna mouth ($r/a \approx$ 1.02) and $r/a \approx$ 0.80. The condition of *adiabatic electrons* has been fully verified occurring for all the considered solutions of Eq. 1. This equation is thus suitable for modelling the PIs produced by LH waves for the considered plasma parameters.

The solutions of the parametric dispersion relation are shown in the Table 1. The first row contains the plasma electron density and temperature that typically occur in the layer close to the antenna mouth. The density value: $n_e \approx 0.2\cdot10^{19}$ m^{-3} is of about three times higher than the cut-off plasma density, so that the condition of propagation for the slow electron plasma wave is satisfied.

Table 1.

	Density ($\cdot10^{19}$ m^{-3})	Electron temperature (eV)	$n_\perp$	$\mathrm{Re}(\omega)$	$\gamma_{max} \equiv \mathrm{Im}(\omega)$	Comments
1	0.2001	10	37,510000	1,989550E-3	4,040080E-3	
2	0.2001	100	65,228000	5,423600E-4	-1,773800E-1	Stable solutions for T_e≥25 eV
3	0.2001	500				
4	1.0000	10	101,388000	1,991600E-3	1,818790E-3	
5	1.0000	100	154,006000	5,878400E-4	-1,771110E-1	Stable solutions for T_e≥23 eV
6	1.0000	500				
7	5.0000	10	216,075998	1,974270E-3	5,158820E-4	
8	5.0000	100	314,883996	5,823370E-4	-1,758230E-1	Stable solutions

						for $T_e \geq 20$ eV
9	5.0000	500				

Higher values of plasma densities are shown in the first column, which correspond to the values occurring in more internal radial regions of the outer radial half of plasma for typical LHCD experiments of FTU with high plasma density. In the second column, different values of the electron temperature are shown. The value shown in the first row corresponds to the temperature measured by Langmuir probes in a layer close to the antenna-plasma interface. In the second and third rows, two higher values of temperature have been assumed in order to study the effect of electron temperature on the growth rate. A similar check of the effect of the electron temperature has been performed for the other values of plasma densities (shown in the column1). This study should be important for assessing the parameters that could be utilised as possible control tool of the growth rate and, consequently, the broadening of the LH wave spectrum for enabling propagation into the plasma core. In this regard, a PI analysis was performed before for parameters of long-lasting internal transport barriers sustained by LHCD during H-mode phase of JET [12,13]. As result, the electron temperature was found to have a determinant role in diminishing the growth rate and consequently the LH spectral broadening and the deposition in the plasma.

In order to perform such a check, different values of electron temperatures have been considered for a fixed density. These values correspond to values of the kinetic profiles measured during FTU experiments performed with the aim of modifying the temperature in the region of the plasma periphery. The 1st, 4th and 8th rows show the solutions of Eq. 1 considering the values measured, respectively, at $r/a \approx 1.02$, $r/a \approx 0.9$ and $r/a \approx 0.80$ during a high-density plasma discharge of FTU (shot number 32164) performed in standard conditions. In this way, a relatively low electron temperature in the plasma periphery and medium plasma density ($n_{e\_av} \approx 1.0 \cdot 10^{20}$ m^{-3}, $n_{e\_0.8} \approx 0.4 \cdot 10^{20}$ m^{-3}) was produced. In these conditions no signature of penetration of the coupled LH power (0.35 MW) into the plasma core was observed to occur. The solutions corresponding to the 1st, 6th and 9th rows correspond to operations with higher temperatures at the plasma edge and also higher plasma densities: $n_{e\_av} \approx 2.0 \cdot 10^{20}$ m^{-3}, $n_{e\_0.8} \approx 0.85 \cdot 10^{20}$ m^{-3} (e.g., FTU plasma discharge 32555). As result, high growth rates are found for lower temperatures, confirming the results of previous analysis of LHCD experiments of JET [14,15]. The 5th column shows the maximum growth rates of the solutions of Eq. 1 obtained considering the $n_\perp$ value of the quasimode as independent variable. For these solutions, the frequency ($\lesssim$15 MHz) lies in the range of ion-sound frequencies, lower than the ion-cyclotron frequency ($\approx$30 MHz). For the highest electron temperature (T_e =500 eV) considered here, only stable solutions of Eq. 1 are found, indicating that ion-sound quasimode-driven PIs occur only in the region of periphery of plasma columm.

3. Comments and conclusions

The growth rate of ion-sound quasimode-driven PI is strongly dependent on the local electron temperature and weakly dependent on the ion temperature [15]. This result indicates that electron dynamics plays a dominant role in determining both the linear (i.e. the antenna coupling of the electron plasma waves, etc.) and non-linear physics (i.e., the mode coupling producing PIs). In all the considered cases, which have considered LH experimental parameters of the regions of the plasma periphery, the solutions of the parametric dispersion relation (Eq. 1) satisfy the condition of adiabatic electrons, $\omega \lesssim k_{//} v_{the}$. It implies that plasma electrons have a dominant dynamics in the instability mechanism and justifies the circumstance that the growth rates are strongly dependent on the local electron temperature (and weakly dependent on the ion temperature). The higher growth rates of the ion-sound quasimode-driven PIs, produced by lower values of plasma temperatures and densities can be explained in terms of the condition of weak turbulence determining the onset of non-linear plasma-wave interactions. The condition of weak turbulence is met easier indeed at low plasma energy densities, for a given RF power density. Conversely, higher temperatures make the plasma more stable.

The occurrence of high growth rates is the necessary condition for broadening the launched LH spectrum. The results shown here confirm that operations with higher electron temperatures in the plasma periphery regions are recommended for enabling, via LH spectral broadening reduction, the penetration of the coupled LH power in the core of reactor grade high density tokamak plasmas.

4. References

[1] Freidberg, J. *Plasma Physics and Fusion Energy*, Cambridge Univ. Press, 2007.

[2] Ikeda, K., Progress in the ITER Physics Basis", *Nucl. Fusion* **47,** (2007) S1–S414

[3] Colas, L., Milanesio, D., Faudot, E., Goniche, M. and Loarte, A. "Estimated RF sheath power fluxes on ITER plasma facing components", *Journ. of Nucl. Materials* 390–391 (2009) 959–962

[4] Brambilla, M., *Kinetic Theory of Plasma Waves*, Oxford Press, 1999

[5] Fisch, N.J., "Confining a tokamak plasma with rf-driven currents", *Phys. Rev. Letters* **41** (1978) 873-876

[6] Bernabei, S., Daughney, C., Efthimion, P., Hooke, W., Hosea, J., Jobes, F., et al., "Lower Hybrid Current Drive in the PLT Tokamak", *Phys. Rev. Letters* **49**, (1982) 1255-1258

[7] Fisch, N.J., "Theory of current drive in plasmas" *Rev. Mod. Physics,* 59, (1987) 175-234

[8] Porkolab, M., Schuss, J.J., Lloyd, B., Takase, Y., Texter, S., Bonoli, P., et al., "Observation of Lower Hybrid Current Drive at High Densities in the Alcator C Tokamak", *Phys. Rev. Letters,* **53** (*1984*) 450

[9] Karney, C.F.F., Fisch, N.J., and Jobes, F. C., "Comparison of the theory and practice of lower hybrid current drive", *Phys. Rev. A* **32** (1985) 2554

[10] Kirov, K.K., Baranov, Y., Goniche, M., Mailloux, J., Mayoral, M.-L., Ongena, J., et al., "LH Wave Absorption and Current Drive Studies by Application of Modulated. LHCD at JET" Proceedings of the 36th European Phys. Society Conf. on Plasma Physics, Sofia, Bulgaria, 2009

[11] Wallace, G., Parker, R.R., Bonoli, P.T., Harvey, R.W., Schmidt, A.E., Smirnov, A.P., et al, "Observation of Lower Hybrid Wave Absorption in the Scrape-Off Layer of a Diverted Tokamak", in Radio Frequency Power in Plasmas, Proceedings of the 18th Topical Conference, Gent, Belgium, editors: V. Bobkov and J.M. Noterdaeme, 24-26 June 2009, American Institute of Physics, 2009, 395-398

[12] R. Cesario et al., "Plasma edge density and lower hybrid current drive in JET", submitted to *Phys. Rev. Letters*

[13] Cesario, R., Cardinali, A., Castaldo, C., Paoletti, F., Mazon, D., and the EFDA-JET Contributors, "Modelling of Lower Hybrid Current Drive in Tokamak Plasmas by including the Spectral Broadening induced by Parametric Instability", *Phys. Rev. Letters* **92** 17 (2004) 175002

[14] Cesario, R., Cardinali, A., Castaldo, C., Paoletti, F., Fundamenski, W., Hacquin, S. and the JET-EFDA workprogramme Contributor, "Spectral broadening of lower hybrid waves produced by parametric instability in current drive experiments on tokamak plasmas", *Nucl. Fusion* **46** (2006) 462-476

[15] Cesario, R. et al., "Current drive at plasma densities required for thermonuclear reactors", *Nature ncomms1052*, August 2010

[16] Cesario, R., and Cardinali, A., "Parametric instabilities excited by ion-sound and ion-cyclotron quasi-modes during lower hybrid heating of tokamak plasmas", *Nucl. Fusion* **29** (1989)

1709-1719

[17] Porkolab, M., " Theory of parametric instability near the lower-hybrid frequency", *Phys. Fluids* **17** (1974) 1432-1442

[18] Liu, C.S., and Tripathi, V.K., "Parametric Instabilities in a magntized plasma", *Physics Reports* **24** (1984) 1709

Theory of Fusion Plasmas: Joint Varenna–Lausanne International Workshop IOP Publishing
Journal of Physics: Conference Series **260** (2010) 012009 doi:10.1088/1742-6596/260/1/012009

Transport physics of ASDEX Upgrade current ramps: experiment and theory

E Fable, C Angioni, S Fietz, J Hobirk, G Pereverzev, the ASDEX Upgrade Team

Max-Planck-Institut für Plasmaphysik, EURATOM Association, 85748 Garching, Germany

E-mail: `emiliano.fable@ipp.mpg.de`

Abstract. Interpretative analysis of experimental data from the ASDEX Upgrade tokamak current ramps–up and ramps–down is carried out to shed light on the properties of confinement and transport in these particular phases of the plasma discharge. It is found that the two ramps are similar in parameters evolution, but are not symmetric with respect to the current level, for several reasons that will be elucidated. Theory–based energy transport modeling allows to understand the underlying transport processes at play, although there are limitations in describing the edge part of the plasma at very low currents, in particular during ramp–up. Possible reasons for this discrepancy are discussed. Finally, the relevant turbulence (linear) regime is identified simultaneously with calculations of quasi–linear particle transport, showing that a broad range of modes, from Trapped Electron Modes (TEM) to Ion Temperature Gradient (ITG) modes, is explored during both the ramp–up and ramp–down.

1. Introduction

Understanding the physics of transport in tokamak plasma current ramps at the beginning and at the end of the discharge is crucial to be able to predict the evolution of important global parameters like the internal inductance, the stored energy, the normalized plasma pressure parameter β_N, and the required loop voltage in ITER [1], in view of defining the best operational space.

Current ramps are characterized by large changes in almost all the relevant plasma parameters as the current is varied. In addition the two ramps are not symmetric at the same current level, as for example the Ohmic heating power (OH) is large and peaked off–axis in the ramp–up, but small and peaked on–axis in the ramp–down. Current ramps are challenging from the theoretical point of view, since a comprehensive description would require the solution of a highly non–linear problem, where the non–linearities are present either at the microscopic level (turbulent fluxes) and at the macroscopic level (coupling of time–evolving parameters in the transport equations). Note that most of the interest is shifted on the electrons, since the electron temperature profile enters as a critical parameter in the internal inductance, in the loop voltage, and in the stored energy in case of OH cases in which the ions play then a minor role for the total stored plasma energy and in β. On the other hand, looking at the ion behavior is instructive to crosscheck the validity of the transport model employed and to understand the ion transport channel in more details.

In this work we approach the problem from both the experimental side, considering discharges from the ASDEX Upgrade tokamak, and from the theory side, with theory–based electron energy

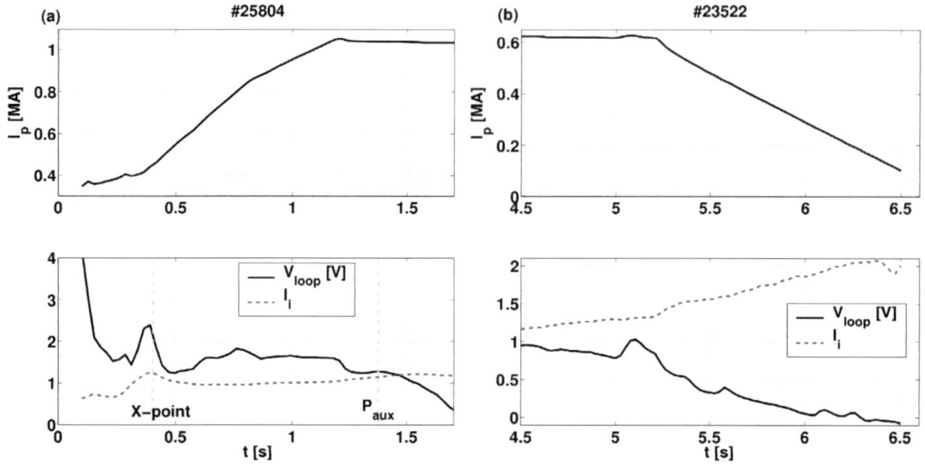

Figure 1. Time evolution of the plasma current $I_{\rm p}$ (top), the loop voltage $V_{\rm loop}$ and the internal inductance $l_{\rm i}$ (bottom), for RU #25804 (a) and for RD #23522 (b). "X–point" and "$P_{\rm aux}$" indicate the times where the X–point forms and where an eventual auxiliary heating is applied.

transport modeling and turbulence calculations to identify the turbulent regime and compare predictions of quasi–linear particle transport with the experiment.

2. Transport phenomenology of ASDEX Upgrade current ramps

We start by considering a set of OH heated and inductively–driven current ramps from the ASDEX Upgrade shots database. In figure 1 we show the time evolution of the total plasma current $I_{\rm p}$, the LCFS loop voltage $V_{\rm loop}$, and the internal inductance $l_{\rm i}$ [1]. The ramp–up (RU) of 1 MA discharge #25804 is shown in figure 1(a), while the ramp–down (RD) of 0.6 MA discharge #23522 in figure 1(b). With vertical dash–dotted lines we show the times at which the X–point is formed and auxiliary heating (NBI in this case) $P_{\rm aux}$ is turned on.

In the RU case, $V_{\rm loop}$ is found to be quite constant at ≈ 1.5 Volts once the breakdown phase is decayed and the X–point is formed. The bump observed before the X–point formation is due to a rapid plasma volume expansion, which perturb both the plasma edge temperature through adiabatic cooling and the voltage itself due to inductive effects (Faraday's law). The presence of an X–point has a large effect on both plasma conductivity and energy transport since it allows the edge electron temperature to reach higher values than in the limited configuration, impacting on plasma conductivity on one side, and on the other side reducing the OH power, thus decreasing transport. As illustrated in figure 1, $l_{\rm i}$ is observed to have a trend which is common to most RUs: it starts from low values at the beginning of the RU, due to an almost flat, or even reversed, q profile, it grows towards a peak, and finally attains a stationary value of $l_{\rm i} \approx 1$.

Looking now at the RD case, we see that it behaves differently: $V_{\rm loop}$ drops almost to zero, whereas $l_{\rm i}$ increases monotonically up to large values of $l_{\rm i} \approx 2$.

In figure 2(a-d), we show radial profiles of the electron temperature $T_{\rm e}$ (a,c) and density $n_{\rm e}$ (b,d) for the same RU (a,b) and RD (c,d) cases, at three time slices. The employed radial coordinate $\rho_{\rm V}$ is defined as $\rho_{\rm V} = \sqrt{V/V_{\rm b}}$, where V is the plasma volume enclosed by the flux surface under consideration. $T_{\rm e}$ attains very low edge values at the beginning of the RU, before the formation of the X–point, and displays a steep gradient, whereas at later times it becomes hotter in the edge and broader. This is observed to be a general behaviour of $T_{\rm e}$ to lose peaking as the ramp proceeds. $n_{\rm e}$ is observed to initially have some degree of peaking, which at later

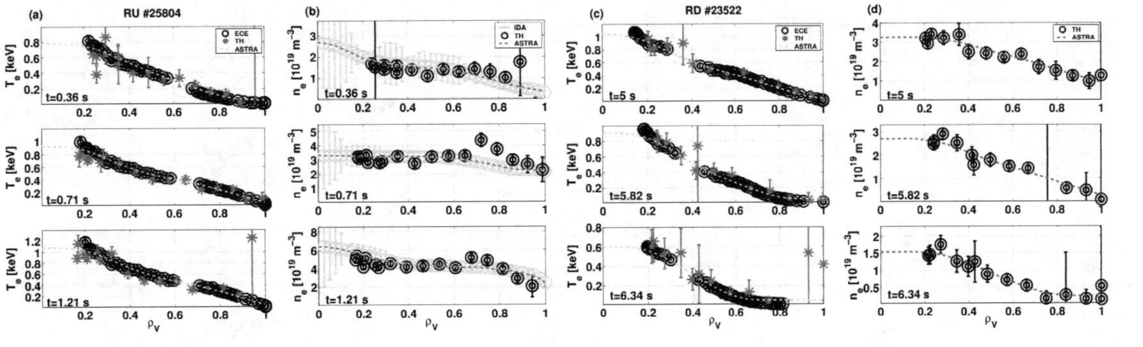

Figure 2. T_e (a,c) and n_e (b,d) profiles for RU #25804 (a,b) and RD #23522 (c,d) shown at different times: for the T_e we show measurements from ECE (empty circles), Thompson scattering (stars) together with a spline fit (dashed lines); for n_e we show the measurements from Thompson scattering (empty circles) together with the spline fit (dashed lines). For #25804, IDA [2] n_e measurements are available and are shown in solid lines.

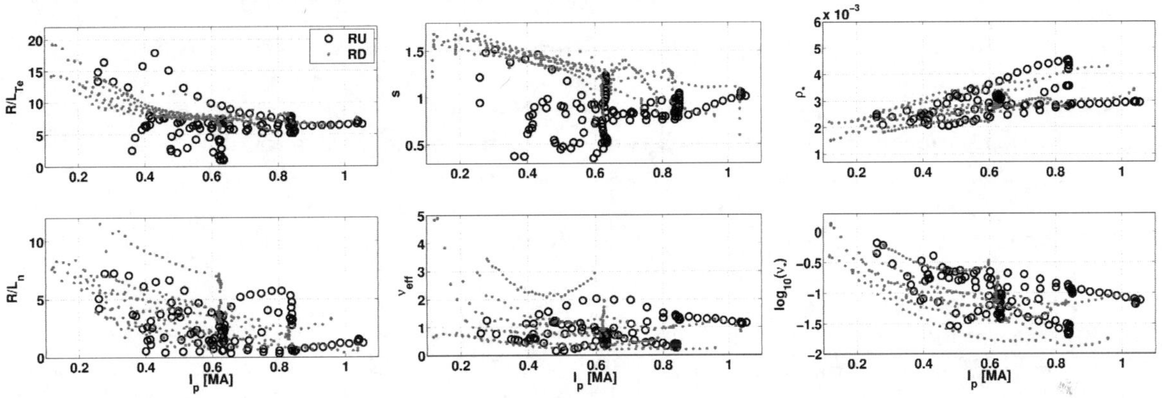

Figure 3. Database points for several plasma parameters plotted versus the plasma current I_p. Open circles belong to ramp–up ('RU') cases, while dots belong to ramp–down ('RD') cases. Radially–dependent quantities are shown averaged over the interval $0.35 < \rho_V < 0.6$.

times will be more or less lost, depending on the evolution of the other parameters, in particular T_e. Note also that the behaviour of n_e can be influenced in case of strong gas fueling at the edge. In the RD an analogous situation emerges for T_e, which decreases at the edge and acquires a larger peaking as the current is decreased. For n_e the behaviour is more striking: it strongly peaks as the ramp proceeds, much more than what it does in the RU at the same current. This difference in n_e behaviour between the two ramps will be discussed later in the context of particle transport simulations.

In figure 3 we plot several parameters averaged in the radial interval $0.35 < \rho_V < 0.65$, for all the studied RU cases (empty circles) and for the RD cases (dots) versus plasma current I_p. The behaviour of the parameters with respect to the current appears to be almost symmetric for some parameters ($R/L_{Te}, R/L_n, \rho_*, \nu_*$), but as said before this is more of a coincidence than of a fundamental property. In particular the local magnetic shear, calculated from a magnetic equilibrium reconstructed with ASTRA [3] and the SPIDER code [4], behaves very differently: it is almost constant during the RU, even being lower at lower currents in that phase, while in the RD it monotonically increases as current decreases. Another interesting parameter is $\hat{\nu} \div \sqrt{\varepsilon}\nu_{ei}qR/v_{th}^{e}$, which represents the impact of collisions on electron banana orbits. High

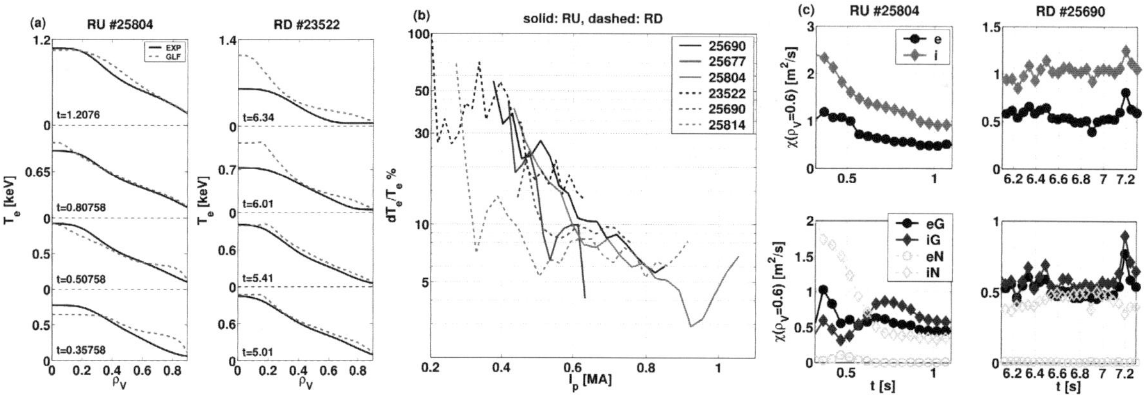

Figure 4. (a) Comparison of experimental (solid lines) and predicted (dashed lines) T_e for RU #25804 (left column) and for RD #23522 (right column); (b) relative error (defined in the text) between predicted and experimental T_e, plotted for several RU (solid lines) and RD (dashed lines) cases versus I_p; (c) time evolution of the energy diffusivities χ evaluated at $\rho_V = 0.6$ for RU #25804 (left column) and for RD #25690 (right column). Top: total diffusivity of electrons (circles) and of ions (diamonds); bottom: diffusivities split in turbulent (eG for electrons, iG for ions) and neoclassical (eN, iN) contributions.

values of this parameter, which are reached at low currents, can change the relative importance on transport of trapped vs passing particles. Note also that the collisionality itself (ν_{eff}) can become quite large at low currents, and the same is also valid for (not shown) the effective charge Z_{eff}, making it necessary to consider collisional effects on turbulence, neoclassical transport, and the impact of impurities on transport.

We have not mentioned the ion temperature profile, since almost no information is provided during the ramps. Therefore the behavior of the ions will be studied semi–empirically by modeling them in some way and checking that global properties like the total plasma energy obtained from measurements is consistent with the modeled ion temperature profiles.

3. Theoretical modeling of electron heat and particle transport

We now address the interpretation of the data with theory–based modeling of heat and particle transport.

3.1. Heat transport

Modeling of electron and ion heat transport is done via combined ASTRA–GLF23[5] simulations, whereas the density is prescribed. ASTRA is employed as the transport modeling framework, for which the internal magnetic equilibrium is reconstructed with the SPIDER code. The GLF23 model provides the turbulence–driven energy diffusivities for $\rho_V \leq 0.9$, outside which radius the profiles are experimentally prescribed. In figure 4(a) we show the comparison between the experimental T_e (solid lines) and the GLF23 predicted T_e (dashed lines) at different times for RU #25804 (left column) and RD #23522 (right column). GLF23 works pretty well either when the RU has already progressed, or when the RD is at the initial stage. At the beginning of the RU it strongly differs in the outer part of the plasma, and the same happens at the end of the RD, although differencies arise depending on the case under study. This emerges clearly when computing the relative difference $dT_e/T_e =< |T_e^{\text{GLF}} - T_e^{\text{EXP}}|/T_e^{\text{EXP}} >$, where $< .. >$ indicates radial averaging, shown in figure 4(b) for all the simulated cases (solid lines for RUs and dashed lines for RDs) plotted versus the plasma current I_p: the tendency of the model is to produce larger deviations at lower currents, with absolute values that can depend on the case. The

Journal of Physics: Conference Series **260** (2010) 012009 doi:10.1088/1742-6596/260/1/012009

reason behind the discrepancy is that the model does not give sufficient diffusivity at the edge, as compared to the experimental estimate. However, a difference exists between RUs and RDs: in RUs most of the OH power is deposited off–axis, due to the loop voltage profile, and so there is a high sensitivity to the details of transport in the outer plasma, resulting in large discrepancies on the T_e profiles. On the other hand, in RDs the OH power is strongly peaked on–axis, leading to an integrated heat flux in the outer plasma which is very small, or even zero if some loss is accumulated on the way (i.e. radiative losses and power exchange to the ions); therefore, details of transport in the edge region are less important. It must be said that, looking at the eigen–frequencies provided by GLF23 in the edge region during the ramps, ITGs are found to be the dominant modes. It is then tempting to conclude that the underestimate by GLF23 of edge transport could be resolved if a TEM–dominated regime takes place. In some sense, an indication that this should be the case when analyzing edge turbulence is given in Ref. [6], although, as mentioned in that reference, one must be careful since edge turbulence has a high degree of non–linearity which can invalidate the quasi–linear approach.

In figure 4(c) we show the time evolution of the heat diffusivites χ evaluated at $\rho_V = 0.6$. In the top subplots we compare the electrons (circles) and the ions (diamonds) total diffusivities: the ions always display a larger diffusivity, moreover both diffusivities strongly decrease in the RU case, while they stay almost constant in the RD case. In the bottom sub–plots we split the diffusivities into the turbulent (eG for electrons, iG for ions) and neoclassical (eN, iN) contributions: in the RU case the neoclassical ion diffusivity dominates at earlier times, thus explaining why the ions are always less confined than the electrons, since a q scaling arising from the neoclassical contribution is well known to go as $\propto q^2$. In the RD case a q scaling seems to be absent, essentially because the variations in q in the core are much less important in the RD than in the RU.

3.2. Particle transport: quasi–linear estimates

Particle transport is modeled via quasi–linear calculations performed with the GS2 code [7], run in linear, electrostatic mode. The magnetic equilibrium is obtained from the Miller parametrization. All profiles are experimentally evaluated, except the ion temperature which in this context is estimated employing the semi–empirical Coppi–Tang transport model [8], which includes fitting coefficients that are varied to match the experimental total plasma energy. For the spectral analysis we employ 12 toroidal modes between $0.08 \leq k_y\rho_i \leq 1.5$ and three species are taken into account: deuterium, electrons, and an impurity species ($A = 24, Z = 12$) with concentration given by the experimental value of $Z_{\rm eff}$ and a normalized density gradient of $0.3R/L_n$. The turbulence–driven part of the electron particle flux must fullfill the following balance equation:

$$\Gamma^{e}_{\rm turb} = -n_e V_W - \frac{1}{\langle|\nabla\rho_V|\rangle V'}\int_0^{\rho_V}\frac{\partial V' n_e}{\partial t}d\rho_V \qquad (1)$$

where V_W is the Ware pinch velocity, V' is the volume derivative over ρ_V. Since $\Gamma^{e}_{\rm turb}$ is a function of the normalized density gradient R/L_n, we perform a scan in this parameter to find the value that satisfies equation (1). We close the problem by assuming a saturated electrostatic potential of the form shown in Ref. [9]. The absolute value of the saturated potential is not needed as in practice we calculate equation (1) normalized to the heat flux Q_e, such that on the left–hand side we have a ratio of turbulent fluxes (negligible electrons neoclassical flux is assumed), while for the right–hand side we use the experimentally estimated power–balance heat flux.

In figure 5(a) we show the results for RU #25804 (left column) and RD #23522 (right column). In the top sub–plots we compare the experimental (circles) and the predicted (lines) value of R/L_n, at $\rho_V = 0.64$ for the RU and $\rho_V = 0.56$ for the RD, as function of time. The

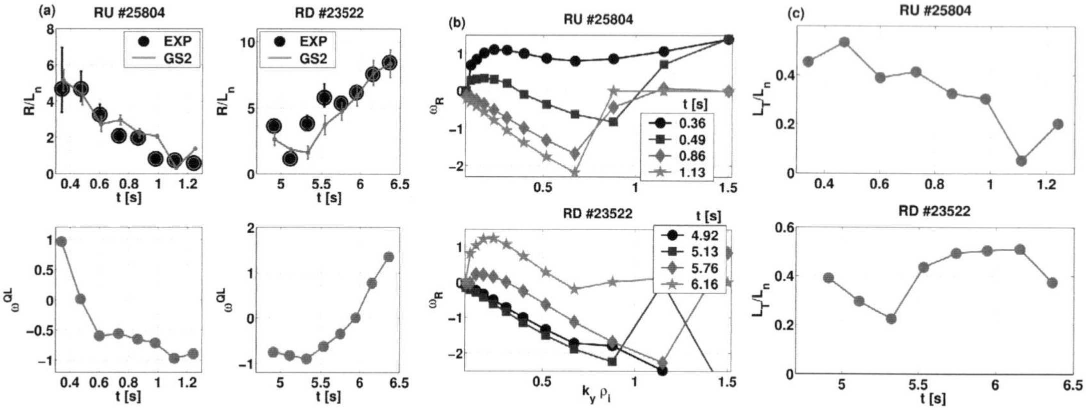

Figure 5. (a) Top: time evolution of predicted (solid lines, with error bars) and experimental (circles, with error bars) $R/L_{\rm n}$ values for RU #25804 (left column) and for RD #23522 (right column). Bottom: time evolution of the spectral average frequency $\omega^{\rm QL}$ (defined in Ref. [10]). $\omega^{\rm QL} > 0$ indicates a TEM and negative $\omega^{\rm QL} < 0$ indicates an ITG; (b) $\omega_{\rm R}$ spectrum for the two cases shown at different times (time values in the two legends); (c) time evolution of the predicted ratio $L_{\rm Te}/L_{\rm n}$.

two radial positions are chosen in the high gradients region not affected by sawteeth. The agreement is pretty good and shows that the code correctly gives the decay (build–up) of the density peaking as the RU (RD) proceeds. The relevant turbulent regime is shown in the bottom sub–plots, where we clearly see a transition from TEM–dominated turbulence ($\omega^{\rm QL} > 0$) at the beginning of the RU, to an ITG–dominated regime ($\omega^{\rm QL} < 0$) towards the flat–top, and viceversa for the RD. The frequency spectrum is shown in figure 5(b), for the RU (top) and for the RD (bottom), at four time slices (in the legend). The observed agreement in $R/L_{\rm n}$ between code results and experimental data and the behaviour in frequency domain are found to be quite sensitive on the input parameters, of which the ion normalized gradient $R/L_{\rm Ti}$ is one of the most important. In fact, if $R/L_{\rm Ti}$ is, say, reduced, then the RU case would have more TEM, eventually leading to lower $R/L_{\rm n}$ values. This has indeed been checked and for other cases (not shown) $R/L_{\rm Ti}$ had to be readjusted to obtain good agreement. Luckily the readjustment involved does not require unrealistic values for $R/L_{\rm Ti}$. In the future, discharges are planned where ion profile measurements will be undertaken during the ramps. Finally, in figure 5(c), we plot the time evolution of the ratio $L_{\rm Te}/L_{\rm n}$ predicted by the code, which shows that this parameter tends to be maximum when the turbulent regime is between the TEMs and the ITGs, at a value of $L_{\rm Te}/L_{\rm n} \approx 0.5$, as expected from a previous study [10].

4. Conclusions

In this work we have analyzed the properties of energy and particle transport and confinement observed during current ramp–up and ramp–down phases in ASDEX Upgrade. Parameters variation point out asymmetries between RUs and RDs. Electrons kinetic profiles are found to become more peaked at lower currents. As regards energy confinement, a clear improvement in time is observed in RU, while no significant change is observed in RD. Theory–based transport modeling is in agreement with the observed trends for confinement, although there are severe limitations on the description of edge transport. GLF23 predicts for these cases a marginal role for a q–scaling of turbulent origin. On the other hand ion neoclassical transport is found to be relevant. Quasi–linear calculations of particle transport show good agreement between predicted and experimental values of $R/L_{\rm n}$, in trend and in amplitude. The relevant turbulent regime is found to change from a TEM–dominated regime at low $I_{\rm p}$ to an ITG–dominated regime at higher I_p. Accordingly, the ratio L_{Te}/L_n follows, being maximal when the regime is closer to TEM, as expected from previous studies [10].

Acknowledgments
The Authors thank M. Kotschenreuther and W. Dorland for having provided the source of the GS2 code, and R. E. Waltz *et al.* for the source of the GLF23 code. The GS2 simulations have been run on the Linux cluster at the RZG in Garching b.M.

References

[1] Y. Gribov *et al.*, *Nucl. Fusion* **47**, S385 (2007)
[2] R. Fischer *et al.*, *Fus. Sci. and Tech.*, to be published
[3] G. V. Pereverzev *et al.*, IPP Report 5/42 (August 1991)
[4] A. A. Ivanov *et al.*, 32nd EPS Conf. on Plasma Phys., ECA Vol.29C, P-5.063 (2005)
[5] R. E. Waltz *et al.*, *Phys. Plasmas* **4**, 2482 (1997)
[6] B. Scott, *Phys. Plasmas* **7**, 1845 (2000)
[7] M. Kotschenreuther *et al.*, *Comput. Phys. Commun.* **88**, 128 (1995)
[8] B. Coppi and W. M. Tang, *Phys. Fluids* **31**, 2683 (1988)
[9] C. Bourdelle *et al.*, *Phys. Plasmas* **14**, 112501 (2007)
[10] E. Fable *et al.*, *Plasma Phys. Control. Fusion* **52**, 015007 (2010)

Journal of Physics: Conference Series **260** (2010) 012010 doi:10.1088/1742-6596/260/1/012010

Entropy production rate in tokamak plasmas with helical magnetic perturbations

X Garbet, J Abiteboul, Y Sarazin, A Smolyakov*, S Allfrey, V Grandgirard, P Ghendrih, G Latu, A Strugarek **

CEA, IRFM, F-13108 St. Paul-lez-Durance cedex, France.

* Department of Physics and Engineering Physics, University of Saskatchewan, 116 Science Place, Saskatoon, SK S7N 5E2 Canada.

** also CEA, SAP, CEA-Saclay, France

Abstract. This paper presents a variational principle based on an extremum of entropy production rate allowing the calculation of the full neoclassical equilibrium (particles, heat and momentum) in tokamaks with an arbitrary helical perturbation.

1. Introduction

Determining the neoclassical equilibrium in tokamaks in presence of a helical perturbation is a question of great interest. Indeed several processes break toroidal axisymmetry of a tokamak, namely field ripple due to finite number of coils [1, 2, 3, 4, 5, 6, 7, 8], MHD instabilities, or externally applied magnetic perturbations [9, 10, 11, 12]. One important point is the calculation of toroidal and poloidal flows, since sheared flows back react on turbulence and MHD instabilities. This question has been much debated in the 90's [5, 8]).

A variational principle based on an extremum of entropy production rate is proposed here to calculate the full neoclassical equilibrium (particles, heat and momentum). This method has been applied with success to the case of ripple induced transport [13, 14]. It is extended in the present work to an arbitrary helical perturbation. This approach provides an efficient means to discriminate between various regimes, depending on collisionality and the amplitude of the helical perturbation. It appears that fluxes and damping rates are usually not negligible. Implications will be discussed, with applications to recent observations on Tore Supra [15].

The remainder of this paper is organized as follows. General properties of a variational principle based on the entropy production rate are described in section 2. The section 3 is dedicated to the application to neoclassical transport in a tokamak in presence of a helical magnetic perturbation. A conclusion follows.

2. Entropy production rate in a tokamak with helical perturbation

2.1. Distribution function

A general expression of the entropy production rate for a non turbulent tokamak plasma in presence of an helical perturbation can be found in ref.[14]. The analysis is restricted to the case of a simple circular concentric equilibrium. The system of coordinates is labeled (r, θ, φ), where r is the minor radius, θ the poloidal angle, and φ the toroidal angle. Moreover a single ion species is considered. The equilibrium Hamiltonian is $H = \frac{1}{2}mv_\parallel^2 + \mu B + e\phi$, where m is the mass, $v_{\parallel}$ the parallel velocity, μ the magnetic moment, B a reference magnetic field, and ϕ the mean electric potential. The perturbed Hamiltonian reads

$$\delta H = -\mu B\epsilon \cos\theta + \mu B\delta \cos\left(N\varphi + M\theta\right) \tag{1}$$

where $\epsilon = \frac{r}{R}$ and $\delta(r)$ is the amplitude of the helical perturbation (it is assumed that it does not depend on the poloidal angle). We call $M_h = M + Nq$ the helical wave number. Particles are trapped in the helical perturbation when the magnetic fields exhibit local extrema along the field lines. If $|M_h| >> 1$, this occurs when $Y = \alpha\left|\sin\theta\right| < 1$, where $\alpha = \frac{\epsilon}{|M_h|\delta}$. This condition defines for each minor radius r a domain in θ for which there exists locally trapped particles. One can define an effective ripple amplitude, which is the depth of the magnetic well along the field lines between successive minima and maxima [3], namely

$$2\delta_{eff} = \frac{B_{max}}{B_{min}} - 1 = 2\delta\left(\sqrt{1-Y^2} - Y\arccos Y\right) \tag{2}$$

If $|M_h| << 1$, two kinds of trapped particles exist: toroidally trapped, and helically trapped particles. The distribution function is a function of motion invariants and is close to a local Maxwellian, namely

$$F_M = \frac{\mathcal{N}}{(2\pi mT)^{3/2}}\exp\left(-\frac{H}{T}\right)\left(1+\frac{mWv_{\parallel}}{T}\right) \tag{3}$$

The function $\mathcal{N}$ is defined as $\mathcal{N} = n\exp\left(e\phi/T\right)$. The density n, potential ϕ and temperature T are functions of the toroidal kinetic momentum $e\psi + mRv_{\parallel}$ (ψ is the poloidal magnetic flux normalised to 2π), while W is a function of the total energy H that ensures finite mean parallel velocity and parallel thermal flux. Once the condition $\frac{RT}{e}\partial_{\psi}\Xi + W = V_T$ is fulfilled, where

$$\partial_{\psi}\Xi = \partial_{\psi}\ln n + \frac{e}{T}\partial_{\psi}\phi + \left(\frac{E}{T} - \frac{3}{2}\right)\partial_{\psi}\ln T \tag{4}$$

and V_T is the toroidal velocity (strictly speaking the parallel velocity), the distribution function Eq.(3) is an exact solution of the Fokker-Planck equation for the unperturbed problem. Multiplying Eq.(3) by $v_{\parallel}^2/v_T^2$, where $v_T = \sqrt{T/m}$ is the thermal velocity, and integrating over the velocity space yields the conventional force balance equation in the large aspect ratio limit [13]

$$-\partial_r\phi + V_pB - V_TB_p = \frac{\partial_r p}{ne} \tag{5}$$

where $V_p = \int d^3\mathbf{v}F_M\frac{v_{\parallel}^2}{v_T^2}W$ is the poloidal velocity, and p the pressure. The entropy production rate can be constructed with the help of the equations derived in [13, 14]. Fluxes and thermodynamical forces can be properly defined from the resonant production rate using the following expression

$$\dot{S}_{res} = -\frac{1}{2}\int dVn\left(\frac{\Gamma}{n}\frac{\partial_r n}{n} + \frac{\mathcal{M}}{nmv_T}\frac{V_T}{v_T} + \frac{Q}{nT}\frac{\partial_r T}{T}\right) \tag{6}$$

where Γ and Q are the particle and heat fluxes, and $\mathcal{M}$ is the rate of dissipated momentum due to toroidal collisional damping. Fluxes are therefore functional derivatives of the resonant entropy production rate.

3. Neoclassical transport tokamaks with an helical perturbation

3.1. Total entropy production rate

In this section, we restrict the analysis to a single ion species. Also the analysis is restricted to the case where transit frequencies associated to $E \times B$ and curvature drifts are smaller than collisional frequencies. Hence this analysis holds for small values of the radial electric field. The collisionality is measured with the conventional parameter for neoclassical theory $\nu^* = \frac{\nu q R}{v_T \epsilon^{3/2}} \ll 1$, where ν is the collision frequency of the considered species. The collision frequency is defined here as $\nu = \frac{4\sqrt{\pi}}{3} \frac{e^4}{(4\pi\epsilon_0)^2} \frac{n}{m_i^2 v_T^3} \ln \Lambda$. The normalized collision frequency is $\bar{\nu}(v) = \frac{3}{4}\sqrt{2\pi}(\Phi(v) - G(v))/v^3$, where $\Phi(v) = \frac{2}{\sqrt{\pi}} \int_0^v dx \exp(-x^2)$ and $G(v) = (\Phi(v) - v\Phi'(v))/2v^2$. All results are expressed as functions of ν^*, δ/ϵ, and wave numbers. With these conventions, the collisionality parameter related to the helical perturbation reads $\nu^*_{hel} = \frac{\nu^*}{|M_h|} \left(\frac{\epsilon}{\delta}\right)^{3/2}$. Also we define an effective collisionality $\nu^*_{eff} = \max(1, M_h^2)\nu^*$. The total resonant entropy production reads

$$\dot{S}_{res} = \dot{S}_{hel} + \dot{S}'_{tor} + \dot{S}_{tor} + \dot{S}'_{hel} \tag{7}$$

where

$$\begin{aligned} \dot{S}_{hel} &= \frac{1}{2}\sqrt{\frac{\pi}{2}} \int dV n \frac{M^2}{|M_h|} \left(\frac{\delta}{\epsilon}\right)^2 \frac{qRv_D^2}{v_T} \int_0^{+\infty} du e^{-u} u^2 \\ &\left(G_0'(\alpha) + G_0(\alpha) \min\left(1, \frac{4}{\pi} \mathcal{I} \nu^*_{hel} \frac{\bar{\nu}}{u^{1/2}}\right)\right) \\ &\left(\partial_r \ln \mathcal{N} + \left(u - \frac{3}{2}\right) \partial_r \ln T + \frac{eB_p}{T} \frac{M_h}{M} V_T\right)^2 \end{aligned} \tag{8}$$

$$\dot{S}_{tor} = \frac{1}{2}\sqrt{\frac{\pi}{2}} \int dV n \frac{qRv_D^2}{v_T} \int_0^{+\infty} du e^{-u} u^{3/2} \bar{\nu}(u) \tag{9}$$

$$\min\left(1, \frac{4}{\pi} \mathcal{I} \nu^* \frac{\bar{\nu}}{u^{1/2}}\right) \left(\partial_r \ln \mathcal{N} + \left(u - \frac{3}{2}\right) \partial_r \ln T + \frac{eB_p}{T} V_T\right)^2 \tag{10}$$

where $\mathcal{I} = 1.38$. The entropy production rate $\dot{S}'_{tor}$ associated to the effect of vertical drift on helically trapped particles reads when $\nu^*_{hel} \ll 1$

$$\begin{aligned} \dot{S}'_{tor} &= K_{hel} \int dV G_1(\alpha) n \left(\frac{Nq}{M_h}\right)^2 \left(\frac{\delta}{\epsilon}\right)^{3/2} \frac{1}{\nu^*} \frac{qRv_D^2}{v_T} \\ &\int_0^{+\infty} du e^{-u} u^{5/2} \frac{1}{\bar{\nu}(u)} \left(\partial_r \ln \mathcal{N} + \left(u - \frac{3}{2}\right) \partial_r \ln T\right)^2 \end{aligned} \tag{11}$$

The rate $\dot{S}'_{tor}$ can be neglected for larger collisionality $\nu^*_{hel} \gg 1$. Similarly the production rate $\dot{S}'_{hel}$ due to effect of the helical perturbation on toroidally trapped particles is small when $\nu^* \gg 1$. If $\nu^* \ll 1$, the expression of $\dot{S}'_{hel}$ depends on the effective collision frequency. In the low effective collisionality regime $\nu^*_{eff} \ll 1$

$$\begin{aligned} \dot{S}'_{hel} &= K'_{hel} \int dV n (Nq)^2 \left(\frac{\delta}{\epsilon}\right)^2 \frac{1}{\nu^*} \frac{qRv_D^2}{v_T} \\ &\int_0^{+\infty} du e^{-u} u^{5/2} \frac{1}{\bar{\nu}(u)} \left(\partial_r \ln \mathcal{N} + \left(u - \frac{3}{2}\right) \partial_r \ln T\right)^2 \end{aligned} \tag{12}$$

while for larger collisionality $\nu^*_{eff} \gg 1$ one has

$$\dot{S}'_{hel} = \frac{1}{2}\sqrt{\frac{\pi}{2}} \int dV n \frac{N^2 q^2}{|M_h|} \left(\frac{\delta}{\epsilon}\right)^2 \frac{qRv_D^2}{v_T}$$

$$\int_0^{+\infty} due^{-u}u^2\left(\partial_r \ln \mathcal{N} + \left(u - \frac{3}{2}\right)\partial_r \ln T\right)^2 \tag{13}$$

where $dV = 4\pi^2 Rrdr$, $v_D = \frac{T}{eBR}$ is the thermal curvature drift velocity, $K_{hel} = \left(\frac{2}{\pi}\right)^{3/2}\min\left(\frac{8}{9}, |M_h|\right)$, and $K'_{hel} = \left(\frac{2}{\pi}\right)^{3/2}\min\left(\frac{8}{9}, \frac{1}{|M_h|^3}\right)$. When $\alpha \gg 1$, the form factors G_0, G'_0, G_1 are given by the relations $G_0(\alpha) = \frac{1}{2\pi}\int_{Y<1} d\theta$, $G'_0(\alpha) = \frac{1}{2\pi}\int_{Y>1} d\theta$, and $G_1(\alpha) = \frac{1}{\pi}\int_{Y<1} d\theta \sin^2\theta$. In the opposite limit one gets $G_0 = G_1 = 1$ and $G'_0 = 0$. The integration domains that correspond to the form factors G_0, G'_0, G_1 do not span the whole phase space. The reason is that each contribution to the entropy production rate is associated to a resonant surface in the phase space where the entropy production rate is maximum. Indeed the production rates come from collisional boundary layers near transitions from trapped to passing particles.

The strategy to determine the neoclassical equilibrium is rather straightforward. First the extremum of the entropy production rate with respect to density variations yields the particle flux. It appears readily that the electron flux is smaller than the ion flux by a factor $\sqrt{m_e/m_i} \ll 1$. Thus the ambipolarity constraint reduces to a condition of vanishing ion flux, $\Gamma = 0$. This constraint gives a relation between the radial electric field, toroidal velocity and gradients of density and temperature (the poloidal velocity can be eliminated using the force balance equation). Second the extremum with respect to toroidal velocity provides the damping rate in the toroidal direction and therefore another constraint. Finally, the extremum with respect to the temperature gradient yields the thermal diffusivity. Several situations may occur, depending on whether local trapping occurs or not, and also depending on collisionality. We will call "weak perturbation regime" the situation where local trapping does not occur, and "strong perturbation regime" the case with local trapping. These two limits are detailed in the following sections.

3.2. Weak perturbation

For small amplitudes of the helical perturbation $Y > 1$, no local trapping occurs, so that only banana particles matter. We restrict the analysis here to the banana regime $\nu^* \ll 1$. In that case two regimes are possible: banana-drift and ripple-plateau (by analogy with the particular case of ripple $M = 0$). These names follow the terminology proposed by Yushmanov [3], and corresponds respectively to the "$1/\nu$" regime and superbanana-plateau regime of Shaing [12]. The transition regime, called "$\sqrt{\nu}$ regime", is not addressed here.

3.2.1. Banana-drift regime Two contributions remain in the entropy production rate: the collisional friction of banana particles on passing particles Eq.(10), and a second related to the effect of ripple on banana particles Eq.(12). The ambipolarity condition reads

$$(1 + k_1)\partial_r \ln \mathcal{N} + (3.37 - 0.17k_1)\partial_r \ln T + k_1 \frac{eB_pV_T}{T} = 0 \tag{14}$$

The collisionality parameter k_1 is similar to the one defined in [8]

$$k_1 = \frac{c_\nu}{c_{bd}} \frac{1}{\min\left(\frac{8}{9}, \frac{1}{|M_h|^3}\right) N^2 q^2}\left(\frac{\epsilon}{\delta}\nu^*\right)^2 \tag{15}$$

where $c_\nu = \mathcal{I}\sqrt{\frac{2}{\pi}}\int_0^{+\infty} due^{-u}u^{3/2}\bar{\nu}(u) \approx 1.1$ and $c_{bd} = \left(\frac{2}{\pi}\right)^{3/2}\int_0^{+\infty} due^{-u}u^{5/2}\frac{1}{\bar{\nu}(u)} \approx 7.4$. Hence the high collisionality regime is defined by the condition $\nu^* \gg Nq\left[\min\left(\frac{8}{9}, \frac{1}{|M_h|^3}\right)\right]^{1/2}\frac{\delta}{\epsilon}$. Note that this condition is consistent with the banana-drift condition $\nu^*_{eff} \ll 1$ if $\frac{\delta}{\epsilon} \ll$

$\frac{1}{Nq}\min\left(1, \frac{1}{|M_h|^{1/2}}\right)$. The extremalisation with respect to the toroidal velocity yields the following evolution equation

$$\partial_t V_T = -c_\nu \frac{1}{1+k_1}\sqrt{\epsilon}\nu\left(V_T - 3.54\frac{\partial_r T}{eB_p}\right) \tag{16}$$

When the toroidal velocity has relaxed, i.e. $V_T = 3.54\frac{\partial_r T}{eB_p}$, the poloidal velocity is found by combining this expression Eq.(14) with the force balance equation. It turns out that, whatever the collisionality, it is equal to the conventional neoclassical poloidal velocity

$$V_p = 1.17\frac{\partial_r T}{eB} \tag{17}$$

The force balance equation (or equivalently Eq.(14)) yields the radial electric field

$$\frac{eE_r}{T} = \partial_r \ln n + 3.37\partial_r \ln T \tag{18}$$

Finally the extremalisation with respect to $\partial_r \ln T$ yields the heat flux $Q = -\chi_i n\partial_r T$ where

$$\chi_i = 36.94\min\left(\frac{8}{9}, \frac{1}{|M_h|^3}\right)N^2q^2\frac{\delta^2}{\epsilon^{1/2}}\frac{v_D^2}{\nu}(1+0.246k_1) \tag{19}$$

These results are consistent with the values given by Kovrizhnykh [8].

3.2.2. Ripple-plateau regime Again two contributions remain in the entropy production rate: banana Eq.(10), and ripple-plateau Eq.(13). The ambipolarity condition reads

$$(1+k_2)\partial_r \ln\mathcal{N} + (1.5 - 0.17k_2)\partial_r \ln T + k_2\frac{eB_pV_T}{T} = 0 \tag{20}$$

where the collisionality parameter k_2 is defined as

$$k_2 = \sqrt{\frac{2}{\pi}}c_\nu\frac{|M_h|}{N^2q^2}\left(\frac{\epsilon}{\delta}\right)^2\nu^* \tag{21}$$

Note that the parameter k_2 corresponds to k_2^{-1} in [8]. The weak collisional regime is defined as $\nu^* \ll \frac{N^2q^2}{|M_h|}\left(\frac{\delta}{\epsilon}\right)^2$, which is consistent with the ripple-plateau condition $\nu^*_{eff} > 1$ if $\frac{\delta}{\epsilon} \gg \frac{1}{Nq}\min\left(|M_h|^{1/2}, \frac{1}{|M_h|^{1/2}}\right)$. The extremalisation with respect to the toroidal velocity yields the following evolution equation

$$\partial_t V_T = -c_\nu \frac{1}{1+k_2}\sqrt{\epsilon}\nu\left(V_T - 1.67\frac{\partial_r T}{eB_p}\right) \tag{22}$$

The poloidal velocity is found by combining the relaxed toroidal velocity $V_T = 1.67\frac{\partial_r T}{eB_p}$ with the force balance equation and the ambipolarity condition Eq.(20). Again it is found to be equal to the conventional neoclassical poloidal velocity $V_p = 1.17\frac{\partial_r T}{eB}$ for any collisionality parameter k_2. The force balance equation yields the radial electric field

$$\frac{eE_r}{T} = \partial_r \ln n + 1.5\partial_r \ln T \tag{23}$$

Finally the extremalisation with respect to $\partial_r \ln T$ yields the heat flux $Q = -\chi_i n\partial_r T$ where

$$\chi_i = 3\sqrt{\frac{\pi}{2}}\frac{N^2q^2}{|M_h|}\left(\frac{\delta}{\epsilon}\right)^2\frac{qRv_D^2}{v_T}(1+0.41k_2) \tag{24}$$

3.3. Strong perturbation

In that regime, all 4 contributions must be kept in the entropy production rate. It is assumed that $\nu^* \ll 1$. Several situations must be considered. Let us assume first that helically trapped particles are in the plateau regime $\nu^*_{hel} = \frac{\nu^*}{|M_h|}\left(\frac{\epsilon}{\delta}\right)^{3/2} \gg 1$. In that case, $\dot{S'}_{tor}$ is negligible. Depending whether $|M_h|$ is of order one, or larger than one, the parameter $\nu^*_{eff} = \max(1, M_h^2)\nu^*$ can be smaller or larger than one. If $\nu^*_{eff} \gg 1$, toroidally trapped particles (bananas) are in the ripple plateau regime. In this particular case, $\dot{S}_{hel}$ and $\dot{S'}_{hel}$ are of the same amplitude, if $M \sim Nq$. In that case, the radial electric field satisfies the relation

$$\partial_r \ln \mathcal{N} + \frac{3}{2}\partial_r \ln T + \frac{MM_h}{M^2 + N^2q^2}\frac{eB_pV_T}{T} = 0 \tag{25}$$

and the toroidal friction rate reads

$$\partial_t V_T = -\sqrt{\frac{\pi}{2}}\,|M_h|\,\frac{N^2q^2}{M^2 + N^2q^2}\frac{v_T}{qR}\delta^2 V_T \tag{26}$$

If $\nu^*_{eff} \ll 1$, toroidally trapped particles are in the banana-drift regime so that $\dot{S'}_{hel} \gg \dot{S}_{hel}$. The radial electric field is given by Eq.(18). The toroidal friction rate reads

$$\partial_t V_T = -\sqrt{\frac{\pi}{2}}\,|M_h|\,\frac{v_T}{qR}\delta^2\left(V_T - 1.87\frac{M}{M_h}\frac{\partial_r T}{eB_p}\right) \tag{27}$$

We now investigate the other extreme case where all particles are in the helically trapped regime, i.e. $\nu^*_{hel} \ll 1$, $G_0' = 0$, and $G_0 = G_1 = 1$. The entropy production rate is then dominated by the two contributions proportional to the inverse of the collision frequency, namely $\dot{S'}_{tor}$ and $\dot{S'}_{hel}$. Moreover $\dot{S'}_{tor}$ is larger than $\dot{S'}_{hel}$ provided that $\delta < \epsilon/M_h^4$ if toroidally trapped particles are in the banana-drift regime $\nu^*_{eff} \ll 1$, and provided that $\nu^* \ll (\epsilon/\delta)^{1/2}/|M_h|$ if they are in the ripple-plateau regime $\nu^*_{eff} \gg 1$. These conditions are usually met. The extremum of Eq.(11) with respect to $\partial_r \ln \mathcal{N}$ and $\partial_r \ln T$ then the particle and heat fluxes

$$\begin{pmatrix} \Gamma \\ Q \end{pmatrix} = -nD_{lt}\begin{pmatrix} \partial_r \ln \mathcal{N} + 3.37\partial_r \ln T \\ 5\partial_r T \end{pmatrix} \tag{28}$$

where $D_{lt} = 6.58\min\left(1, \frac{9}{8}|M_h|\right)\frac{N^2q^2}{M_h^2}\delta^{3/2}\frac{v_D^2}{\nu}$. This expression is a generalization of the expression given by Connor et al. [2] for ripple induced fluxes in tokamaks. Since electron ripple losses are small, the ambipolarity constraint $\Gamma = 0$ imposes the value of the radial electric field, which is the Connor value Eq.(18). At this point, one has still to find the ordering between $\dot{S}_{hel}$ and $\dot{S}_{tor}$. One finds easily that $\dot{S}_{hel}/\dot{S}_{tor}$ is of the order of $\left(\frac{M}{M_h}\right)^2\left(\frac{\delta}{\epsilon}\right)^{1/2}$. If $\frac{M}{M_h} \ll \left(\frac{\epsilon}{\delta}\right)^{1/4}$, the banana term $\dot{S}_{tor}$ is dominant. The extremum with respect to the toroidal velocity yields the value $V_T = 3.54\frac{\partial_r T}{eB_p}$ and a damping rate $\nu_\varphi = c_\nu\sqrt{\epsilon}\nu$. The toroidal velocity can be combined with the radial electric field in the force balance equation to find the poloidal velocity $V_p = 1.17\frac{\partial_r T}{eB}$. If $\frac{M}{M_h} \gg \left(\frac{\epsilon}{\delta}\right)^{1/4}$, the term $\dot{S}_{hel}$ is dominant. One finds $V_T = 3.54\frac{M}{M_h}\frac{\partial_r T}{eB_p}$ with the damping rate $\nu_\varphi = c_\nu\sqrt{\delta}\nu$. The toroidal velocity can be combined with the radial electric field in the force balance equation to find the poloidal velocity $V_p \approx \left(3.54\frac{M}{M_h} - 2.54\right)\frac{\partial_r T}{eB}$.

4. Discussion and conclusion

In conclusion, a variational principle based on a principle of extremum of entropy production rate has been developed. This method allows the calculation of the neoclassical equilibrium in tokamaks with arbitrary helical perturbation. In tokamaks, helical perturbations usually come from MHD perturbations or external coils (RMPs). Since the amplitude is small, the relevant regime is banana-drift in the nearly collisionless regime [9, 10, 11, 12]. The regime that is usually retained corresponds to $k_1 \gg 1$ (so-called "$1/\nu$" regime). Nevertheless, it turns out that the weakly collisional regime $k_1 \ll 1$ is not that difficult to reach. In that case the toroidal friction rate is of the order of $\nu\sqrt{\epsilon}$. More interestingly, the diffusion coefficient is not that small in this regime, and exceeds the banana diffusion coefficient. However it remains smaller than a turbulent diffusion coefficient. In ohmic plasmas in the Tore Supra tokamak, the level of density fluctuations, and therefore turbulent transport, has been found to be very small ($\sim 10^{-3}$) inside the $q = 1$ surface. It appears that the measured diffusion coefficient of electrons is larger than the neoclassical banana value. However the impurity flux does agree with the neoclassical value when assuming axisymmetric geometry [15]. This paradox can be resolved if one accounts for the presence of a $(1, 1)$ internal kink mode with a reasonable amplitude. Electron collisional diffusion increases due to the helical perturbation, while the impurity flux is barely affected. Hence ripple induced transport matters for toroidal momentum transport, but also in some cases for particle transport.

Acknowledgements
This work, supported by the European Communities under the contract of Association between EURATOM and CEA, was carried out within the framework of the European Fusion Development Agreement. The views and opinions expressed herein do not necessarily reflect those of the European Commission.

References

[1] Stringer T E 1972 *Nucl. Fusion* **12** 689
[2] Connor J W and Hastie R J 1973 *Nucl. Fusion* **13** 221
[3] Yushmanov P N 1990 *Review of Plasma Physics*, Vol. 16, New York, Consultants Bureau
[4] Shaing K C 1993 *Phys. Fluids B* **5** 3841 .
[5] Mikhailovskii A B 1995 *Plasma Phys. Rep.* **21** 529
[6] Kuvshinov B N, Mikhailovskii A B and Sharapov S E 1995 *Plasma Phys. Rep.* **21** 713
[7] Shaing K C 1996 *Phys. Rev. Lett.* **76** p 4364
[8] Kovrizhnykh L M 1999 *Plasma Phys. Rep.* **25** 760
[9] Shaing K C 2003 *Phys. Plasmas* **10** 1443
[10] Cole A J, Hegna C C, and Callen J D 2007 *Phys. Rev. Letters* **99** 065001
[11] Becoulet M *et al* 2009 *Nucl. Fusion* **49** 085011
[12] Shaing K C, Sabbagh S A and Chu M S 2010 *Nucl. Fusion* **50** 025022
[13] Garbet X *et al* 2009 *Phys. Plasmas* **16** 062503
[14] Garbet X *et al* 2010 Phys. Plasmas **17** 072505
[15] Guirlet R *et al* 2010 Nucl. Fusion **50** 095009

Journal of Physics: Conference Series **260** (2010) 012011 doi:10.1088/1742-6596/260/1/012011

Nonlocal effects in gyrokinetic turbulence simulations using GENE

T Görler[1], X Lapillonne[2], S Brunner[2], J Chowdhury[3], T Dannert[4], F Jenko[1], B F McMillan[2], F Merz[1], D Told[1] and L Villard[2]

[1] Max-Planck-Institut für Plasmaphysik, IPP-EURATOM Association, Garching, Germany
[2] École Polytechnique Fédérale de Lausanne, Centre de Recherches en Physique des Plasmas, Association Euratom-Confédération Suisse, CH-1015 Lausanne, Switzerland
[3] Institute for Plasma Research, Bhat, Gandhinagar - 382 428, India
[4] Rechenzentrum der Max-Planck-Gesellschaft und des Max-Planck-Institutes für Plasmaphysik, EURATOM Association, D-85748, Germany

E-mail: `tobias.goerler@ipp.mpg.de`

Abstract. Modeling of anomalous transport, typically based on gyrokinetic theory, is an essential tool for better understanding and possibly improving the confinement of magnetic fusion plasmas. A well-benchmarked and established implementation of the gyrokinetic Vlasov-Maxwell system of equations can, for instance, be found in the software package GENE. This Eulerian code has recently been enhanced to optionally consider full radial temperature, density, and geometry variations allowing for investigations of nonlocal effects instead of using the radially local flux-tube approximation. First investigations on the importance of finite-size effects are presented and compared with results from the gyrokinetic Lagrangian PIC code ORB5 [S. Jolliet *et al.* 2007 *Comput. Phys. Commun.* **177** 409, B.F. McMillan *et al.* 2008 *Phys. Plasmas* **15** 052308]. Special attention is drawn to a specific nonlocal effect, namely heat flux avalanches.

1. Introduction

The thorough understanding and reliable prediction of the so-called anomalous transport of heat, momentum, and particles across the magnetic surfaces (hereafter referred to as the radial direction) is one of the key physical problems on the way to efficient fusion power plants based on toroidal magnetic confinement. By now, this effect is commonly attributed to small-scale (roughly comparable to the ion or electron gyroradius), low-frequency (much smaller than the ion and electron gyrofrequency) turbulence driven by microinstabilities which extract free energy from the background temperature and density gradients. An appropriate theoretical framework for high-temperature, low-density and thus weakly collisional fusion plasmas is provided by the gyrokinetic approach [1, 2] where fast dynamics (e.g., the particle gyromotion) are eliminated from the full kinetic description but low-frequency physics is kept. In general, the resulting gyrokinetic Vlasov-Maxwell system of equations in five-dimensional phase space can only be solved numerically. Existing implementations can be classified into so-called local and global codes. The former typically employ a specific flux-tube simulation domain [3, 4, 5] which is restricted to a narrow box perpendicular to the magnetic field lines. Consequently, temperature and density profiles and their gradients are just evaluated at the (radial) center position of this domain and periodic boundary conditions allowing for the application of fast and efficient spectral methods are employed. In this case, however, one implicitly assumes a gyro-Bohm transport scaling, i.e. a Bohm scaling reduced by the gyroradius-to-machine-size ratio ρ^* where the latter has to be small. Investigations on the limit of such assumed scalings and applications to small fusion devices thus have to rely on global codes where full radial temperature, density and geometry profiles are considered. Amongst others, an implementation of both approaches can be found in the software package GENE [6, 7, 8, 9] which is a massively parallelized, comprehensive Eulerian δf code. In this paper, we present comparisons of the recently developed global version with the well-established local code version and the global gyrokinetic Lagrangian PIC code ORB5 [10, 11] and hereby study the role of finite-size effects.

2. Code description

In gyrokinetic theory, a gyrocenter distribution function f_σ per species σ with mass m_σ and charge q_σ is evolved in time using an accordingly transformed Vlasov equation (or Boltzmann equation if weak collisions are considered) which reads in advection form [1]

$$\frac{\partial f_\sigma}{\partial t} + \dot{\mathbf{X}} \cdot \nabla f_\sigma + \dot{v}_\parallel \frac{\partial f_\sigma}{\partial v_\parallel} + \dot{\mu} \frac{\partial f_\sigma}{\partial \mu} = 0 \tag{1}$$

where the magnetic moment is an adiabatic invariant fulfilling $\dot{\mu} = 0$. The time derivatives of the gyrocenter coordinate $\mathbf{X}$ and the parallel velocity $v_\parallel$ are given in the low-β limit (small thermal pressure compared to the magnetic pressure) by

$$\dot{\mathbf{X}} = v_\parallel \hat{\mathbf{b}}_0 + \frac{B_0}{B_{0\parallel}^*} \left(\mathbf{v}_{\bar{\chi}} + \mathbf{v}_{\nabla B} + \mathbf{v}_c \right) \qquad \text{and} \qquad \dot{v}_\parallel = -\frac{\dot{\mathbf{X}}}{m_\sigma v_\parallel} \cdot \left(q_\sigma \nabla \bar{\phi}_1 + \frac{q_\sigma}{c} \hat{\mathbf{b}}_0 \dot{\bar{A}}_{1\parallel} + \mu \nabla B_0 \right).$$

Here, B_0 denotes the modulus of the magnetic (background) field vector $\mathbf{B}_0$, $\hat{\mathbf{b}}_0 = \mathbf{B}_0/B_0$ the corresponding unit vector, $B_{0\parallel}^* = \hat{\mathbf{b}}_0 \cdot \mathbf{B}_0^*$ the parallel component of $\mathbf{B}_0^* = \nabla \times (\mathbf{A}_0 + \frac{m_\sigma c}{q_\sigma} v_\parallel \hat{\mathbf{b}}_0)$, and $\bar{\chi}_1 = \bar{\phi}_1 - \frac{v_\parallel}{c} \bar{A}_{1\parallel}$ the gyroaveraged scalar potential in the gyrocenter moving frame with the fluctuating fractions of the electrostatic potential ϕ_1 and the parallel vector field component $A_{1\parallel}$. The total drift velocity consists of the generalized $\mathbf{E} \times \mathbf{B}$ velocity $\mathbf{v}_{\bar{\chi}} = \frac{c}{B_0} \hat{\mathbf{b}}_0 \times \nabla \bar{\chi}_1$, the gradient-B velocity $\mathbf{v}_{\nabla B_0} = \frac{\mu c}{q_\sigma B_0} \hat{\mathbf{b}}_0 \times \nabla B_0$, and the curvature drift velocity $\mathbf{v}_c = \frac{m_\sigma c}{q_\sigma B_0} v_\parallel^2 \hat{\mathbf{b}}_0 \times \frac{\nabla B_0}{B_0}$. Overbars and $\langle \ldots \rangle$ brackets denote gyroaverages being defined as $\bar{\phi}_1(\mathbf{X}) \equiv \mathcal{G}\left[\phi_1(\mathbf{X})\right] \equiv \frac{1}{2\pi} \oint \mathrm{d}\theta \, \phi_1(\mathbf{X} + \mathbf{r}(\theta))$ with the gyroradius vector $\mathbf{r}(\theta)$ being orthogonal to the magnetic field.
In the following, a δf splitting is employed, i.e. the fluctuating part f_1 of the distribution function is considered to be small compared to the stationary background part f_0 – here, chosen to be a local Maxwellian – such that the gyrokinetic Vlasov equation can be rewritten and simplified. Keeping only first order terms in the perturbation parameter expansion, the $\nabla \chi \times \mathbf{B}$ nonlinearity is retained while higher-order terms like the so-called $v_\parallel$-nonlinearity are neglected. This is in line with careful studies in Refs. [12, 13, 14]. The distribution functions for the different species are coupled in the low-β approximation through the gyrokinetic Poisson equation and the parallel component of Ampère's law. The former reads

$$-\nabla_\perp^2 \phi_1 = 4\pi \sum_\sigma \left(q_\sigma \bar{n}_{1\sigma} + n_{0\sigma} \frac{q_\sigma^2}{T_{0\sigma}} \left[\frac{B_0}{T_{0\sigma}} \int \langle \bar{\phi}_1(\mathbf{x} - \mathbf{r}) \rangle e^{-\frac{\mu B_0}{T_{0\sigma}}} \mathrm{d}\mu - \phi_1(\mathbf{x}) \right] \right) \tag{2}$$

where the gyrocenter density is $\bar{n}_{1\sigma} = \frac{2\pi}{m_\sigma} \iint B_{0\parallel}^* \langle f_{1\sigma}(\mathbf{x} - \mathbf{r}) \rangle \mathrm{d}v_\parallel \mathrm{d}\mu$. If adiabatic electrons are assumed, the electron contribution on the right hand side of Eq. (2) is replaced by $n_{1e} = \frac{e n_{0e}}{T_{0e}} (\phi_1 - \langle \phi_1 \rangle_{\mathrm{FS}})$, with $\langle \ldots \rangle_{\mathrm{FS}}$ indicating a flux surface average. Note, that derivatives of equilibrium quantities are here ordered small (as before). The second field equation is

$$-\nabla_\perp^2 A_{1\parallel} = \frac{4\pi^2 B_0}{c} \sum_\sigma q_\sigma \iint \langle f_{1\sigma}(\mathbf{x} - \mathbf{r}) \rangle v_\parallel \mathrm{d}v_\parallel \mathrm{d}\mu. \tag{3}$$

In GENE, Eqs. (1)-(3) are evaluated in the first order perturbation expansion using a field-aligned coordinate system in order to take advantage of the strong anisotropy of plasma turbulence itself. Hence, just a few (on the order of several 10) grid points are required in the direction along the field line where turbulent structures hardly vary. As mentioned earlier, local codes typically employ the flux tube concept where periodic boundary conditions can be taken in both directions (x, y) perpendicular to the magnetic field. In the parallel (z) direction quasi-periodic boundary conditions are implemented which account for the stretching of the simulation box due to magnetic shear. They are also adopted in the global version of GENE. However, in the latter, the radial box size is significantly enhanced to a large fraction of the minor radius and periodic boundary conditions are inapplicable since full radial profiles, e.g. of temperatures and densities are considered. Consequently, (pseudo-) spectral methods can only be employed in the binormal (y) direction. One major consequence is that gyroaverage operators, for instance, cannot be given a simple analytic form as in the (k_x, k_y) Fourier space. Hence, the integration over gyroangles has to be performed using interpolation techniques in real space for the radial direction. In GENE, the latter is realized using finite element interpolation which effectively amounts to a Hermite polynomial interpolation, for details see [9].

Besides hyper diffusion terms and a simple Krook operator, $\mathrm{d}f_{1\sigma}/\mathrm{d}t = -S_{\mathrm{Krook}} f_{1\sigma}$, which can be added close to the radial boundaries to create artificial buffer zones where fluctuations are damped, two further sources/sinks are available in GENE. On the one hand, an artificial Krook-type heat source term similar to the model in Ref. [11],

$$\mathcal{S}_K(x, |v_\parallel|, \mu) = -\gamma_h \left[\langle f_{1\sigma}(\mathbf{X}, |v_\parallel|, \mu)\rangle_{\mathrm{FS}} - \langle f_{0\sigma}(\mathbf{X}, |v_\parallel|, \mu)\rangle_{\mathrm{FS}} \frac{\langle \int d\mathbf{v} \, \langle f_{1\sigma}(\mathbf{X}, |v_\parallel|, \mu)\rangle_{\mathrm{FS}}\rangle_{\mathrm{FS}}}{\langle \int d\mathbf{v} \, \langle f_{0\sigma}(\mathbf{X}, v_\parallel, \mu)\rangle_{\mathrm{FS}}\rangle_{\mathrm{FS}}} \right], \tag{4}$$

with $f_{1\sigma}(\mathbf{X}, |v_\parallel|, \mu) = (f_{1\sigma}(\mathbf{X}, v_\parallel, \mu) + f_{1\sigma}(\mathbf{X}, -v_\parallel, \mu))/2$, is added to the right hand side of the Vlasov equation. Being applied over the whole radial simulation domain, it is designed to fix the temperature profile on average, while conserving the flux-surface averaged density and parallel momentum. Alternatively, a localized heat source model closely following the implementation being described in Ref. [15] can be used. In normalized units, it is added to the right hand side of the Vlasov equation as $\frac{\mathrm{d}f_1}{\mathrm{d}t} = \mathcal{S}_H = \mathcal{S}_0 \mathcal{S}_x \mathcal{S}_E$ with

$$\mathcal{S}_E = \frac{2}{3} \frac{1}{p_{0\sigma}(x)} \left(\frac{v_\parallel^2 + \mu B_0}{T_{0\sigma}(x)/T_{0\sigma}(x_0)} - \frac{3}{2} \right) f_{0\sigma}, \qquad \mathcal{S}_x = \mathcal{S}_{x,in}(x) \Big/ \int \mathrm{d}^3x \, \mathcal{S}_{x,in}(x) J(x, z), \tag{5}$$

and the source amplitude $\mathcal{S}_0$. Here, $J(x, z)$ denotes the configuration space Jacobian and $\mathcal{S}_{x,in}(\hat{x})$ is a user-defined radial source profile.

3. Linear GENE results

Since GENE offers the possibility to run simulations in the global as well as in the local mode, using pseudo-spectral methods in the radial direction in the latter case, it is obvious to take this inherent advantage for two reasons. Firstly, both approaches should produce similar results in the small $\rho^* = \rho_s/a$ (with the Tokamak minor radius a) limit which can be used as a test for the implementation of the (different) numerical schemes. Secondly, it becomes possible to gain some first insights on the range of ρ^* values where finite-size effects become important. In this section, we present corresponding linear simulation results.

Here, the temperature and density profile shapes for the global simulations are either chosen to exhibit peaked logarithmic gradient profiles by considering

$$(T_\sigma, n_\sigma) = (T_{\mathrm{ref}}, n_{\mathrm{ref}}) \exp\left[-\kappa_{(T_\sigma, n_\sigma)} \varepsilon \Delta(T, n) \tanh\left(\frac{(x - x_0)/a}{\Delta(T, n)} \right) \right], \tag{6}$$

or flat top gradient profiles by choosing

$$(T_\sigma, n_\sigma) = (T_{\rm ref}, n_{\rm ref}) \left[\frac{\cosh\left(\frac{(x-x_0)/a+\delta(T,n)}{\Delta(T,n)}\right)}{\cosh\left(\frac{(x-x_0)/a-\delta(T,n)}{\Delta(T,n)}\right)} \right]^{-\kappa(T_\sigma,n_\sigma)\,\varepsilon\,\Delta(T,n)/2} . \quad (7)$$

In definitions (6) and (7), $T_{\rm ref}$ and $n_{\rm ref}$ denote reference temperatures and densities at the center position x_0 (here, $x_0 = 0.5a$). Furthermore, $\delta(T,n)$ and $\Delta(T,n)$ are characteristic gradient profile widths, which will be varied while $\kappa_T = \max(R_0/L_T)$ and $\kappa_n = \max(R_0/L_n)$ denote the maximum temperature and density gradient values and ε is the inverse aspect ratio between minor radius a and major radius R_0. If not stated otherwise, these values will be chosen as in the Cyclone Base Case (CBC) [16] parameter set, i.e. $\kappa_T = 6.96$, $\kappa_n = 2.23$ and $\varepsilon = 0.36$. The shape of the flux surfaces is assumed to be circular and concentric such that x can be identified with the minor radius r of each flux surface (see Ref. [17] for details). The safety factor profile is $q(x/a) = 0.498(x/a)^4 - 0.466(x/a)^3 + 2.373(x/a)^2 + 0.854$ where the center values of $q_0 = q(x_0) = 1.42$ and shear $\hat{s}_0 = \hat{s}(x_0) = 0.8$ match the CBC values. In a first step, the radial simulation box length l_x is kept fixed with respect to the ion gyroradius $\rho_i = \rho_s$. Hence, with decreasing ρ^* parameter, it becomes smaller and smaller compared to the minor radius a so that eventually only a very narrow region about a central flux surface at $x/a = 0.5$ is taken into account. Naturally, such simulations should converge towards those performed with the local code if periodic boundary conditions are assumed. The resulting growth rates using adiabatic electrons with $T_e = T_i$ are presented in Fig. 1. Note that each simulation has been performed at

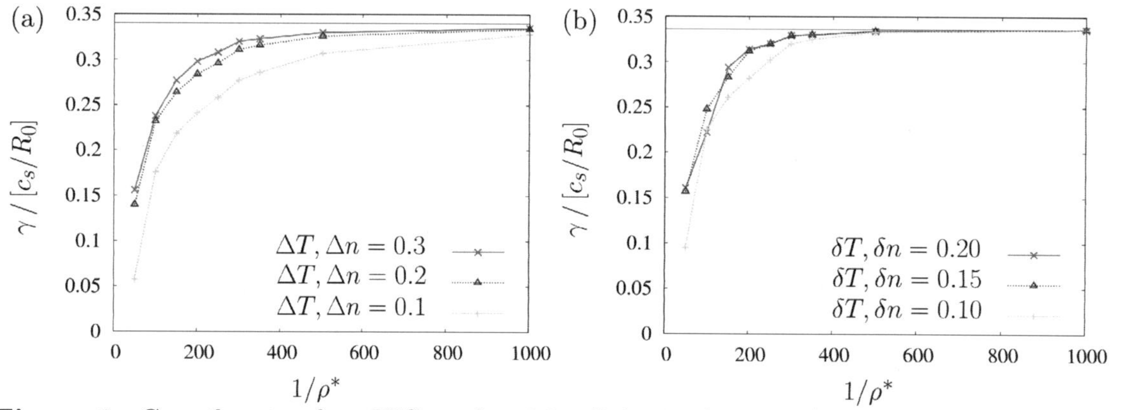

Figure 1. Growth rate of an ITG mode with adiabatic electrons for $k_y\rho_s \approx 0.3$ as function of the inverse ρ^* value using (a) the peaked and (b) the flat temperature and density gradient profiles. The latter are additionally employing $\Delta T, \Delta n = 0.025$ as second characteristic width. Here, the radial simulation box length is kept fixed with respect to the gyroradius. The local code result considering the maximum gradient at x_0 is shown as thin, black line.

the wave number being closest to $k_y\rho_s = 0.3$. The exact value cannot be chosen in general since the box size in the y direction is constrained by the torus dimensions. For details on this issue and on the chosen numerical parameters, see Ref. [18]. The resulting deviations tend to be larger with increasing ρ^* values which explains the jagged behavior in this region. All in all, a very good agreement with the local result can be observed for different profiles and widths at small ρ^*. However, with increasing value of ρ^*, the global code results start to deviate from the local results depending on the profile shape. Here, the broader logarithmic gradient profile shape, Eq. (7) sticks longer to the local result than the narrow one, Eq. (6), which can be attributed to a larger effective (radial) drive region. Further simulation results using a different physical scenario –

namely, an ITG-Kinetic Ballooning (Alfvénic ITG) Mode transition with gyrokinetically treated electrons (with true proton-electron mass ratio) and electromagnetic fluctuations – are shown in Fig. 2. For these runs, the peaked logarithmic gradient profile, Eq. (6), has been employed

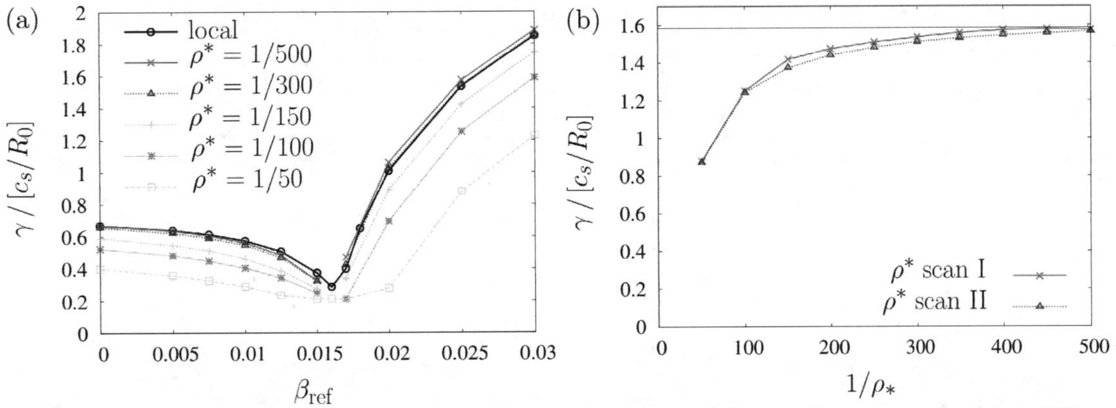

Figure 2. (a) Growth rate at $k_y\rho_s \approx 0.284$ as function of β_{ref} and parametrized by different values of ρ^* together with the local code result. (b) Growth rate at the same binormal wave number but fixed $\beta_{\text{ref}} = 2.5\%$ as function of the inverse ρ^* value. Here, the radial simulation box is kept fixed with respect to (I) the gyroradius and (II) the minor radius. The local code result using the maximum gradients is again shown as thin, black line. The ideal MHD ballooning mode threshold approximately evaluates to $\beta_{\text{ref}} \approx 1.6\%$.

with fixed $\Delta T, \Delta n = 0.3$. The left plot still considers l_x/ρ_s =const. and contains the linear growth rate for different values of ρ^* and $\beta_{\text{ref}} = 8\pi p_{\text{ref}}/B_{\text{ref}}^2$ with reference pressure p_{ref} and magnetic field B_{ref}. Obviously, the previously observed convergence behavior seems to hold even for a wide range of β_{ref} values since the global growth rates do well agree with the local ones for $\rho^* \lesssim 1/300$. A comparison with a different kind of ρ^* scan is presented in fig. 2(b) for a fixed β_{ref} value of 2.5%. Here, the box size is kept fixed with respect to the minor radius a so that more and more radial grid points are required to resolve the mode structures with decreasing gyroradius-to-machine-size ratio. In this case, only minor deviations, i.e. a slightly slower convergence towards the local limit, can be observed. However, if the mode structures exhibit a significantly large ballooning angle – i.e. a finite k_x wave number, respectively – the growth rates in the small ρ^* limit might appear to be systematically below the local code results (see, e.g., Ref. [18]). As is discussed in Ref. [19, 20] in more detail, this can be linked to the radial mode structure in relation to the radial width of the linearly unstable region.

4. Nonlinear GENE and ORB5 results

Since the prediction of heat and particle diffusivities is the true motivation for performing gyrokinetic simulations, it is now most interesting to study the nonlinear physics and the underlying transport scaling. Although this task has already been tackled in the past [21, 22], no coherent picture has evolved so far due to inconsistent findings. In the following, we present results of nonlinear gyrokinetic simulations using adiabatic electrons with $T_e = T_i$ for different values of ρ^* gained from GENE and the Lagrangian PIC code ORB5 [10, 11]. Here, the Krook-type heat sources have been applied in order to approximately maintain the average profiles

$$R_0 \frac{\mathrm{d}\ln(T_i, n)}{\mathrm{d}x} = \kappa_{(T_i,n)} \left[1 - \cosh^{-2}\left(\frac{x - x_0 - \delta/2}{a\Delta}\right) - \cosh^{-2}\left(\frac{x - x_0 + \delta/2}{a\Delta}\right)\right] \quad (8)$$

taken for $|x - x_0| \le \delta/2$ and zero otherwise with $\delta = 0.8a$, $\Delta = 0.04$, $x_0 = a/2$, and the maximum logarithmic gradient lengths $\kappa_n = 2.2$ and $\kappa_{T_i} = 7.1$ and 7.5. The latter is indeed set to these two different initial values in order to allow for a linear interpolation of the time-averaged heat diffusivities at the CBC value of $\kappa_{T_i} = 6.96$ in the quasi-stationary saturation phase where the initial temperature profiles are relaxed by some amount but still clearly above the nonlinear threshold. Again, the flux surfaces are assumed to be circular concentric with a safety factor profile of $q(x) = 0.85 - 0.01x/a + 2.28(x/a)^2 - 0.09(x/a)^3 + 0.22(x/a)^4$. Further details, e.g. on the numerical parameters, can be found in Refs. [20, 23]. The resulting ρ^* dependencies of the ion heat diffusivity measured in units of $\chi_{\rm GB} = \rho_s^2 c_s/a$ are shown in Fig. 3(b). First of all, both codes

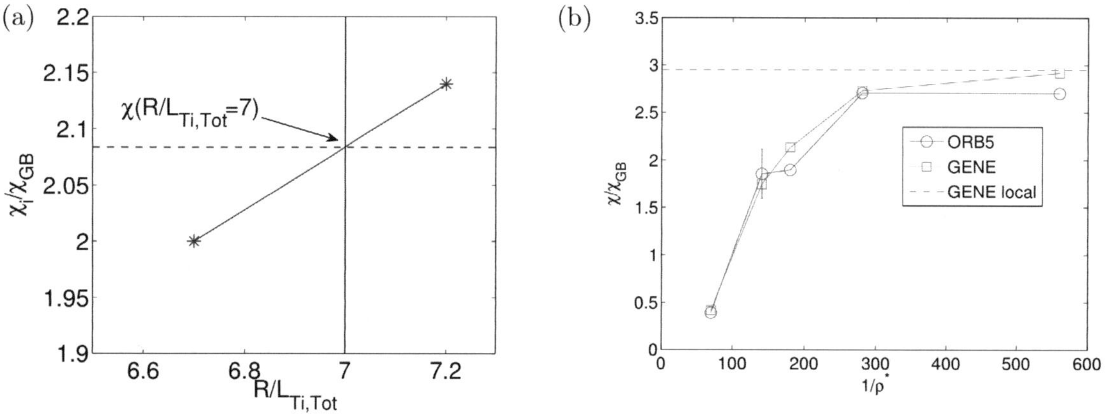

Figure 3. (a) The ion heat diffusivity for the nominal CBC logarithmic gradient $\kappa_T \approx 7$ is extracted using a linear interpolation of the time-averaged heat fluxes and gradients taken from two simulations with different initial temperature profiles. (b) Using this method, the GENE and ORB5 results are plotted for a series of ρ^* values and compared to the ion heat diffusivity of a local GENE simulation. The error bar at $\rho^* = 1/180$ is estimated using three ORB5 simulations with different initial conditions.

show excellent agreement though they are based on completely different numerical methods and are thus potentially subject to different types of discretization errors. Within the error bars, which have exemplarily been evaluated at $\rho^* = 1/180$ using three different initial conditions with ORB5, both codes approach the local GENE (flux tube) result at about $\rho^* \lesssim 1/500$ such that gyro-Bohm scaling would hold for large devices as ITER. Comparing with earlier results using similar parameters, the asymptotic value agrees quantitatively with the largest GTC run of Ref. [21]. Qualitatively, these results concur with the conclusion of Ref. [22] that the global results converge towards the local results in the $\rho^* \to 0$ limit. The exact value in the latter publication is interestingly quite close to the GENE result using an $\hat{s} - \alpha$ magnetic equilibrium model which differs in the treatment of small inverse aspect ratio terms [17]. Hence, the current investigations imply that small but decisive differences in the equilibrium models might be a very likely reason for earlier disagreement in ρ^* scalings. However, when comparing with experiments, it should be noted that profile shapes, for instance, might have a strong influence on the scaling as has already been observed above in linear investigations and in Ref. [20, 22]. More complete physics, e.g., a gyrokinetic treatment of electrons or electromagnetic fluctuations, might furthermore alter those results such that dedicated simulations are called for.

5. Avalanches in local and global simulations

As the anomalous transport scaling appears to be gyro-Bohm like at small ρ^* but non-gyro-Bohm like at $\rho^* \gtrsim 1/300$, the question as to which mechanisms are responsible for this transition arises. An often favored candidate are so-called avalanches, i.e. ballistically propagating structures in various observables as, for instance, in the heat fluxes [24, 25, 26]. The latter shall be considered in the following. As can be seen in Fig. 4 such structures are indeed present in nonlocal, collisionless GENE simulations of adiabatic electron ITG modes using CBC-like parameters and they appear to be independent on whether a gradient or flux driven mode is chosen. Another

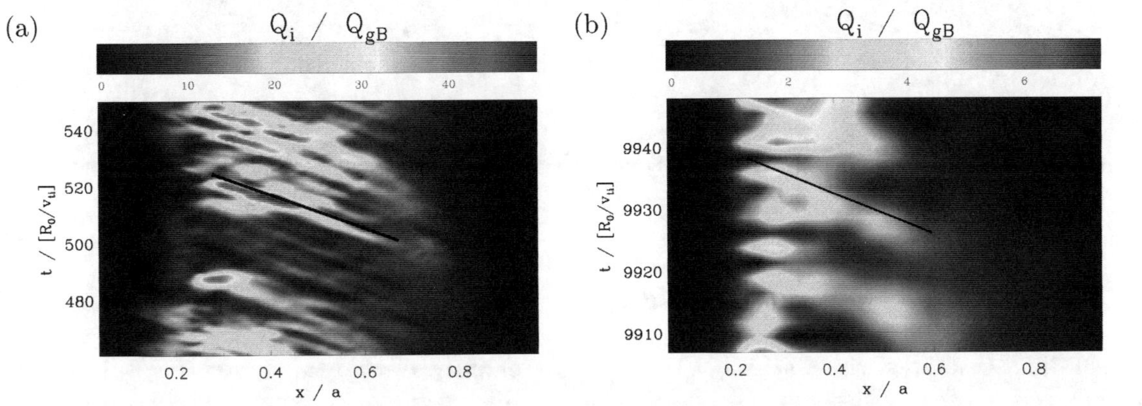

Figure 4. Flux surface averaged heat flux in units of $Q_{\mathrm{gB}} = p_0 c_s \rho_s^2 / R_0^2$ vs. time and radial coordinate. Plot (a) shows the result of a gradient driven simulation using the Krook type heat source at $\rho^* = 1/140$ while (b) contains the result of a low-resolution, "flux-driven" (by the localized physical heat source) simulation at $\rho^* = 1/80$. Black lines are added to emphasize the ballistic propagation of heat flux structures.

interesting feature is that both simulations being presented exhibit similar avalanche propagation velocities of about 2 in units of $\rho_s / R_0 c_s$ although they are performed at significantly different values of ρ^* and differing average temperature gradients – in the first case about 6.8 – 6.9 while in the second about 6.6. For gradient driven simulations of adiabatic electron ITG modes, it has recently be stated that avalanches can be dominated by local physics instead of being governed (as often presumed) by global dynamics since characteristic avalanche parameters appeared to be closely correlated with the turbulent eddy space and time scales and not the machine size [26]. At least, for the simulations at hand the same argument seems to hold for flux-driven simulations. Based on these findings, avalanches should also be visible in local simulations. A corresponding result for the CBC parameters is shown in Fig. 5. Clearly, avalanche-like structures can be observed on this microscopic ($\rho^* \to 0$) scale, as well. However, due to the periodic boundary conditions and the constant temperature and density profiles, they do not exhibit a certain propagation direction but are restricted in their radial extent by large scale zonal flows, see Fig. 5(b). In line with the statements in Ref. [26], the avalanche propagation velocity is again close to $2\rho_s / R_0 c_s$ and corresponding radial correlation lengths being extracted by hand from Fig. 4(a) and Fig. 5 are very similar (about 105 and $85\rho_s$, respectively). Since in general, variations e.g. in the q-profile may affect particle orbits and wave-particle decorrelation physics, further investigations are required to determine the parameters space where the above findings still hold.

6. Conclusion

The new extension of the grid-based gyrokinetic turbulence code GENE by radial variations of the background profiles and metric coefficients has been introduced and tested for various linear and nonlinear scenarios. In this context, it was found that finite-size effects seem to be less important for large fusion devices, where gyro-Bohm scaling can be assumed for the given parameters. However, significant deviations from the local results are observed consistently at $\rho^* \gtrsim 1/300$ by two different gyrokinetic codes and explanations are provided why earlier studies appeared to be

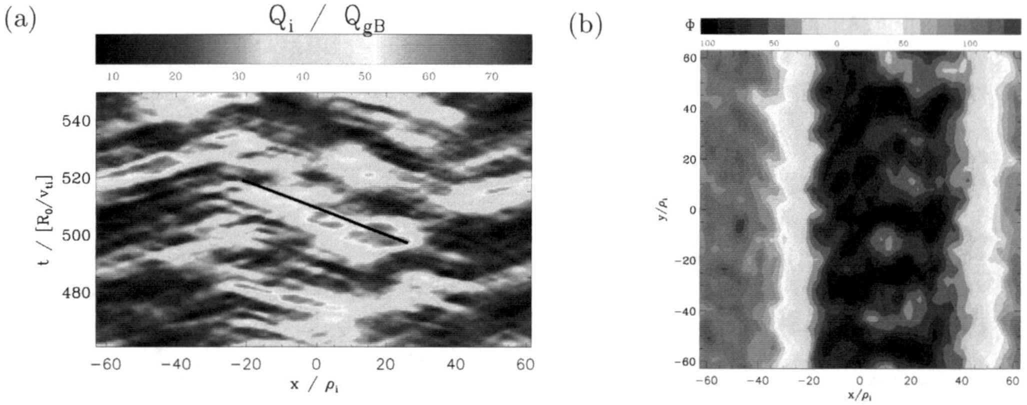

Figure 5. (a) Flux surface averaged heat flux vs. time and radial coordinate for a local GENE simulation using CBC parameters. The black line is added to emphasize the ballistic propagation of heat flux structures. (b) A snapshot ($t = 500\,R_0/c_s$) of the electrostatic potential at outboard midplane is shown to visualize the strong zonal flows which seem to "confine" the radial avalanche extent.

inconclusive. Heat flux avalanches have been investigated in three different scenarios and earlier claims on a (partially) local nature of these ballistically propagating structures are confirmed.

[1] Brizard A J and Hahm T S 2007 *Reviews of Modern Physics* **79** 421 (pages 48)
[2] Garbet X, Idomura Y, Villard L and Watanabe T H 2010 *Nuclear Fusion* **50** 043002
[3] Roberts K V and Taylor J B 1965 *Physics of Fluids* **8** 315–322
[4] Cowley S C, Kulsrud R M and Sudan R 1991 *Physics of Fluids B* **3** 2767–2782
[5] Beer M A, Cowley S C and Hammett G W 1995 *Phys. Plasmas* **2** 2687–2700
[6] Jenko F, Dorland W, Kotschenreuther M and Rogers B N 2000 *Phys. Plasmas* **7** 1904–1910
[7] Dannert T and Jenko F 2005 *Phys. Plasmas* **12** 072309
[8] Merz F 2008 *Gyrokinetic Simulation of Multimode Plasma Turbulence* Ph.D. thesis Westfälische Wilhelms-Universität Münster
[9] Görler T *et al* 2010 *J. Comput. Phys.* Submitted
[10] Jolliet S *et al* 2007 *Comput. Phys. Commun.* **177** 409–425
[11] McMillan B F, Jolliet S, Tran T M, Villard L, Bottino A and Angelino P 2008 *Phys. Plasmas* **15** 052308
[12] Candy J, Waltz R E, Parker S E and Chen Y 2006 *Phys. Plasmas* **13** 074501
[13] Idomura Y, Ida M, Tokuda S and Villard L 2007 *J. Comput. Phys.* **226** 244–262
[14] Jolliet S *et al* 2009 *Phys. Plasmas* **16** 072309
[15] Sarazin Y *et al* 2010 *Nucl. Fusion* **50** 054004
[16] Dimits A M *et al* 2000 *Phys. Plasmas* **7** 969–983
[17] Lapillonne X *et al* 2009 *Phys. Plasmas* **16** 032308
[18] Görler T 2009 *Multiscale Effects in Plasma Microturbulence* Ph.D. thesis Universität Ulm
[19] Villard L *et al* 2010 *Plasma Phys. Controlled Fusion* Accepted for publication
[20] McMillan B *et al* 2010 *Phys. Rev. Lett.* Accepted for publication
[21] Lin Z, Ethier S, Hahm T S and Tang W M 2002 *Phys. Rev. Lett.* **88** 195004
[22] Candy J, Waltz R E and Dorland W 2004 *Phys. Plasmas* **11** L25–L28
[23] Lapillonne X 2010 *Microturbulence in electron internal transport barriers in the TCV Tokamak and global effects* Ph.D. thesis École Polytechnique Fédérale de Lausanne
[24] Sarazin Y, Garbet X, Ghendrih P and Benkadda S 2000 *Phys. Plasmas* **7** 1085–1088
[25] Candy J and Waltz R E 2003 *J. Comput. Phys.* **186** 545–581
[26] McMillan B F *et al* 2009 *Phys. Plasmas* **16** 022310

Journal of Physics: Conference Series **260** (2010) 012012 doi:10.1088/1742-6596/260/1/012012

Barriers to field line transport in 3D magnetic configurations

D Grasso[1,2], D Borgogno[2], F Pegoraro[3], TJ Schep[4]

[1] Istituto dei Sistemi Complessi - CNR, Roma, Italy
[2] Dipartimento di Energetica, Politecnico di Torino, Italy
[3] Phys. Dept. Pisa University, Pisa, CNISM, Italy
[4] Phys. Dept. Eindhoven University of Technology, The Netherlands

E-mail: daniela.grasso@infm.polito.it

Abstract. The transport properties of magnetized plasma configurations are studied that arise from a one-dimensional, current layer that is unstable to reconnecting modes. These magnetic configurations are partially stochastic. It is shown that ridges in the Finite Time Lyapunov Exponents (FTLE) distribution are aligned with the invariant manifolds related to the lines of uniform hyperbolicity. It is shown that these ridges form approximate Lagrangian Coherent Structures (LCS) and act as barriers to the transport of magnetic field lines.

1. Introduction

In this paper we study the spatial topology of 3D magnetic configurations with a strong background magnetic field. Such configurations will be a mixture of volumes where the field is regular and the field lines form periodic or quasi-periodic trajectories, and volumes where the field is irregular and the trajectories are chaotic. The standard tool for analyzing such structures is the method of Poincaré sections in which the field equations are integrated for sufficiently large values of the parameter along the field lines that plays the role of an effective-time and one starts from many initial conditions. Although this technique is rather powerful, it has some serious limitations. In the first place it is limited to periodic or quasi-periodic systems and cannot deal with fully 3D systems. In the second place, Poincaré plots will not tell you where and to what degree a system becomes chaotic so that it does not provide information on the transport process that characterize the transition from local to global chaos.

The magnetic field is represented by

$$\mathbf{B} = B_0\mathbf{e}_z + \mathbf{e}_z \times \nabla\Psi, \tag{1}$$

where the strong background field B_0 is constant and normalized to unity, and $\Psi = \Psi(x, y, z, t)$ is the poloidal magnetic flux function, which varies in time under a reconnection process.

At each time t the trajectory $\mathbf{x}(z; z_0, \mathbf{x}_0)$ of the magnetic field passing through $x(z_0) = \mathbf{x}_0$, obeys the Hamiltonian system

$$\frac{dx}{dz} = -\frac{\partial\Psi}{\partial y}, \qquad \frac{dy}{dz} = \frac{\partial\Psi}{\partial x}. \tag{2}$$

These equations describe a dynamical system moving in field line time z in the $(x,\ y)$ plane in the presence of a flow field with components $(-\Psi_y, \Psi_x)$.

The system is integrable if the flux function does not depend on the field-line-time, i.e., the axial coordinate z. Then, the elliptic and hyperbolic points of (2) are the solutions for which its right-hand vanishes.
At a hyperbolic point (X-point) lines emanate, the separatrices, that form the stable and unstable manifolds associated with the hyperbolic point. These manifolds, curves in the $(x,\ y)$ plane in the 2D context, are invariant surfaces that separate topologically different type of field lines, i.e., they separate regions with closed field lines from regions with open field lines.
Perturbations that are z-dependent will destroy the integrability. The position of an X-point, i.e. a point where the gradient of the flux function vanishes, will also be z-dependent. The path that is traced out by such a point is, however, not a solution of Eq.(2). This means that in 2D motions with arbitrary z-dependence, the trajectory of the "instantaneous" X-point of the field is not a hyperbolic trajectory and no invariant surfaces are related with it! Nevertheless, there exist hyperbolic trajectories in 3D phase-space (x, y, z) that are the generalizations of the 2D hyperbolic points. Such a trajectory *is* a solution of the equations of motion and, thus, a field line. It has the property that all neighboring field lines approach this trajectory exponentially either forward or backward in 'time' z ([1, 2, 3, 4]). It also means that the largest associated Lyapounov exponent is real and positive. The invariant manifolds associated with these trajectories, are the generalizations of the 2D separatrix surfaces, and play an essential role in the topology of magnetic fields.
The stable and unstable manifolds cannot intersect themselves, but they can intersect each other. The intersections of the unstable and stable manifolds of the same hyperbolic trajectory form homoclinic tangles, which determine the local transport. Intersections between a stable (unstable) and an unstable (stable) manifold belonging to different hyperbolic trajectories form heteroclinic tangles. These govern the global transport between nearby chaotic layers. The transition between regular and chaotic configurations was studied in [5] in terms of the field-line-time it takes for heteroclinic intersections to appear.

Although the invariant manifolds contain essential information on the transport properties of the system, a drawback of this method is that these invariant surfaces become so densely folded that they are impossible to trace for sufficiently long times such that their intersections become numerically visible. A consequence is that it is impossible to quantify transport on this basis. Therefore, it might be rewarding to settle for a less exact method and to define transport on the basis of approximate, asymptotic properties of the system.
In this article, we base our investigations on the properties of the field of the Finite Time Lyapunov Exponent (FTLE). The largest positive FTLE measures the exponential separation between two neighboring field lines after a given interval of field-line-time. As will be discussed, this field exhibits clear ridges that form approximate Lagrangian coherent structures (LCS). It is argued that these LCS's play a decisive role in the transition from local to global chaos. During this phase, these hidden coherent structures in the developing chaotic sea, attract field lines and act as barriers to transport of magnetic field lines. The properties of transport between regions separated by such barriers are the subject of this paper. As far as we know this is the first time that the FTLE properties are used to describe chaotic magnetic fields in the context of fusion plasmas.
The paper is organized as follows. Section 2 describes the model and the chaotic system that is generated by a collisionless reconnection event; section 3 describes the LCS we find in this chaotic configuration; section 4 shows why these LCS can be interpreted as transport barriers. We conclude in section 5.

2. 3D chaotic magnetic configuration

We generate a chaotic magnetic field by means of a collisionless reconnection process, extensively described in [6]. We recall here its main features. The two-fluid model derived in [8] is adopted. In this model the reconnection of magnetic field lines is permitted by considering the effect of the electron inertia in Ohm's law, which takes into account also electron pressure gradient effects. The model is valid in the presence of a strong guide field. The model equations are solved numerically in a $3D$-periodic slab geometry starting from a static equilibrium configuration with a one-dimensional shear magnetic field. The spontaneous reconnection process is induced by the onset of multiple helicity perturbations in a configuration with large values of the linear stability parameter Δ'. We consider a configuration with background toroidal and poloidal magnetic fields such that a resonant mode is excited at each of two neighboring surfaces.
The magnetic flux function consists of an equilibrium part $\psi_{eq}(x)$ and a wave-like contribution $\psi(x, y, z; t)$,

$$\Psi(x, y, z, t) = \psi_{eq}(x) + \psi(x, y, z; t), \tag{3}$$

where $\psi(x, y, z; t)$ may be written as a sum over perturbations with different helicities

$$\psi(x, y, z; t) = \Sigma_i \psi_i(x, k_{yi}y + k_{zi}z, t). \tag{4}$$

with $k_{yi} = 2\pi m_i/L_y$, $k_{zi} = 2\pi n_i/L_z$.
The surfaces $x = x_{si}$ where the modes are resonant are characterized by $\mathbf{B}_{eq} \cdot \nabla\Psi = 0$, which yields

$$\frac{d\psi_{eq}(x)}{dx} = -\frac{\partial\psi_i/\partial z}{\partial\psi_i/\partial y} = -\frac{k_{zi}}{k_{yi}}. \tag{5}$$

where $k_{yi} = 2\pi m_i/L_y$, $k_{zi} = 2\pi n_i/L_z$ with $L_y = 4\pi, L_z = 32\pi$.
The numerical simulations were carried out in a triple-periodic slab, with size $L_x = 2\pi$, $L_y = 4\pi$, $L_z = 32\pi$, starting from an equilibrium configuration with magnetic flux function $\psi_{eq} = 0.19\cos(x)$. The initial perturbation consists of two unstable contributions, ψ_1 and ψ_2, with different helicities

$$\psi(x, y, z; t) = \hat{\psi}_1(x, t)\ \exp(ik_{y1}y + ik_{z1}z) + \hat{\psi}_2(x, t)\ \exp(ik_{y2}y + ik_{z2}z). \tag{6}$$

The functions $\hat{\psi}_{1,2}(x, t)$ are chosen so as to approximate the analytic solutions of the linearized dynamical equations. The wave numbers (m_i, n_i) of the two components of the perturbation are $(1, 0)$, for $i = 1$, and $(1, 1)$, for $i = 2$. The amplitude $\hat{\psi}_1$ is of order 10^{-4} and is ten times bigger than $\hat{\psi}_2$.
In the small amplitude linear phase, when the two helicities evolve independently from each other, each mode induces a magnetic island chain around its resonant surface $x = x_{si}$, where $\mathbf{B}_{eq} \cdot \nabla\Psi = 0$. For the case we present here $x_{s1} = 0, \pi$ and $x_{s2} = 0.71, \pi - 0.71$. We remark that resonant surfaces with $x_s > \pi/2$ are simply due to the periodicity of the magnetic equilibrium ψ_{eq}. Since we are interested in the interaction between different helicities, hereafter we we will focus on the magnetic field structure in the reduced interval $-\pi/2 < x < \pi/2$.
When the magnetic islands are sufficiently large to interact with each other, the nonlinear phase of the process enters. Modes with different helicities and higher order modes of the same helicities of the initial perturbation are generated. At this stage the magnetic field topology exhibits volumes where field lines are stochastic and whose extension tends to spread during the evolution of the reconnection process [6].
Here we will not deal with the dynamic evolution of the magnetic field, but we will focus on investigating the transport properties of magnetic field lines at a particular fixed value in real time, chosen just before the transition to global chaos.

3. FTLE ridges and LCS

The solution to (2) for the maximum value of the distance $||\delta\mathbf{x}(z)||$ between two neighboring field lines can be written as

$$max\ ||\delta\mathbf{x}(z)|| = e^{\sigma(z,z_0,\mathbf{x}_0)|z-z_0|}\ ||\delta\mathbf{x}(z_0)|| \tag{7}$$

where $||\delta\mathbf{x}(z_0)||$ is the initial distance at field-line-time z_0 and σ is defined by

$$\sigma(z,z_0,\mathbf{x}_0) = \frac{1}{|z-z_0|}\ln\sqrt{\lambda(z,z_0,\mathbf{x}_0)}, \tag{8}$$

where $|z-z_0|$ is the length of the effective time at which the FTLE is computed and λ is the largest positive eigenvalue of the tangent map associated with (2)[11]. The square root of λ is the factor by which a perturbation is maximally stretched.

In the infinite time limit $|z-z_0| \to \infty$ the standard Lyapunov coefficient is obtained. While this standard Lyapunov coefficient is constant along each field line, the FTLE is not. In [10] it is shown that

$$\frac{d\sigma(z,z_0,\mathbf{x}_0)}{dz_0} = \mathcal{O}(\frac{1}{|z-z_0|}), \tag{9}$$

so that at long field-line-times, the FTLE becomes a constant.

Within the context of FTLE theory, Lagrangian Coherent Structures may be defined as second-derivative ridges of the scalar FTLE field and following [10] we define a ridge as a curve in the $(x,\ y)$ plane such that the gradient in the FTLE-field is along the curve and such that the second order derivative, given by the Hessian

$$\Sigma = \frac{d^2\sigma(\mathbf{x})}{d\mathbf{x}^2} \tag{10}$$

in the direction perpendicular to the curve is minimal.

In order to extract the ridges in the FTLE field from the magnetic data obtained by the numerical simulations of 3D reconnection processes we developed a new computational algorithm. This solver starts with the computation of the Lyapunov exponents for a set of magnetic field lines at different finite effective-time z according to the method described in [12]. Finite Time Lyapunov exponents for magnetic field line trajectories obtained from both forward and backward time integration of eqs.(2) have to be computed. Forward-time integration, in fact, allows to reveal the repelling branches of the Lagrangian coherent structures, e.g., the unstable manifolds, while in order to locate the attracting patterns, e.g., the stable manifolds, a backward-time integration has to be carried out.

In order to simplify the numerical analysis, controlling in particular the computational time of the FTLE field, we will use approximate descriptions of the Hamiltonian function ψ. As shown in [5], only the higher amplitude modes obtained from the Fourier decomposition of the original data need to be considered, in order to approximate the Hamiltonian accurately. This truncation results in a magnetic flux function with 20 spectral components. For the evolution times considered here it turns out that these latter modes have the helicity of either of the two original perturbations.

As stated in Ref.[13] when ridges are computed using the Hessian of the scalar field, noise amplification can become an issue, especially when chaotic "velocity fields" are taken into account, as in the case of the magnetic fields we considered in this paper. A large collection of criteria addressing ridge filtering is available in technical literature [13]. Here, in order to remove numerical noise effects, we have chosen a natural, easy to implement, criterion which prescribes a minimum height of the ridge. It has be shown that in the case of finite Lyapunov exponent

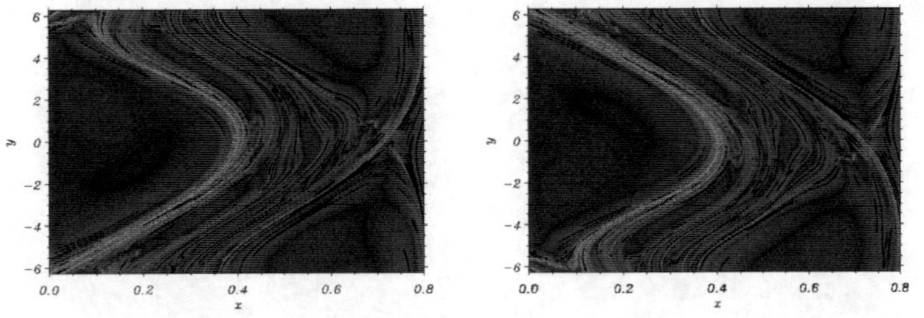

Figure 1. Contour plots of the FTLE computed at $z = 16L_z$ forward (left frame) and backward (right frame) for a set of magnetic field, initially distributed over a uniform 8000×16000 mesh on the domain $0 < x < 0.8$, $-2\pi < y < 2\pi$ of the $z = 0$ section. Superimposed are the ridges extracted by the corresponding FTLE fields.

ridges this leads to significant, consistent, and reliable visualizations [14].
In fig. 1 we show the contour of the FTLE fields evaluated for the dynamical time of the reconnection process we are focusing on and that refers to the local-global chaos transition. The left frame refers to the forward-time integration of the magnetic field line equations, while the right frame refers to the backward-time integration. These FTLE fields have been calculated for a set of $1.28 \cdot 10^8$ magnetic field lines initially distributed at $z = 0$ over a uniform 8000×16000 mesh on the domain $0 < x < 0.8$, $-2\pi < y < 2\pi$.

Figure 1 shows the results after 16 iterations along the toroidal direction. This choice is motivated by a comparison of the results obtained for different integration times. In particular we have carried out simulations up to 20 iterations. Indeed, the essential structures are pretty well represented already after 12 integration times. Superimposed on the contour the corresponding ridges that have been extracted from the FTLE fields are shown in black. These ridges have already been proved in [7] to act as barriers to magnetic field line transport and in the next section we analyze their properties.

4. LCS as transport barriers

In this section we focus on the transport properties of magnetic field lines in the presence of the LCS identified in the previous section. We recall that within the context of a reconnecting system transport of field lines is related with heteroclinic intersections of unstable/stable manifolds with stable/unstable manifolds belonging to different hyperbolic lines, as extensively discussed in [5]. Unfortunately, it is numerically impossible to follow these manifolds sufficiently long in z in order to obtain any realistic estimate of transport based upon heteroclinic intersections. Nevertheless, a comparison of the ridges with the stable and unstable manifolds associated with hyperbolic trajectories show that ridges play the role of magnetic field line transport barriers. This is evident in fig. 2 where we show that the ridges of fig. 1 tend to be aligned with and largely coincide with the initial pieces (*i.e.* evaluated for a few L_z) of the branches of specific stable and unstable manifolds.

In particular, the ridges are shown together with the stable and unstable manifolds associated to the hyperbolic points $(0, -2\pi)$ (green), $(0.51, 0)$ (magenta), $(0.617, 0))$ (light blue), $(0.51827, -1)$ (dark blue), $(0.71, 0)$ (red). The stable and unstable manifolds have been calculated with a Contour Dynamics code [5, 15] only for a few L_z periods. These manifolds are characterized by different frequencies. In particular, we verified that corresponding hyperbolic points have the following periodicities along z: $0L_z$ (green manifold), $5L_z$ (magenta manifold), $3L_z$ (dark blue manifold), $4L_z$ (light blue manifold) and $1L_z$ (red manifold). Since magnetic field lines can not cross these invariant manifolds and since the ridges tend to coincide with the

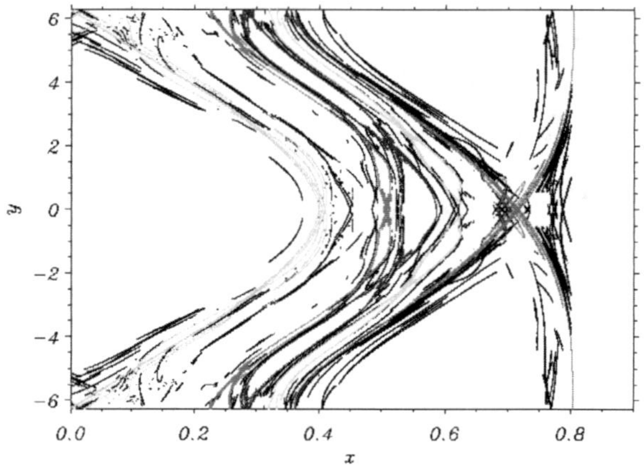

Figure 2. Ridges extracted from the FTLE distribution shown in fig.1 (black curves) with superimposed the corresponding stable and unstable manifolds.

initial part of these manifolds it is expected that these FTLE ridges form barriers with respect to magnetic field line transport.

To quantify how robust these barriers are with respect to the magnetic field line penetration we compare the magnetic topology and the ridges distribution at different effective times in fig. 3. Here the Poincarè plots, obtained for a sample of magnetic field lines initially distributed around a circle centered at the hyperbolic point $(0, -2\pi)$ are overplotted together with the FTLE ridges. We observe that up to $z = 2000L_Z$ the magnetic field lines remain confined by the barrier associated with the magenta manifold. Then up to $z = 7000L_Z$ the confining structure is identified by the barriers associated to the dark blue manifold. After that the relevant confining barriers become the ones associated with the light blue and red manifolds respectively for $z = 12000L_z$ and $z = 20000L_z$. This step-like behavior, where the magnetic field lines remain trapped inside the ridges for finite effective time intervals, is a clear sign of the efficiency of the ridges as barriers.

5. Conclusions

We analyzed the transport properties of a partially chaotic system generated by a collisionless reconnection process, when the transition from local to global chaos occurs. The Lagrangian Coherent Structures that influence the transport of magnetic field lines in this transition phase have been identified by means of the ridges of the Finite Time Lyapunov Exponents. These ridges have been shown to play the role of transport barriers by showing that they partly coincide with the branches of specific stable and unstable manifolds. We observe that the transport of magnetic field line has a step-like behavior. Field lines are at first trapped by the barrier corresponding to the first LCS structures that they encounter in their motion. Then, once they pass this barrier, they spread in a larger volume surrounded by a new barrier and stay there for a second field line time interval. Further in effective time they proceed in the same way until they cross all the barriers and spread over all the chaotic volume.

Acknowledgments

This work was partly supported by the Euratom Communities under the e contract of Association between EURATOM/ENEA. The views and opinions expressed herein do not necessarily reflect those of the European Commission.

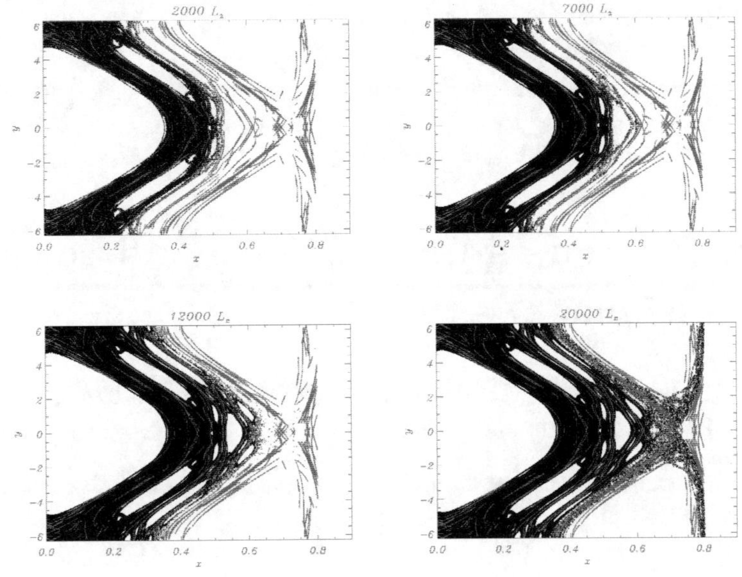

Figure 3. Poincarè maps obtained integrating the field line equations for four different z intervals: up to 2000 toroidal periods (left frame top row), up to 7000 toroidal periods (right frame top row), up to 12000 toroidal periods (left frame bottom row) and up to 20000 toroidal periods (right frame bottom row). Superimposed are the relevant barriers.

References

[1] K. Ide, D. Small, S. Wiggins, *Nonlin. Process.Geophys.*, **9**, 237 (2002).
[2] Ning Ju, D. Small, S. Wiggins, *Int. J. of Bifurcation and Chaos*, **13**, 1449 (2003).
[3] G. Haller, *Chaos*, **10**, 99 (2000).
[4] Ya.B. Pesin, Chapter 7 in *Dynamical Systems II*, Ya. G. Sinai (Ed), Springer Verlag 1989.
[5] D. Borgogno, D. Grasso, F. Pegoraro, T.J. Schep, *Phys. Plasmas.*, **15**, 102308 (2008).
[6] D. Borgogno, D. Grasso, F. Porcelli, F. Califano, F. Pegoraro, F. Farina, *Phys. Plasmas.*, **12**, 032309 (2005).
[7] D. Borgogno, D. Grasso, F. Pegoraro, T.J. Schep, *to be submitted*, (2010).
[8] T.J. Schep, F. Pegoraro, B.N. Kuvshinov, *Phys. Plasmas.*, **1**, 2843 (1994).
[9] G. Haller, Physica D **149**, 248-277 (2001).
[10] S.C. Shadden, F. Lekien, J.E. Marsden, *Phys. D*, **212**, 271-304 (2005); Manikandan Mathur, George Haller, Thomas Peacock, José E. Ruppert-Felsot, Harry L. Swinney, Phys.Rev.Lett. **98**, 144502 (2007).
[11] G. Benettin *et al.*, *Nuovo Cimento* **50 b**, 211 (1979).
[12] G. Rangarajan, S. Habib, R.D. Ryne, *Phys. Rev. Lett.*, **80**, 17 (1998).
[13] F. Sadlo, R. Peikert, *IEEE TRANSACTIONS ON VISUALIZATION AND COMPUTER GRAPHICS* **13**, 6, 1456-1463 (2007).
[14] F. Sadlo and R. Peikert, *Topology-based Methods in Visualization, Mathematics + Visualization*, accepted for publication, 2007.
[15] P.W.C. Vosbeek and R.M.M.Matthey, J.Comput.Phys. **133** 222 (1997).

The impact of energetic particles and rotation on tokamak plasmas

M Hole[1], **K G McClements**[2], **G Dennis**[1], **M Fitzgerald**[1] **and R Akers**[2]

[1] Research School of Physical Sciences and Engineering, Australian National University, 0200 ACT, Australia

[2] EURATOM/CCFE Fusion Association, Culham Science Centre, Abingdon, Oxfordshire, OX14 3DB, UK

E-mail: `matthew.hole@anu.edu.au`

Abstract.

We discuss two contributions that elucidate the impact of energetic particles and rotation on tokamak plasmas: `FLOW-M` (M. J. Hole and G. Dennis, Plasma Phys. Control. Fusion 51, 035014, 2009), a generalisation of the ideal MHD flow code FLOW to multiple quasi-neutral fluids, and recent work on steady poloidal and toroidal bulk flows in tokamak plasmas [K. G. McClements and M.J. Hole, Phys. Plasmas **17**, 082509 (2010)]. Hole and Dennis have generalized ideal MHD to consider multiple quasi-neutral fluids, each in thermal equilibrium and each thermally insulated from each other such that no population mixing occurs. Kinetically, such a model may be able to approximate the ion or electron distribution function in regions of velocity phase space with a large number of particles, at the expense of more weakly populated phase space, which may have uncharacteristically high temperature and hence pressure. As magnetic equilibrium effects increase with the increase in pressure, this work constitutes an upper limit to the effect of energetic particles. McClements and Hole have examined the effects of poloidal and toroidal flows on tokamak plasma equilibria in the MHD limit. Transonic poloidal flows, of the order of the sound speed multiplied by the ratio of poloidal magnetic field to total field B_θ/B, can cause the (normally elliptic) Grad-Shafranov (G-S) equation to become hyperbolic in part of the solution domain. The discontinuity in variables produced by this transition indicates a breakdown in the validity of the MHD model in tokamak plasmas. It is pointed out that the range of poloidal flows for which the G-S equation is hyperbolic increases with plasma beta and B_θ/B, thereby complicating the problem of determining spherical tokamak plasma equilibria with transonic poloidal flows. When the assumption of isentropic flux surfaces is replaced with the more tokamak-relevant one of isothermal flux surfaces, a simple expression can be obtained for the variation of density on a flux surface when poloidal and toroidal flows are simultaneously present. Combined with Thomson scattering measurements of density and temperature, this expression could be used to infer information on poloidal and toroidal flows on the high field side of a tokamak plasma, where direct measurements of flows are not generally possible.

1. Introduction

Neutral beam heating produces two lowest-order corrections to toroidal magnetically confined plasmas: toroidal and poloidal plasma flows, and anisotropy. Bulk plasma motion can be incorporated into ideal MHD by including flow terms in the equations of continuity, momentum and the ideal form of Ohm's law. [1, 2] When combined with a suitable closure condition, and the assumption of toroidal symmetry, the steady-state form of the equations yield a flow-modified

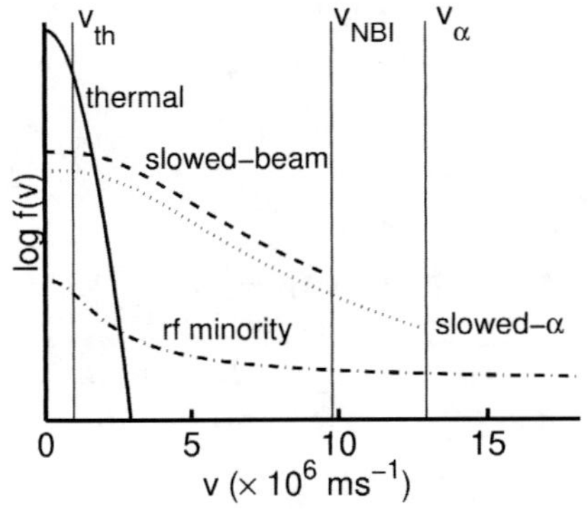

Figure 1. Illustrative 1D distribution function showing thermal, neutral beam slowing-down, α-particle slowing-down, and radio frequency minority heated populations. Also shown are the birth speeds of the beam ions and α-particles.

Grad-Shafranov equation. The effects of anisotropy are typically studied through the use of a single-fluid model with different parallel and perpendicular pressures. [3] A more recent approach that captures both bulk flows and anisotropy involves the use of energetically-resolved fluids that separate the thermal and beam populations. The effects of these energetic components increase with flow speed and anisotropy. If the poloidal flows become "transonic", in the sense that they approach the sound speed multiplied by the ratio of poloidal magnetic field to total field B_θ/B, they cause the (normally elliptic) flow modified Grad-Shafranov equation to become hyperbolic in part of the solution domain. [1] Radial discontinuities in density and pressure can form at the interface between the elliptic and hyperbolic regions.

The aim of this paper is to illustrate some of the magnetic configuration changes that can be induced by the presence of energetic particles and rapid rotation. In Sec. II we introduce a multiple fluid model and describe a code that solves the model equations, presenting results for a spherical tokamak equilibrium with both toroidal and poloidal flows. In Sec. III we describe the effects of transonic poloidal flows on plasma equilibria, showing that the Grad-Shafranov equation is hyperbolic for a much greater range of flow Mach numbers in high performance spherical tokamaks than in conventional tokamaks. In Sec. IV we compute the anisotropy of a fast-ion distribution of a Mega Ampére Spherical Tokamak (MAST) discharge from a TRANSP [4] simulation. Our concluding remarks appear in Sec. V.

2. Multiple Fluid Equilibria

Rather than attempt to solve the equilibrium field given a distribution function for the electron and ions, we use the energetic fluid model of Hole and Dennis [5] to resolve different quasi-neutral fluids, and solve for the field configuration of a multiple fluid plasma in an axisymmetric toroidal configuration. Each fluid represents a different energetic component of the distribution function, with ions paired with electrons within each fluid to ensure quasi-neutrality. Figure 1 shows an illustrative one-dimensional distribution function in velocity space for a plasma comprising a thermal population, beam slowing-down & fusion α-particle slowing-down populations, and a minority radio-frequency heated population. Also shown are the birth speeds of the beam ions and α-particles.

The model used to describe the plasma is given by Eqs. (1)-(7) of Hole and Dennis [5]. That is, each quasi-fluid, labeled by subscript α, is described by ideal MHD, and governed in steady state by:

$$\nabla \cdot (\rho_\alpha \mathbf{v}_\alpha) = 0, \tag{1}$$

$$\begin{aligned}
\rho_\alpha (\mathbf{v}_\alpha \cdot \nabla) \mathbf{v}_\alpha &= -\nabla p_\alpha + \mathbf{J}_\alpha \times \mathbf{B}, & (2)\\
\mathbf{E} + \mathbf{v}_\alpha \times \mathbf{B} &= 0, & (3)\\
\nabla \cdot \mathbf{B} &= 0, & (4)\\
\nabla \times \mathbf{E} &= 0, & (5)\\
\nabla \times \mathbf{B}_\alpha &= \mu_0 \mathbf{J}_\alpha. & (6)
\end{aligned}$$

Here, γ_α is the ratio of specific heats and μ_0 the permeability of free-space, and the unknowns are: ρ_α, the mass-density; $\mathbf{v}_\alpha$, the fluid velocity; p_α, the isotropic partial pressure; $\mathbf{J}_\alpha$, the current density; $\mathbf{E}$, the electric field; $\mathbf{B}$, the total magnetic field; and $\mathbf{B}_\alpha$, the magnetic field produced by the current $\mathbf{J}_\alpha$. The set of equations is completed by a closure equation for the pressure. Highly collisional plasmas obey the standard steady state form of the ideal MHD adiabatic equation [6]

$$\mathbf{v}_\alpha \cdot \nabla \left(\frac{p_\alpha}{\rho_\alpha^{\gamma_\alpha}} \right) = 0, \qquad (7)$$

Equations (1)-(6), together with Eq. (7) define the relationship between these unknowns for each fluid in the plasma.

To simplify this system it is convenient to introduce the poloidal flux functions $\Omega_\alpha(\Psi)$, a toroidal rotation frequency; $f_\alpha(\Psi)$, a modified toroidal flux function and $\sigma_\alpha(\Psi)$, an adiabatic flux function. These functions are defined by

$$\begin{aligned}
\Omega_\alpha(\Psi) &\equiv \frac{v_{\alpha,\phi}}{R} - \frac{B_\phi}{\rho_\alpha R} \zeta'_\alpha(\Psi), & (8)\\
f_\alpha(\Psi) &\equiv B_{\alpha,\phi} R - \mu_0 v_{\alpha,\phi} R \zeta'_\alpha, & (9)\\
\sigma_\alpha(\Psi) &\equiv \frac{p_\alpha}{\rho_\alpha^{\gamma_\alpha}}. & (10)
\end{aligned}$$

As $\rho_\alpha \mathbf{v}_\alpha$ is divergence free [see Eq. (1)] it can be defined by

$$\rho_\alpha \mathbf{v}_\alpha = \frac{1}{R} \nabla \zeta_\alpha(\Psi) \times \hat{\mathbf{e}}_\phi + \rho_\alpha v_{\alpha,\phi} \hat{\mathbf{e}}_\phi, \qquad (11)$$

such that $\zeta(\Psi)$ is the poloidal mass flux of the fluid. The set of equations (1) - (7) can then be reduced to a Bernoulli relation for each fluid

$$H_\alpha(\Psi) = \frac{1}{2} v_\alpha^2 + \frac{\gamma_\alpha}{\gamma_\alpha - 1} \sigma_\alpha(\Psi) \rho_\alpha^{\gamma_\alpha - 1} - \Omega_\alpha(\Psi) v_{\alpha,\phi} R, \qquad (12)$$

together with the generalised Grad-Shafranov equation

$$\begin{aligned}
&\frac{1}{\mu_0} \nabla \cdot \left[\left(1 - \mu_0 \sum_\alpha \frac{1}{\rho_\alpha} \zeta_\alpha'^2 \right) \frac{\nabla \Psi}{R^2} \right] = - \sum_\alpha \rho_\alpha H'_\alpha - R^2 \sum_\alpha \rho_\alpha \Omega_\alpha \Omega'_\alpha - \\
&B_\phi R \left(\frac{1}{\mu_0} \frac{f'}{R^2} + \sum_\alpha (\Omega_\alpha \zeta'_\alpha)' \right) + \sum_\alpha \frac{\rho_\alpha^{\gamma_\alpha}}{\gamma_\alpha - 1} \sigma'_\alpha - \sum_\alpha \frac{1}{R^2} \frac{\zeta'_\alpha \zeta''_\alpha}{\rho_\alpha} \left[|\nabla \Psi|^2 + (B_\phi R)^2 \right]. \qquad (13)
\end{aligned}$$

Hole and Dennis have modified the 2D flow equilibrium code `FLOW` [6] to solve the coupled set of N algebraic Bernoulli equations and the nonlinear PDE, Eq. (13). The new code, `FLOW-M`, takes as user input profiles the quasi-toroidal magnetic field $B_0(\Psi)$, together with the following variables for each fluid: the mass density $D_\alpha(\Psi)$; the pressure $P_\alpha(\Psi)$; the poloidal sonic Mach number $M_{\alpha,\theta}(\Psi)$ (defined as the poloidal flow normalised to the sound speed c_s and multiplied by the ratio of B to poloidal field B_θ); and the toroidal sonic Mach number $M_{\alpha,\phi}(\Psi)$. The

Function	Definition
$f(\Psi)$	$R_0 B_0(\Psi)$
$\zeta'_\alpha(\Psi)$	$\sqrt{\gamma_\alpha P_\alpha(\Psi) D_\alpha(\Psi)}\frac{M_{\alpha,\theta}(\Psi)}{B_0(\Psi)}$
$\Omega_\alpha(\Psi)$	$\sqrt{\gamma_\alpha \frac{P_\alpha(\Psi)}{D_\alpha(\Psi)}}\frac{M_{\alpha,\phi}(\Psi)-M_{\alpha,\theta}(\Psi)}{R_0}$
$H_\alpha(\Psi)$	$\gamma_\alpha \frac{P_\alpha(\Psi)}{D_\alpha(\Psi)}\left(\frac{1}{\gamma_\alpha - 1} + M_{\alpha,\theta}(\Psi) M_{\alpha,\phi}(\Psi) - \frac{1}{2} M^2_{\alpha,\phi}(\Psi)\right)$
$\sigma_\alpha(\Psi)$	$\frac{P_\alpha(\Psi)}{[D_\alpha(\Psi)]^{\gamma_\alpha}}$

Table 1. Relationship between FLOW-M constraint functions and quasi-variable user input profiles, with R_0 the radial position of the geometric axis.

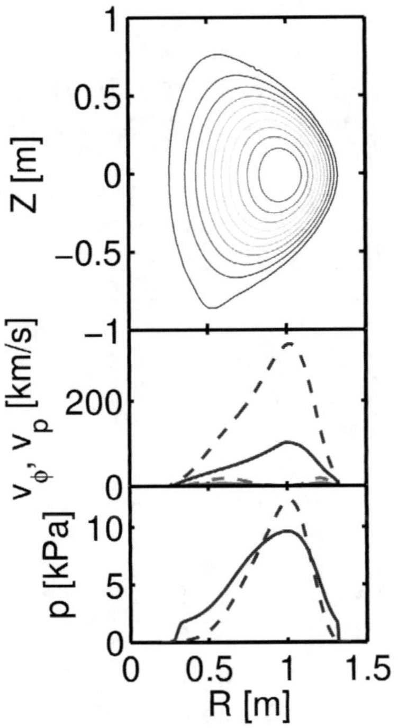

Figure 2. A MAST-like multi-fluid equilibrium with toroidal and poloidal rotation. The plasma boundary and thermal pressure profile is based on MAST discharge #7085 at 290ms, [7] and the flow profiles parametrized. The middle panel shows the toroidal (blue) and poloidal (red) rotation profiles for the thermal (solid) and energetic (dashed) fluids. The lower panel shows the pressure profile for the thermal (solid) and energetic (dashed) fluids.

relationship between FLOW-M constraint functions and quasi-variable user input profiles is shown in Table 1.

Figure 2 shows the equilibrium for a two quasi-neutral fluid plasma with toroidal and poloidal flow; with similar parameters to those of MAST plasmas. The boundary profile and thermal pressure profile $P_0(\Psi)$ were taken from discharge #7085 at 290ms [7], with the energetic pressure profile chosen to be of the form $P_1(\Psi) \propto (1-\psi_n)^2$. The toroidal and poloidal flow profiles for both thermal and energetic species were parametrized as

$$M_\phi(\Psi) = M_{\phi,max}(1-\psi_n)^{1/2} \tag{14}$$

$$M_\theta(\Psi) = 4M_{\theta,max}(1-\psi_n)\psi_n \tag{15}$$

with $M_{0,\phi,max} = 0.30$, $M_{1,\phi,max} = 0.70$ and $M_{0,\theta,max} = 0.1$, $M_{0,\phi,max} = 0.25$. These rotation profiles are representative of MAST plasmas. [8]

The impact of these energetic quasi-neutral fluids on the Shafranov shift, the central safety factor, and the poloidal flux surfaces can be assessed by independently changing $M_{1,\phi,max}$,

$M_{1,\theta,max}$ and $p_1(0)$. As the toroidal rotation $M_{1,\phi,max}$ is increased from 0 to 1.3 the Shafranov shift increases by 60% due to the increased centrifugal force, and the central safety factor decreases from 1.7 to 0.8. The latter occurs because the toroidal current decreases with the toroidal Mach number and the magnetic axis moves outboard. As $M_{1,\theta,max}$ is increased from 0 to 0.4, the central safety factor increases from 1.3 to 1.6: this results from an increase of toroidal current with increasing $M_{1,\theta}$. Finally, as $p_1(0)$ is increased from 0 kPa to 8 kPa the Shafranov shift increases by $\approx$ 30%, and central safety factor drops from 2.5 to 1.0 as the magnetic axis moves outboard. In all cases however, the change in poloidal flux is less than 2%. We conclude that although changing the energetic profiles has little impact on the flux surface geometry and magnitude of the poloidal flux, it does have a significant effect on the Shafranov shift, and more importantly, the central safety factor.

3. Poloidal Rotation

It has long been recognised that "transonic" poloidal flows $v_\theta \approx c_s B_\theta / B$ can cause the normally elliptic Grad-Shafranov equation to become hyperbolic in part of the solution domain. The effect of this modification is to produce radial contact discontinuities in plasma properties such as density. The fact that internal transport barriers in tokamaks, which are characterised by sharp transitions in plasma profiles, are often associated with high poloidal flows suggests a possible link with elliptic/hyperbolic transitions in MHD equilibria, although it has been shown that no such transitions occur in the case of transonic flows when either two-fluid or kinetic effects are taken into account [3, 9]. McClements and Hole [10] have recently sought to clarify some of the outstanding issues raised by the presence of poloidal and toroidal flows in tokamaks, in particular with regard to the effect of poloidal flows on density profiles, the dependence of the elliptic/hyperbolic transition on the choice of closure condition, and whether there is evidence for such transitions occurring in existing tokamak experiments.

In a single-fluid model with isentropic flux surfaces the Bernoulli relation can be used to express total mass density ρ in terms of $\nabla\Psi$. The terms in the Grad-Shafranov equation involving second order derivatives of Ψ can then be written as

$$\left(1 - \mu_0 \frac{F'^2}{\rho}\right)\left(A_{RR}\frac{\partial^2\Psi}{\partial R^2} + A_{RZ}\frac{\partial^2\Psi}{\partial R\partial Z} + A_{ZZ}\frac{\partial^2\Psi}{\partial Z^2}\right), \tag{16}$$

where $F = F(\Psi)$ is defined such that poloidal momentum density is $\nabla F \times \nabla\Psi$ and A_{RR}, A_{RZ}, A_{ZZ} depend on the flow components together with c_s, the local Alfvén speed and the local poloidal Alfvén speed. Equation (16) indicates that the Grad-Shafranov equation is elliptic (hyperbolic) if the quantity $D = A_{RZ}^2 - 4A_{RR}A_{ZZ}$ is negative (positive); it can be shown that

$$D = -4\frac{1 - M_{s\theta}^2(1+\beta)}{1 - M_{s\theta}^2(1+\beta) + \beta M_{s\theta}^4(B_\theta/B)^2}, \tag{17}$$

where $\beta = \gamma\mu_0 p/B^2$, γ being the single-fluid ratio of specific heats.

Equation (17) indicates that in the usual conventional tokamak ordering ($\beta \ll 1$, $B_\theta \ll B$) the flow-modified Grad-Shafranov equation is hyperbolic only for a very narrow range of values of $M_{s\theta}$ [6]. However, on the low field side of high performance spherical tokamak plasmas B_θ/B and β can both be of order unity; in this case the flow-modified Grad-Shafranov equation is hyperbolic at lower $M_{s\theta}$ and for a greater range of values of this parameter. Figure 3, which illustrates this point, shows D versus $M_{s\theta}$ for two pairs of values of β and B_θ/B corresponding to typical conditions in (a) a conventional tokamak, and (b) a high performance spherical tokamak.

When toroidal and poloidal flows v_ϕ, v_θ are both present there exists a flux function Ω such that [2]

$$\Omega = \frac{v_\varphi}{R} - \frac{v_\theta}{R}\frac{B_\phi}{B_\theta}, \tag{18}$$

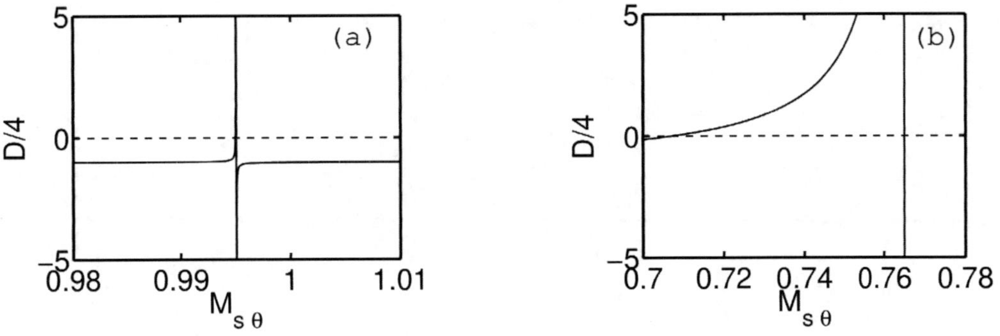

Figure 3. Discriminant D of the second order derivatives of the Grad-Shafranov equation plotted versus poloidal sonic Mach number $M_{s\theta}$ for (a) $\beta = 0.01, B_\theta/B_\phi = 0.1$, and (b) $\beta = 1, B_\theta/B_\phi = 1$. Negative (positive) values of D indicate that the Grad-Shafranov equation is elliptic (hyperbolic).

and the Bernoulli relation for isothermal flux surfaces can be written as

$$H(\Psi) = \frac{2T(\Psi)}{m_i} \ln\left(\frac{\rho}{\rho_0}\right) + \frac{v_\phi^2 + v_\theta^2}{2} - \Omega R v_\phi, \tag{19}$$

where H is a flux function, T is the single-fluid temperature, m_i is ion mass, and ρ_0 is an arbitrary constant. Eliminating v_ϕ from Eqs. (18) and (19) we obtain the following expression for particle density $n \approx \rho/m_i$:

$$n = n_1(\Psi) \exp\left[\frac{\Omega^2(\Psi)}{2v_i^2(\Psi)}\left(R^2 - R_0^2\right) - \frac{1}{2}M_{s\theta}^2\right], \tag{20}$$

where n_1 is a flux function and $v_i = (2T/m_i)^{1/2}$. The quantities Ω, v_i and $M_{s\theta}$ can be measured on the low field side of the plasma using neutral beam diagnostics (charge exchange and motional Stark effect). Because of beam attenuation and injection geometry, such measurements are generally not possible on the high field side. Thomson scattering data can be used to infer n_e and electron temperature T_e across the midplane; measurements of inboard/outboard density asymmetry, combined with Eq. (20), can then yield $M_{s\theta}$ on the high field side. Moreover Eq. (18) can be used to determine v_ϕ on the high field side, provided that $B_\theta^2 \ll B^2$. The above expressions could thus provide a more complete picture of global tokamak dynamics.

4. Pressure anisotropy of fast ions

The inclusion of pressure anisotropy is also known to have an effect on flow equilibrium. [3] For instance, the characteristic poloidal flow speeds at which the MHD picture is hyperbolic is changed for large anisotropy. The centrifugal shift can also be increased or diminished/reversed with larger perpendicular or parallel temperature, respectively.

In preliminary working we have also computed pressure anisotropy using the TRANSP [4] NBI fast ion distribution output for MAST # 18696. Figure 4 shows fast ion density and parallel mean flow speed as well as the computed parallel and perpendicular pressures. For this discharge $p_\perp/p_\parallel \approx 1.7$. Work is underway to establish how this level of anisotropy changes the equilibrium.

Journal of Physics: Conference Series **260** (2010) 012013 doi:10.1088/1742-6596/260/1/012013

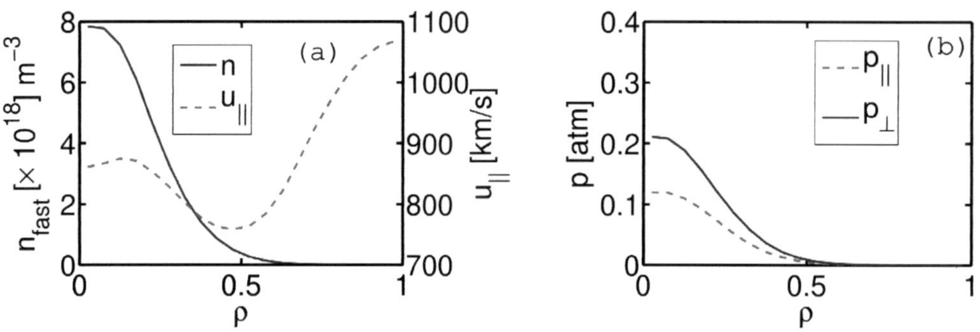

Figure 4. Profiles of (a) fast ion density and parallel mean flow speed and (b) parallel and perpendicular pressures, computed from a TRANSP distribution function for MAST discharge #18696. The horizontal axis is normalised radius.

5. Conclusions

In this paper we have explored the effect of energetic particles and rotation on tokamak plasmas. A multiple energetically resolved fluid model was described, and related to an illustrative distribution function for a plasma with neutral beam, α, and radio frequency heating. Using the new model, and its accompanying numeric solver `FLOW-M`, we have computed a MAST-like equilibrium comprising a slowly rotating thermal population and a rapidly rotating neutral beam population, broadly consistent with MAST data. We have also summarised earlier parameter scans that explore how the plasma configuration changes with varying fast ion flow speed and pressure. These show that although the poloidal flux has only a weak dependence on the energetic fluid, the dependency with the safety factor is strong. Resolving the different energetic components of the distribution function may thus be important for assessing MHD stability, even of the thermal fluid.

As the poloidal flow approaches transonic speeds, we have also re-examined the change in behaviour of the normally elliptic flow-modified Grad-Shafranov equation, which becomes hyperbolic in part of the solution domain. Associated with the spatial transitions from elliptic to hyperbolic behaviour are radial discontinuities in the plasma parameters, such as density. By computing the discriminant of the second order derivatives in the flow-modified Grad-Shafranov equation for both large aspect ratio tokamaks and the spherical tokamak, we have shown that the transition from elliptic to hyperbolic regimes occurs at lower values of the poloidal Mach number, and for a wider range. In the framework of the MHD model such transitions imply the presence of discontinuities in variables such as density, and so the profiles have gradients that are not consistent with the MHD ordering of length scales. One would thus expect discontinuities in MHD equilibria to be replaced with smooth gradients (for example, on length scales of the order of the ion skin depth or Larmor radius) in two-fluid or kinetic models [3, 9]. Indeed it has been shown explicitly that no transition to hyperbolic behaviour occurs at transonic values of the poloidal flow when either two-fluid [9] or kinetic [3] effects are taken into account. Moreover there appears to be no clear experimental evidence of density discontinuities of this type associated with elliptic/hyperbolic transitions occurring in present-day tokamak plasmas [10]. Finally, we report on the calculation of the pressure anisotropy of a TRANSP simulation of a MAST discharge, showing that the effects of anisotropy in MAST plasmas may be quite significant. In ongoing work we plan to compute the impact of this level of anisotropy on the plasma.

Acknowledgments

This work was partly funded by the United Kingdom Engineering and Physical Sciences Research Council under grant EP/G003955, the European Communities under the contract of Association between EURATOM and CCFE, and the Australian Research Council through grant FT0991899. The views and opinions expressed herein do not necessarily reflect those of the European Commission.

References

[1] E. Hameiri. The equilibrium and stability of rotating plasmas. *Phys. Fluids*, 26(1):230–237, January 1983.

[2] K. G. McClements and A. Thyagaraja. Azimuthally symmetric magnetohydrodynamic and two-fluid equilibira with arbitrary flows. *Mon. Not. R. Astron. Soc.*, 323:733–742, 2001.

[3] R. Iacono, A. Bondeson, F. Troyon, and R. Gruber. Axisymmetric toroidal equilibrium with flow and pressure. *Phys. Fluids B*, 2(8):1794–1803, 1990.

[4] R. J. Hawryluk. An Empirical Approach to Plasma Transport. In B. Coppi, editor, *Physics of plamas close to thermonuclear conditions*, volume 1, pages 19–46, CEC Brussels, 1980.

[5] M. J. Hole and G. Dennis. Energetically resolved multiple-fluid equilibrium of tokamak plasmas. *Plas. Phys. Con. Fus.*, 51:035014, 2009.

[6] L. Guazzotto, R. Betti, J. Manickam, and S. Kaye. Numerical study of tokamak equilibria with arbitrary flow. *Phys. Plasmas*, 11(2):604–614, February 2004.

[7] M.J. Hole, R.J. Akers, L.C. Appel, R.J. Buttery, C. Brickley, N.J. Conway, M. Gryaznevich, T.C. Hender, O.J. Kwon, M. Valovic, S. Medvedev, A. Patel, S. Saarelma, D. Taylor, H.R. Wilson, and the MAST Team. Ideal MHD stability of the mega-ampere spherical tokamak. *Plasma Phys. Controlled Fusion*, 47:581, 2005.

[8] A. R. Field, J. McCone, N. J. Conway, M. Dunstan, S. Newton, and M. Wisse. Comparison of measured poloidal rotation in MAST spherical tokamak plasmas with neo-classical predictions. *Plas. Phys. Con. Fus.*, 51:105002, 2009.

[9] J. P. Goedbloed. Variational principles for stationary one- and two- fluid equilibria of axisymmetric laboratory and astrophysical plasmas. *Phys. Plasmas*, 11:L81–L84, 2004.

[10] K. G. McClements and M. J. Hole. On steady poloidal and toroidal flows in tokamak plasmas. *Phys. Plas.*, 17:082509, 2010.

Theory of Fusion Plasmas: Joint Varenna–Lausanne International Workshop IOP Publishing
Journal of Physics: Conference Series **260** (2010) 012014 doi:10.1088/1742-6596/260/1/012014

Global linear gyrokinetic simulations for LHD including collisions

K Kauffmann[1], R Kleiber[1], R Hatzky[2], M Borchardt[1]

[1] Max-Planck-Institut für Plasmaphysik, EURATOM-Association, D-17491 Greifswald, Germany

[2] Max-Planck-Institut für Plasmaphysik, EURATOM-Association, D-85748 Garching, Germany

E-mail: karla.kauffmann@ipp.mpg.de

Abstract. The code EUTERPE uses a Particle-In-Cell (PIC) method to solve the gyrokinetic equation globally (full radius, full flux surface) for three-dimensional equilibria calculated with VMEC. Recently this code has been extended to include multiple kinetic species and electromagnetic effects. Additionally, a pitch-angle scattering operator has been implemented in order to include collisional effects in the simulation of instabilities and to be able to simulate neoclassical transport. As a first application of this extended code we study the effects of collisions on electrostatic ion-temperature-gradient (ITG) instabilities in LHD.

1. Introduction

Particle and heat transport in fusion devices often exceed the neoclassical prediction. This anomalous transport is produced by turbulence which is probably caused by microinstabilities such as ion and electron-temperature-gradient (ITG/ETG) and trapped-electron-mode (TEM) instabilities, the later ones known for being strongly influenced by collisions. In stellarators, neoclassical transport can be dominant in the core, therefore, collisions gain additional importance. It is also relevant to study the interplay between neoclassical transport and these instabilities for stellarator configurations.

The aim of this paper is twofold: First, to verify the collisional code by running simulations for a cylinder and a tokamak. Collisional effects are rather diverse for different geometries: Slab ITGs are much more sensitive to collisions than toroidal ITGs [1, 2]. These simulations provide the background for the second aim, the physical application, i.e. to investigate how ITG modes react to collisions in a stellarator. Stellarators have the complexity of being three-dimensional and ITG modes thus can be destabilized by a mixture of curvature and slab drive. The influence of collisions on ITG instabilities is then unclear and may depend on the details of the configuration. Thus, in this work and as a first step we study the influence of collisions on ITG instabilities for the LHD stellarator.

Journal of Physics: Conference Series **260** (2010) 012014 doi:10.1088/1742-6596/260/1/012014

2. Collisional gyrokinetic model

EUTERPE [3, 4] is a global code that uses the PIC method with a 3D potential grid to solve the set of gyrokinetic equations. The distribution function of each species evolves according to

$$\frac{\mathrm{d}f}{\mathrm{d}t} = \frac{\partial f}{\partial t} + \dot{\mathbf{R}}\frac{\partial f}{\partial \mathbf{R}} + \dot{v}_{\|}\frac{\partial f}{\partial v_{\|}} = C(f) \tag{1}$$

where $C(f)$ is a collision operator. The guiding center trajectories (characteristics) for the *collisionless* case are:

$$\dot{\mathbf{R}} = v_{\|}\hat{\mathbf{b}} + \frac{1}{B}\left(\frac{\mu B + v_{\|}^2}{\Omega_{\mathrm{i}}}\hat{\mathbf{b}} \times \nabla B + \frac{v_{\|}^2}{\Omega_{\mathrm{i}}}(\nabla \times \mathbf{B})_{\perp}\right) \tag{2}$$

$$\dot{v}_{\|} = -\mu\left(\hat{\mathbf{b}} + \frac{v_{\|}}{B\Omega_{\mathrm{i}}}(\nabla \times \mathbf{B})_{\perp}\right) \cdot \nabla B, \tag{3}$$

where $\mu \overset{\mathrm{def}}{=} \frac{v_{\perp}^2}{2B}$ and $\hat{\mathbf{b}} \overset{\mathrm{def}}{=} \frac{\mathbf{B}}{B}$. The gyroaverage is defined as

$$\nabla\langle\phi\rangle = \frac{1}{2\pi}\int_0^{2\pi} \nabla_{\mathbf{x}}\,\phi\,|_{\mathbf{x}=\mathbf{R}+\vec{\rho}}\;\mathrm{d}\alpha \tag{4}$$

with $\vec{\rho}$ being the vector for the Larmor radius and α the gyroangle. Quasineutrality, and the assumption of adiabatic electrons, closes the system by giving the field equation

$$-\nabla_{\perp} \cdot \left(\frac{m_i n_0}{q_i B^2}\nabla_{\perp}\phi\right) + \frac{e n_0}{k_{\mathrm{B}}T}\phi = \langle n\rangle = \int f(\mathbf{R}, v_{\|}, \mu)\,\delta(\mathbf{R} + \vec{\rho} - \mathbf{x})\,\mathrm{d}\mathbf{R}\,\mathrm{d}\mathbf{v}\,. \tag{5}$$

where i is the ion species. As boundary conditions for the field equation $\phi = 0$ has been used at the outer boundary, while natural boundary conditions were used at the center. Since we are interested only in the linear development, a phase factor, which allows the treatment of small scale modes, has been introduced: i.e. ϕ is replaced by $\tilde{\phi}\exp\left[i(M_0\theta + N_0\varphi)\right]$ and only the time evolution of $\tilde{\phi}$ needs to be simulated. For the present study, the pitch-angle (λ) scattering collision operator was used to model self-collisions:

$$C_a(f_a) = \frac{\nu_a}{2}\frac{\partial}{\partial \xi}\left(1 - \xi^2\right)\frac{\partial f_a}{\partial \xi}, \tag{6}$$

where $\xi = \cos\lambda$ and ν_a is the velocity independent collision frequency for each species a, which is a free parameter. Note that this particular operator does not conserve momentum. The spatial diffusion which arises due to the gyroaverage of the collision operator [5] was neglected.

In EUTERPE, the collisions have been implemented using the scheme proposed in [6]. In one time step, the velocity is modified in two phases: First, it evolves according to Eq. (3), giving the parallel and perpendicular velocity, and then these incoming velocities are changed by a random amount, using the following scheme:

$$v_{\mathrm{in}} = \sqrt{v_{\|}^2 + v_{\perp}^2} \tag{7}$$

$$\xi_{\mathrm{in}} = \cos\left[\operatorname{atan}\left(\frac{v_{\perp}}{v_{\|}}\right)\right] \tag{8}$$

$$\xi_{\mathrm{out}} = \sin\Delta\theta\sin\varphi\sqrt{1 - \xi_{\mathrm{in}}^2} + \xi_{\mathrm{in}}\cos\Delta\theta \tag{9}$$

$$v_{\|,\mathrm{out}} = v_{\mathrm{in}}\,\xi_{\mathrm{out}} \tag{10}$$

$$v_{\perp,\mathrm{out}} = \sqrt{v_{\mathrm{in}}^2 - v_{\|,\mathrm{out}}^2} \tag{11}$$

where φ is a random number from a uniform distribution between 0 and 2π and $\Delta\theta = 2R\sqrt{\nu\Delta t}$, where R is a random number from a normal distribution with expectation value zero and variance one.

The code uses the δf method [7] in which, for each species, the distribution function is separated into a background equilibrium part f_0 (here assumed to be a local Maxwellian) and a perturbation part δf, where $\delta f \ll f_0$. Replacing this in the original Eq. (1) we obtain

$$\frac{\mathrm{d}\delta f}{\mathrm{d}t} = C(f_0) + C(\delta f) - \frac{\mathrm{d}f_0}{\mathrm{d}t}. \tag{12}$$

This allows us to compute the time evolution of the perturbation only, which reduces the noise as compared with the full-f method. The particle discretization of δf is performed by

$$\delta f = \sum_{j=1}^{N} w_j(t)\, \delta^3(\mathbf{R} - \mathbf{R_j})\, \delta(v_\parallel - v_{\parallel j})\, \delta(\mu - \mu_j)/J_\mathrm{B}, \tag{13}$$

where $w_j(t)$ is the weight of each marker j and J_B is the phase space Jacobian. With collisions, however, the weight evolution of the Monte-Carlo markers cannot be solved directly by the method of characteristics, thus, it is imperative to use a generalized collisional δf scheme, known as the two-weight scheme [8, 9, 10]. Due to the stochastic nature of collisions, we can no longer calculate the "characteristic line" along which the convective derivative of the weight is to be evaluated. This problem can be overcome by not defining the weight in terms of the local δf, but rather treating it as an additional dimension of phase space. By taking the moments of a new kinetic equation in the extended phase space, it is possible to choose the weights to be consistent with the original δf equation. By doing so, one finds that the evaluation of the marker density becomes difficult, unless a second weight is added to the extended phase space.

For the simulation of the ITG instability we are not interested in the effect of neoclassical transport. Thus, the zeroth order part of the source term $\mathrm{d}f_0/\mathrm{d}t$, coming from the unperturbed orbits, has been neglected. For a tokamak configuration this term vanishes if a canonical Maxwellian is used. This is due to the fact that the toroidal momentum and the energy are constants of motion and a true equilibrium distribution function can be constructed. Since stellarators do not posses continuous symmetries, there is no other conserved quantity, apart from the energy and magnetic moment. This is not sufficient to define a true collisionless equilibrium with temperature or density profile. So the simplest choice is to use a Maxwellian as a function of energy and flux surface label and to artificially neglect the zeroth order of the source term. For the pitch angle scattering operator used here we also have $C(f_0) = 0$. Therefore, the second weight does not evolve, which is equivalent to having a one-weight scheme.

3. EUTERPE simulation results

Once modified to simulate collisions, results from EUTERPE were tested. As a benchmark, the neoclassical radial particle flux was calculated [11] and compared with the analytical calculations, as well as with the Drift Kinetic Equation Solver (DKES) code [12]. It is important to note that in this case the zeroth order part of the source term must be kept since it drives the transport.

With collisions now implemented, we want to investigate its effects on instabilities. As a first test case, we vary the collision frequency to observe how it influences the growth rate of the ITG mode. We study this for three different geometries: cylinder, tokamak and stellarator. For all these cases we have a single species plasma and assume adiabatic electrons. Infinite $\eta_i = \kappa_\mathrm{T}/\kappa_\mathrm{n}$ was chosen with $\kappa_\mathrm{n} = 0.0$ and $\kappa_\mathrm{T} = 3.5$ constant across the plasma. Density and temperature profiles are assumed to be tanh-like with κ_n and κ_T representing the maximum of the logarithmic gradient of the density and the temperature respectively.

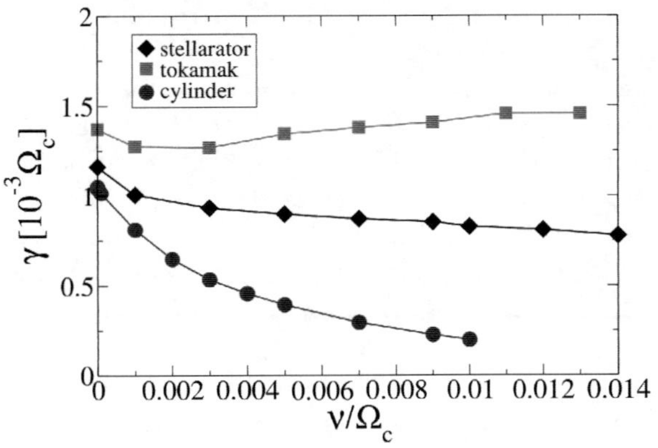

Figure 1. Growth rate dependence on collisionality for different geometries. Here, $\kappa_{\rm n} = 0.0$ and $\kappa_{\rm T} = 3.5$.

We perform a scan from $\nu = 0$ to values near $\nu = 0.01\,\Omega_{\rm c}$ (where $\Omega_{\rm c} = q_i B_0/m_i$ is the ion gyrofrequency using a reference magnetic field B_0). Larger values of ν have no physical meaning since for such a high collisionality, gyrokinetic theory itself is no longer valid.

For the tokamak configuration, an aspect ratio $A = 5$ was used with minor radius $a = 0.8$ m. An equivalent geometry was used for the cylinder. For the stellarator, the LHD configuration was used, with $R = 3.75$ m and $\beta = 1.5\,\%$. In all stellarator cases studied, the following phase factor and filter values were used: $M_0 = -37$, $N_0 = -33$, $\Delta m = 50$, $\Delta n = 4$. The grid size is $64 \times 128 \times 128$ in s, θ, φ and particle number is 16 million. It is important to note that the stellarator results cannot be quantitatively compared with the tokamak and cylindrical results, since they were computed using different physical parameters. They are plotted together for the purpose of qualitative analysis only.

Figure 1 shows the growth rate γ as a function of the collision frequency ν for the three different configurations. It is possible to observe that for the slab ITG modes in cylindrical geometry, the growth rate decreases with increasing collisionality (this is in agreement with [2]). However, if conservation of parallel momentum is added, the growth rate should slightly increase for higher collision rates [1]. In contrast to the slab ITG mode, the toroidal ITG mode (tokamak) is not significantly affected by collisionality. In the stellarator case, the ITG growth rate decreases, although not as strongly as in the cylinder case.

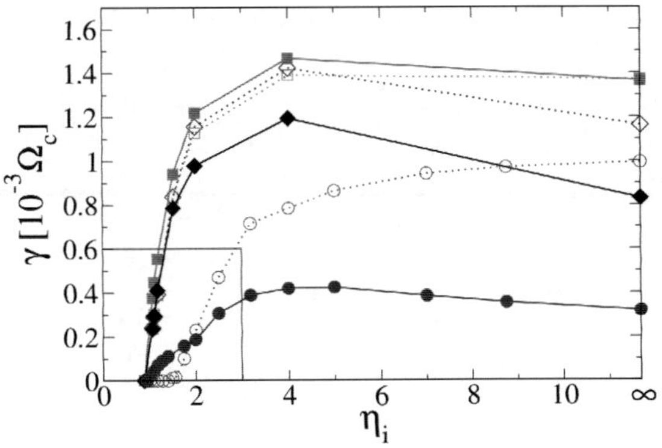

Figure 2. Growth rate dependence on η_i for a cylinder (●), a tokamak (■) and a stellarator (♦) configuration. Dotted lines and open symbols indicate the collisionless case. Solid lines and filled symbols indicate the collisional case ($\nu = 0.01\,\Omega_c$). The black box indicates the area enlarged and displayed in Figure 3.

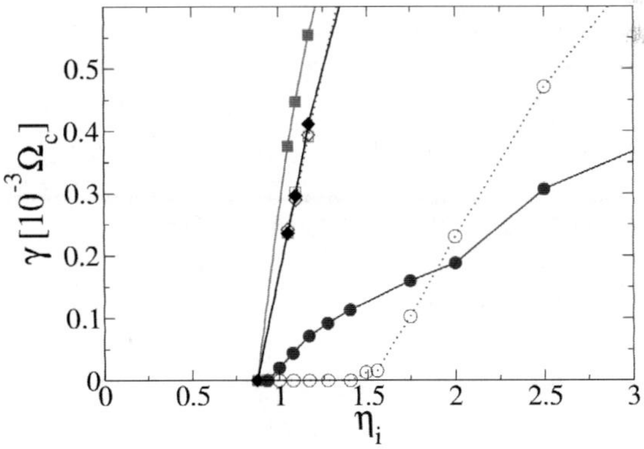

Figure 3. Growth rate dependence on η_i (zoom). Collisions only affect the instability threshold in the case of slab ITG modes.

The growth rate dependence on η_i is displayed in Figures 2 and 3. The η_i variation is performed by varying the value of $\kappa_{\rm n}$ in the range of $[0\,,4]$ and leaving the inverse of the temperature length scale constant $\kappa_{\rm T} = 3.5$. It can be noticed (Figure 3) that the instability threshold value in cylindrical geometry is reduced by collisions. This is due to the collisional coupling between parallel and perpendicular temperatures [2], allowing the system to access more degrees of freedom. In the "collisionless" plasma limit, there is no coupling, thus the threshold for slab ITG modes is higher. In the opposite case, in which the plasma is highly collisional, the threshold drops as a consequence of the additional degrees of freedom of the system and the parallel and perpendicular temperatures are now strongly coupled.

In the tokamak case, collisions do not have a significant effect on the threshold (Figure 3), which is in agreement with theoretical results [2]. Similarly for the stellarator configuration, collisions do not modify the onset of the instability. Nevertheless, for sufficiently large values of η_i, a decrease of the growth rate is observed with increasing η_i (see Figure 2) in both collisionless and collisional cases, phenomena which are not present in the tokamak case. In the collisional case it is even more pronounced.

To investigate further about the nature of the modes, we looked at the spatial structure of the potential. In Figures 4-7, the real part of the potential $\mathcal{R}e(\phi)$ for the collisional and collisionless cases at the beginning and the middle of a period is shown. Interestingly, it is found that the mode is not localized at the low field side but at the bottom of the device. Comparing the structure of the potential at the beginning and middle of a period for the collisional and non collisional case, shows that, despite the strong helical twist of LHD, the modes are nearly axisymmetric and only slightly modified by the variation of the equilibrium with φ.

When collisions are included, we find that the structures once present in the collisionless case, become wider and more irregular. It is also observed that some of the structures are slanted in comparison with their collisionless counterparts.

An additional effect of collisions can be seen in the Fourier spectrum of the electrostatic potential (see Figure 8). Without collisions the mode has a significant coupling between the m and n components of the perturbation. Despite the pronounced helical structure of LHD, the toroidal (n) sidebands are relatively small. As collisionality increases, another mode appears with almost the same growth rate, and we can see the interplay between them as they evolve in time. Above a critical value of the collision frequency, only the second mode remains. This mode shows a much weaker coupling in m. Also, for different values of $\kappa_{\rm n}$ we found that the coupling can be larger. It can be seen from the figure that the modes are well localized in Fourier space. Nevertheless, a high grid resolution and consequently many Fourier modes are necessary to prevent the growth of spurious unresolved modes, located at the edge of the filter, which can otherwise dominate the simulation.

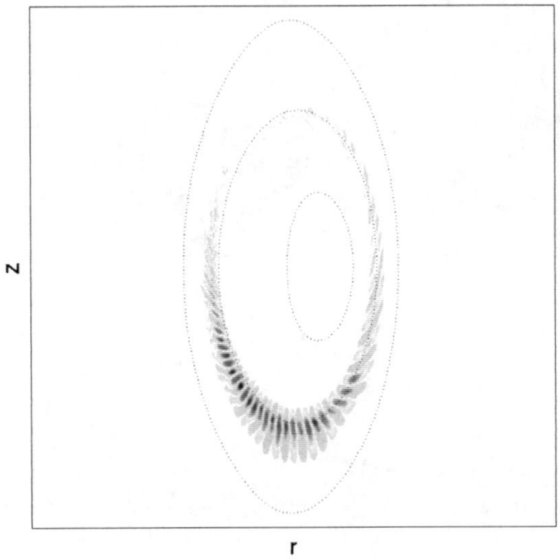

Figure 4. $\mathcal{R}e(\phi)$ for the collisionless ($\nu = 0.0$) linear ITG mode in LHD. Cross section at the beginning of a period ($\varphi = 0$). The dashed lines represent the flux surfaces $s = 0.1, 0.5, 1.0$

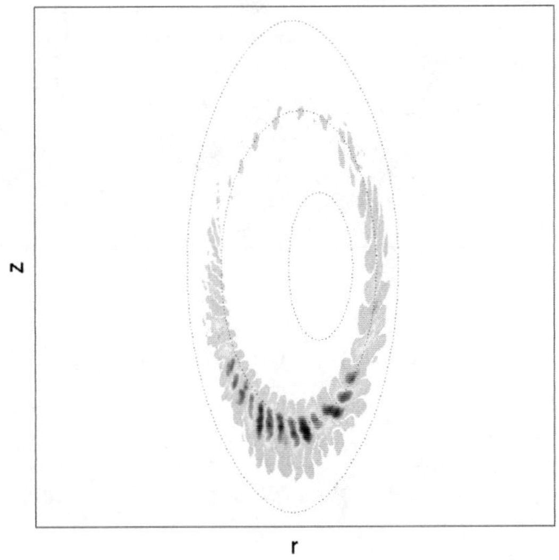

Figure 5. $\mathcal{R}e(\phi)$ for the collisional ($\nu = 0.01$) linear ITG mode in LHD. Cross section at the beginning of a period ($\varphi = 0$).

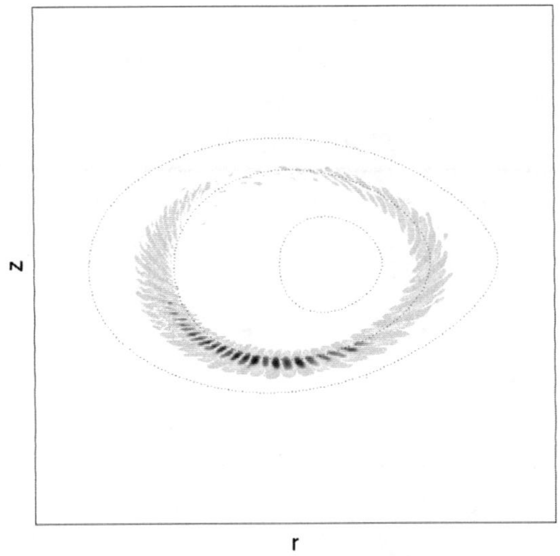

Figure 6. $\mathcal{R}e(\phi)$ for the collisionless ($\nu = 0.0$) linear ITG mode in LHD. Cross section at the middle of a period ($\varphi = \frac{1}{2}\frac{2\pi}{10}$).

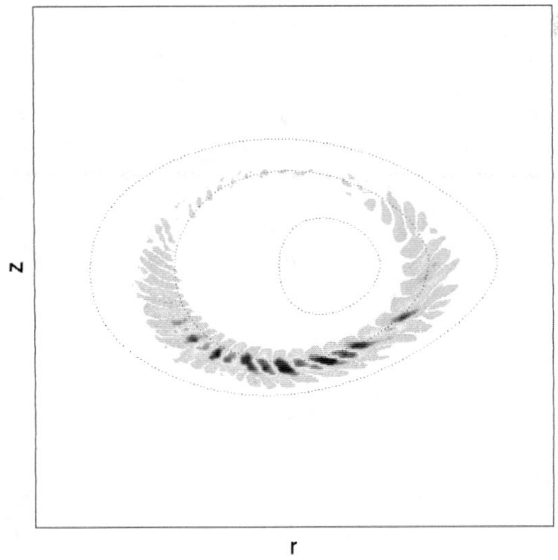

Figure 7. $\mathcal{R}e(\phi)$ for the collisional ($\nu = 0.01$) linear ITG mode in LHD. Cross section at the middle of a period ($\varphi = \frac{1}{2}\frac{2\pi}{10}$).

Journal of Physics: Conference Series **260** (2010) 012014 doi:10.1088/1742-6596/260/1/012014

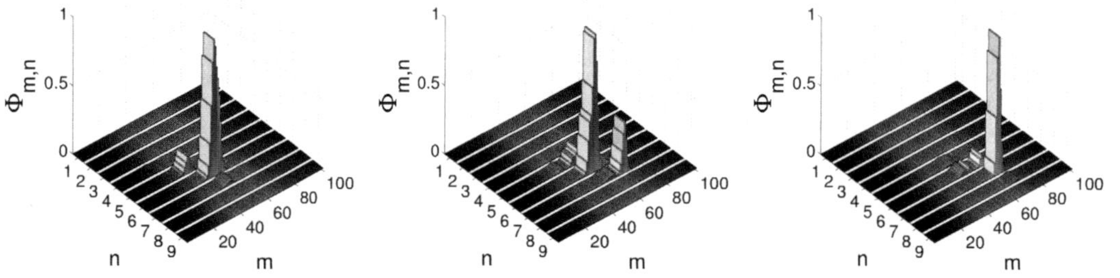

Figure 8. Fourier spectra of the ITG modes for LHD configuration. Shown here with increasing collisionality: a) $\nu = 0$, b) $\nu = 7.0 \times 10^{-3}\,\Omega_c$, c) $\nu = 1.0 \times 10^{-2}\,\Omega_c$.

4. Conclusions

The gyrokinetic code EUTERPE has been extended to include collisional effects. As a first test and application we have investigated the influence of collisions on electrostatic ITG modes in LHD for different values of η_i. For this configuration, we found a dependency of the growth rate on the collision frequency which is weaker than the one for a cylinder but stronger than the one for a tokamak. It is yet to be investigated whether this may change for other equilibrium configurations in which the ITG mode is more slab like than the one found in LHD. Since TEM modes are expected to be more sensitive to collisions, it is important to simulate such modes also for a stellarator. A study in this direction is under way.

References

[1] Chang Z and Callen J D 1992 *Phys. Fluids B* **4** 1182
[2] Dimits A M and Cohen B I 1994 *Physical Review E* **49** 709
[3] Jost G, Tran T M, Cooper W A, Villard L and Appert K 2001 *Phys. Plasmas* **8** 3321
[4] Kornilov V, Kleiber R, Hatzky R, Villard L and Jost G 2004 *Phys. Plasmas* **11** 3196
[5] Xu X Q and Rosenbluth M N 1991 *Phys. Fluids B* **3** 627
[6] Takizuka T and Abe H 1977 *J. Comput. Phys.* **25** 205
[7] Kotschenreuther M 1988 *Bull. Am. Phys. Soc.* **33** 2107
[8] Chen Y and White R B 1997 *Phys. Plasmas* **4** 3591
[9] Brunner S, Valeo E and Krommes J A 1999 *Phys. Plasmas* **6** 4504
[10] Wang W X, Nakajima N, Okamoto M and Murakami S 1999 *Plasma Phys. Contr. Fusion* **41** 1091
[11] Kauffmann K, Kleiber R and Hatzky R 2010 Collisional effects on global gyrokinetic particle-in-cell simulations of ITG and TEM instabilities in tokamaks. Poster presented at: 37th *Conference on Plasma Physics*; Dublin, Ireland.
[12] van Rij W I and Hirshman S P 1989 *Phys. Fluids B* **1** 563

Electron response in gyrofluid simulations of magnetic reconnection

A Perona[1,3], D Grasso[2,1] and D Borgogno[1]

[1] Burning Plasma Research Group, Dip. di Energetica, Politecnico di Torino, Italy
[2] Istituto dei Sistemi Complessi-CNR, Roma, Italy

E-mail: anna.perona@polito.it

Abstract. We investigate the behaviour of an electron population during the evolution of a spontaneous collisionless magnetic reconnection event, reproduced by a two-dimensional version of a Hamiltonian gyrofluid model. This model can be applied in electromagnetic problems where nonlinear effects are important, such as the sawtooth crash in the core of fusion devices. The electron dynamics in the presence of the reconnection fields is reconstructed by means of a test particle code, based on a relativistic Hamiltonian guiding centre formulation of the electron motion. The moments of the electron distribution function and the effects determined by the peculiar structure of the parallel electric field on the electron temperature in the reconnection region are reconstructed and compared with the corresponding quantities evolved by the fluid reconnection model.

1. Introduction

One of the main features of magnetic reconnection is that a local relaxation of the topological magnetic structure is accompanied by a local fast release of magnetic energy, that can be converted into heating of the plasma and particle acceleration. Thermal electrons can be accelerated by the strong electric fields formed during magnetic reconnection, thus becoming suprathermal or even relativistic [1]. Electron acceleration following sawtooth crashes and disruptions has been observed in laboratory tokamak experiments [2], [3], as well as during solar flares [4] and in diffusion regions of the Earth's magnetotail [5]. The theoretical and numerical investigation of the electron acceleration mechanism has considered various models of magnetic reconnection in different parameter regimes [6], [7], [8], [9], [10]. As these works have pointed out, the structure and, when considered, the evolution of the reconnection electric field play a significant role in determining the energy and the amount of energetic electrons generated.
In the rarefied, high temperature plasmas that characterize many configurations of interest, collisions are so rare that they do not represent an efficient mechanisms for breaking the magnetic field lines. Moreover, the electron-ion collision time can be longer than the observed relaxation time as is the case, for instance, of sawtooth crashes occurring in Tokamak esperiments.
In a fluid representation of magnetic reconnection in such collisionless plasmas, the finite electron mass can account for reconnection on so fast relaxation times (see [11] and references therein).

[3] Present address: Dipartimento di Energetica, Politecnico di Torino, corso Duca degli Abruzzi 24, 10138 Torino, Italy

In our work, we reconstruct the response of a population of thermal electrons to the fields characterizing an inertial magnetic reconnection process where a strong guide field is present. This process is described by a gyrofluid model where finite ion Larmor radius effects are retained. The resulting electric field exhibits a structure considerably different with respect to the electric field arising in the two-fluid, cold-ion model that represents the limit of the gyrofluid model for $\rho_i \to 0$ [12], [13]. In the cold-ion limit, the parallel electric field has a monopole structure, peaked in the X-point region, that significantly accelerates the electrons during the nonlinear phase of the process, although even at this stage neither highly energetic nor relativistic tails are observed in the electron distribution function [14]. Aim of our investigation is to determine whether a more significant amount of energetic electrons is generated by the finite-ρ_i electric field and how the electron distribution function and the kinetic moments are distorted with respect to the corresponding fluid quantities.
The gyrofluid model of magnetic reconnection and the equations reproducing the dynamics of the electron guiding centers in the presence of the reconnection fields are introduced in Section 2 and in Section 3, respectively. The results of the numerical simulations performed with the code where these equations have been implemented are presented in Section 4. Finally, conclusions are drawn in Section 5.

2. The gyrofluid reconnection model

In our investigation, we follow the evolution of the reconnection fields evolved by a Padé approximant version of a Hamiltonian gyrofluid model introduced in Ref. [15]. This model, that has been investigated in Ref. [16], can describe both the "inertial" ($\beta_e \ll m_e/m_i$) and the "kinetic" ($\beta_e \gg m_e/m_i$) regimes of the Alfvén wave (β_e is the ratio of the electron kinetic pressure to the magnetic pressure) and extends the models adopted in Refs. [17] and [18] by additionally evolving the ion guiding center density. In particular, we focus on a two-dimensional configuration, where the effects related to the curvature of the magnetic field lines and the dependence of the fields on z are neglected. The density of the ion guiding centers, n_i, the electron density, n_e, and the magnetic flux, $\psi = A_z$, are evolved according to the equations

$$\frac{\partial n_i}{\partial t} + [\Phi, n_i] = 0\ , \tag{1}$$

$$\frac{\partial n_e}{\partial t} + [\phi, n_e] - [\psi, \nabla^2\psi] = 0\ , \tag{2}$$

$$\frac{\partial(\psi - d_e^2\nabla^2\psi)}{\partial t} + [\phi, \psi - d_2^2\nabla^2\psi] + \rho_s^2[\psi, n_e] = 0\ , \tag{3}$$

where A_z is the $z-$component of the vector potential, $\rho_s = \sqrt{T_e m_i/(eB^2)}$ is the sound Larmor radius, related to the electron temperature, T_e, $d_e = \sqrt{m_e/(\mu_0 n e^2)}$ is the electrons skin depth, ϕ is the electrostatic potential, $\Phi = 1/(1-\rho_i^2\nabla^2)\phi$ is the gyroaveraged electrostatic potential, $\rho_i = \sqrt{T_i/T_e}\rho_s$ is the ion Larmor radius, T_i, m_i and m_e represent the ion temperature, the ion and the electron masses, respectively, and $[A, B] = \mathbf{e}_z \cdot \nabla A \times \nabla B$. Moreover, the electron density is related to the ion density and to the electrostatic potential

$$n_e = \frac{n_i}{1 - \frac{\rho_i^2}{2}} + \frac{\nabla^2}{1 - \rho_i^2\nabla^2}\phi\ . \tag{4}$$

Equations (1)-(4) have been normalized according to

$$(t, x, d_e, n_i, n_e, \psi, \phi) = (\frac{v_A}{L}\hat{t}, \frac{\hat{x}}{L}, \frac{\hat{d}_e}{L}, \frac{L}{\hat{d}_i}\frac{\hat{n}_i}{n_0}, \frac{L}{\hat{d}_i}\frac{\hat{n}_e}{n_0}, \frac{\hat{A}_z}{BL}, \frac{\hat{\rho}_s^2}{L^2}\frac{L}{d_i}\frac{e\hat{\phi}}{T_e}) \tag{5}$$

where $v_A = L/\tau_A$, L is the macroscopic scale length and $\tau_A = \sqrt{nM}L/B_{y0}$ is the Alfvén time. The fluid fields characterizing the magnetic reconnection process are calculated considering an equilibrium magnetic flux function, $\psi_{eq} = 0.1/\cosh^2 x$, unstable to perturbations around the rational surface located at $x = 0$ and chosen in such a way as to avoid too early cross-talking between magnetic islands. In particular, the analytical equilibrium flux function is approximated up to 11 modes by a truncated Fourier series, since we impose double periodic boundary conditions. The ion and the electron densities are assumed to be uniform at the equilibrium, and the temperatures for both the ions and the electrons are taken to be constant throughout the reconnection process. The initial perturbation is applied on the ion density, $\tilde{n_i} \propto \cos(x+y) - \cos(x-y)$.

3. The electron equations

The dynamics of the electron guiding centers in the presence of the reconnection fields has been reconstructed in the same geometry of the gyrofluid model introduced in Sec. 2. In order to take into account the eventuality of a very large increase in the electron velocity, we adopted a relativistic Hamiltonian formulation [19]. The unperturbed relativistic guiding-centre phase-space Lagrangian is expressed in terms of the guiding-centre coordinates as $L \equiv (p_{||}\mathbf{b} - e\mathbf{A}) \cdot \dot{\mathbf{R}} - \frac{m}{e}\mu\dot{\theta} - W$, where $\mathbf{b} = \mathbf{B}/B$, $p_{||} \equiv \mu_0\gamma\mathbf{v}\cdot\mathbf{b}$, $\gamma \equiv \sqrt{1 + (p_{||}^2 + p_\perp^2)/(m_0^2c^2)}$, $m \equiv m_0\gamma$, $\mathbf{R}$ denotes the guiding-centre position, $(p_{||}, \mu, \theta)$ are the guiding-centre momentum coordinates, and (t, W) are the guiding-centre time and energy coordinates. The usual procedure yields the equations of motion from the relativistic Hamiltonian, i.e.

$$\dot{y} = \partial H/\partial p_y\ , \qquad \dot{z} = \partial H/\partial p_z\ , \tag{6}$$

$$\dot{p_y} = -\partial H/\partial y\ , \qquad \dot{p_z} = -\partial H/\partial z\ , \tag{7}$$

where $H = \sqrt{p_{||}^2c^2 + 2\mu B m_0 c^2 + m_0^2c^4} - e\Phi$, $\mu = p_\perp^2/(2m_0B)$ is the magnetic momentum, $W = e\Phi$ is the energy, c is the speed of light, and the conjugate momenta to the y and z coordinates are given by

$$p_y = p_{||}b_y - eA_y\ , \qquad p_z = p_{||}b_z - eA_z\ . \tag{8}$$

The resulting set of equations describes the evolution in time of the $(x, y, p_{||})$ guiding-centre coordinates [14]

$$\dot{x} = \frac{1}{(b_yg_z - b_zg_y)}\{b_y[\dot{p}_z + e(\frac{\partial\tilde{\psi}}{\partial t} + \frac{\partial\tilde{\psi}}{\partial y}\dot{y}) - p_{||}(\frac{\partial b_z}{\partial t} + \frac{\partial b_z}{\partial y}\dot{y})] - b_z[\dot{p}_y - p_{||}(\frac{\partial b_y}{\partial t} + \frac{\partial b_y}{\partial y}\dot{y})]\}\ , \tag{9}$$

$$\dot{y} = \frac{p_{||}^2\frac{\partial b_z}{\partial x} - p_{||}e\frac{\partial A_z}{\partial x} - mb_z(\mu\frac{\partial B}{\partial x} - e\frac{\partial\Phi}{\partial x})}{m(g_zb_y - g_yb_z)}\ , \tag{10}$$

$$\dot{p}_{||} = \frac{1}{(b_yg_z - b_zg_y)}\{g_z\dot{p}_y - g_y\dot{p}_z + g_y[p_{||}(\frac{\partial b_z}{\partial t} + \frac{\partial b_z}{\partial y}\dot{y}) - e(\frac{\partial\tilde{\psi}}{\partial t} + \frac{\partial\tilde{\psi}}{\partial y}\dot{y})] - g_zp_{||}(\frac{\partial b_y}{\partial t} + \frac{\partial b_y}{\partial y}\dot{y})\}\ , \tag{11}$$

where the two quantities g_y and g_z are defined as

$$g_y = (p_{||}\frac{\partial b_y}{\partial x} - e\frac{\partial A_y}{\partial x})\ , \qquad g_z = (p_{||}\frac{\partial b_z}{\partial x} - e\frac{\partial A_z}{\partial x})\ .$$

and the components of the vector potential can be expressed in terms of the magnetic field components as $A_x = 0$, $A_y = B_0x$, and $A_z = (\psi_{eq} + \tilde{\psi})$.

3.1. Kinetic moments in the δf approach

The behaviour of the electron guiding-centers has been reconstructed by means of a test particle code, where Eqs. (9)-(11) have been implemented and a Maxwellian distribution function has been assumed for the electrons at the equilibrium. We adopted a δf technique, which has the advantage of reducing the number of particles required in order to adequately fill the four-dimensional phase space $(x, y, p_{||}, p_{\perp})$ while resolving small fluctuation in the electron distribution function [20]. The δf algorithm assigns to each particle (*marker*), whose dynamics is described by the equations introduced in Sec. 3, a *weight*, expressing the number of particles represented by the marker itself. The kinetic momens are then derived in each cell (labelled by i) of the real space by summing the contribution expressed by the weight of the markers, w_j. This yields to the following expressions for the perturbed density

$$(\tilde{n_e})_i = \sum_{j\in i}(w_j \Delta\Gamma_j^{(c)})/(\Delta\Gamma_i^{(p)}) \,, \tag{12}$$

and the current density

$$\tilde{J} = \sum_{j\in i}(e p_{||j} w_j \Delta\Gamma_j^{(c)})/(m\gamma\Delta\Gamma_i^{(p)}) \,, \tag{13}$$

where $\Gamma_j^{(c)}$ represents the volume in canonical space occupied by marker number j, and $\Gamma_i^{(p)}$ is the volume of the $i-th$ cell in the real space. By the same way, the electron temperature can be obtained as $T_e \equiv p_e/(n_0 + \tilde{n}_e)$, where p_e is the trace of the pressure tensor.

4. Numerical results

Fluid simulations were performed using a grid of 1024×128 points. Periodic boundary conditions along x and y have been imposed both to the fluid and to the test particle codes. In order to carry out a comparison with the cold-ion limit of the reconnection model, we ran two fluid simulations characterized by different values of the (normalized) parameter $\rho_i = 0.2, 0, 4$, while the values of the two parameters d_e and ρ_s have been kept fixed and chosen in such a way to correspond to typical fusion devices parameters. Thus, assuming a characteristic length $L = 10^{-2}$ m, simulations were performed for the (un-normalized) values $d_e = 0.002$ m and $\rho_s = 0.004$ m, corresponding to an equilibrium density $n_0 = 0.710^{19}$ m^{-3} and an electron thermal temperature, $T_e \approx 1.53$ keV, respectively. According to the characteristic evolution time derived from the linear growth rate calculated in the simulations, which is in agreement with the asymptotic formula $\gamma \approx 2(2d_e\rho_\tau^2/\pi)^{1/3}$ $(\rho_\tau^2 = \rho_i^2 + \rho_s^2)$ [21], the fluid fields are adequately sampled by saving the output data each 7 Alfvén times for the $\rho_i = 0.2$ simulation and each 5 Alfvén times for the $\rho_i = 0.4$ simulation. With the amplitude of the perturbation adopted in the simulations (10^{-6}), the time required in order to reach the nonlinear phase varies between 110 τ_A for $\rho_i = 0$ and 50 τ_A for $\rho_i = 0.4$, while the saturated phase of the process is attained after 270 τ_A when $\rho_i = 0$ and 190 τ_A when $\rho_i = 0.4$.

The perturbed magnetic flux and the electrostatic potential calculated by the reconnection code are then read by the electron code, which evolves the spatial position, the parallel velocity and the weight of the markers. The results of the electron simulations presented in this paper have been obtained by loading 1×10^7 particles.

A first comparison between the moments provided by the electron simulations and the corresponding fluid quantities confirm a good agreement of the linear growth rate, as shown for the current density in fig. 1.

Similarly to the case $\rho_i = 0$, in the linear phase of the process the parallel electric field provided by the fluid simulations, $E_{||} = -\nabla_{||}\tilde{\Phi} - \partial\tilde{\psi}/\partial t$, exhibits a structure peaked in the X-point region also when $\rho_i > 0$. Differences arise during the nonlinear phase. As shown in fig. 2, when $\rho_i = 0$ the electric field has a positive peak still located at the X-point, whereas it is negative elsewhere. In the case $\rho_i \neq 0$, instead, the electric field structure exhibits highly positive values along fine spirals inside the magnetic island as well as all along the magnetic island separatrix (fig. 3). This difference suggests that the amount of thermal electrons that could be accelerated by the parallel electric field in the latter case could be larger than the number of accelerated particles in the case $\rho_i = 0$. In the cold-ion limit, in fact, we found that electrons crossing the X-point region start becoming significantly accelerated by the electric field at the end of the nonlinear stage, although this acceleration is not so important as to produce either highly energetic tails or relativistic ones in the electron distribution function [14]. Also for the two considered values of $\rho_i \neq 0$, no energetic electrons are observed during the linear phase and the *parallel* electron temperature, reconstructed according to the procedure described in Sec. 3.1, remains constant until the beginning of the nonlinear phase. At this step, hotter regions appear along the separatrix close to the X-point first, and then elongate along the separatrix, according to the positive electric field, as plotted in figs. 4 and 5. When the process approaches saturation, significantly accelerated electrons are visible also in the denser regions inside the magnetic island. Such a behaviour of the temperature can be related to the positive structures of the electric field that characterize also the inner regions at this step of the process, as shown in figs. 6 and 7. This evolution of the electron temperature, that increases in $\approx 50\%$ in the hottest regions of the magnetic island both for $\rho_i = 0.2$ and $\rho_i = 0.4$, corresponds to a different deformation of the electron distribution function for the two values of ρ_i, drawn in figs. 8 and 9. While the bulk of the electron distribution is considerably smoothed in both cases, the amount of thermal electrons that increase their parallel velocity is larger when $\rho_i = 0.4$, thus giving a more significant contribution to the high energy slopes of the distribution. Although the maximum value of the parallel electric field is similar for both values of ρ_i m, in fact, in the case $\rho_i = 0.4$ this value is nearly uniform along the fine positive spirals inside the island, as well as along the island separatrix, while it characterizes only the region surrounding the X-point when $\rho_i = 0.2$, as can be seen in fig. 10. Therefore, in the latter case only electrons crossing the X-point region undergo the same relevant acceleration that affects all electrons streaming along the highly positive electric field structures when $\rho_i = 0.4$.

5. Conclusions

We have numerically investigated the response of a thermal electron population to the fields developed during collisionless reconnection in plasma regimes where the effects of the electron and ion temperatures are important. In particular, the presence of a finite ion temperature affects the evolution of the electric field parallel to the magnetic field lines during the nonlinear stage of the reconnection process, when highly positive fine structures develop not only along the magnetic separatrix, but in all the area enclosed by the magnetic island. Since this feature becomes more distinct for higher values of the ion temperature, and in order to compare the effects of such structures with those related to the cold-ion limit of the reconnection model, we considered two different values of the ion Larmor radius, corresponding to a ratio between the ion and the electron temperatures equal to 1 and 1/4, respectively. Similarly to the cold-ion case, also for $\rho_i \neq 0$ a visible electron acceleration can be observed after the beginning of the nonlinear phase, when the regions characterized by a larger density and by positive values of the electric field become rapidly hotter than the surrounding areas. This effect is more pronounced towards the saturated stage, when the fine positive structures of the electric field accelerate also the electrons streaming inside the magnetic island. As a result, the distribution function of the electron parallel velocities, initially set as a Maxwellian, is considerably deformed, with a larger amount of energetic electrons produced for the larger value of the ion temperature. Furthermore, the electron temperature, set by the (constant) electron skin depth at the beginning of the simulations, varies up to the 50% of its initial value in the hottest regions of the magnetic island. Although a more refined reconstruction of the electron temperature behaviour would be

Journal of Physics: Conference Series **260** (2010) 012015 doi:10.1088/1742-6596/260/1/012015

obtained only if the kinetic code would feed back its quantities to the fluid reconnection code, these results show that the isothermal assumption is respected during the linear phase of the reconnection process, while the higher the ion temperature is, the most it is violated during the nonlinear phase.

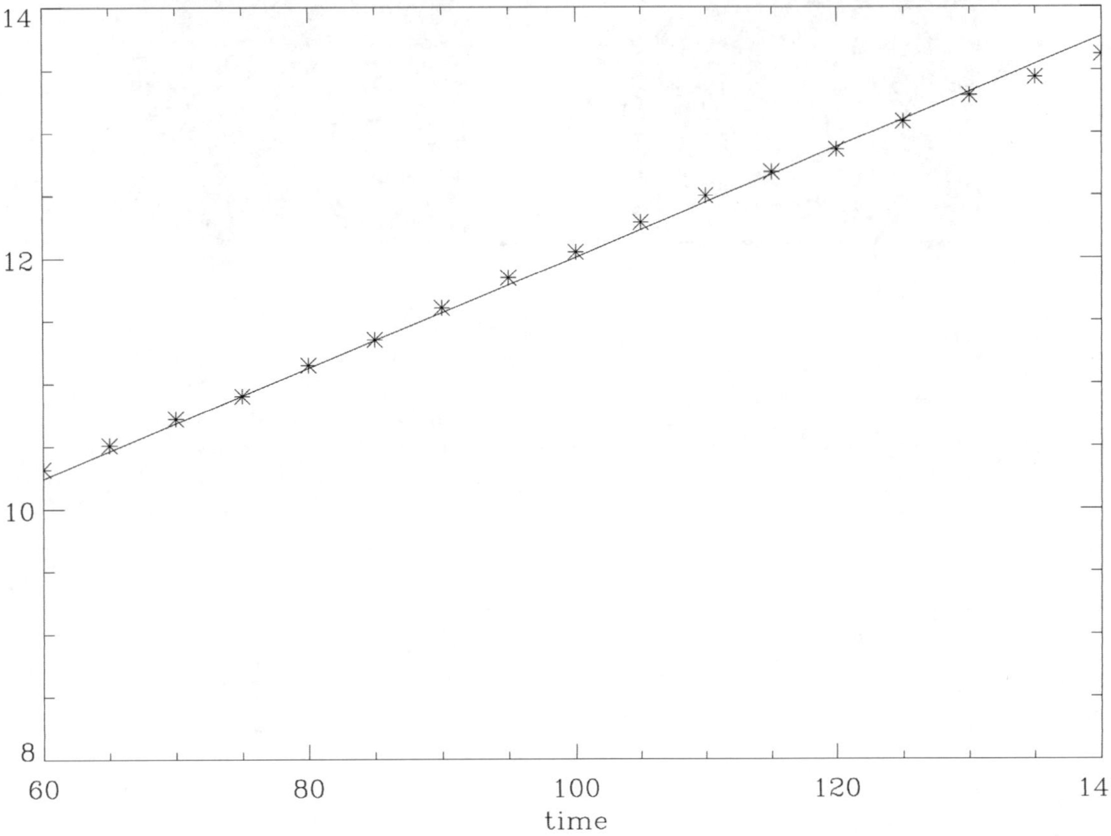

Figure 1. Growth in time of the fluid current density (continuous line) and of the kinetic current density (stars) at the $X - point$ during the linear phase of the reconnection process.

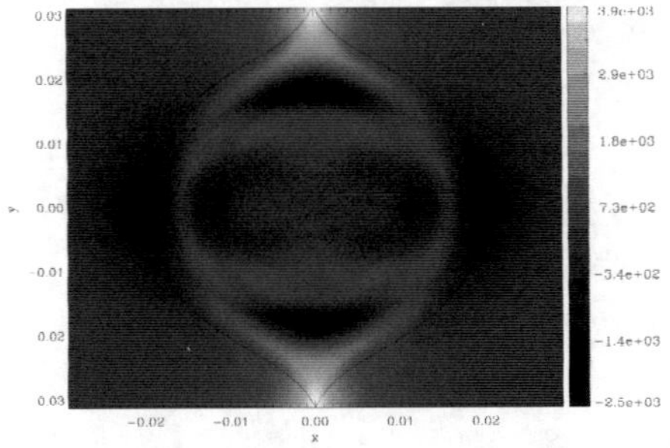

Figure 2. Contour plot of the parallel electric field for $\rho_i = 0$ m.

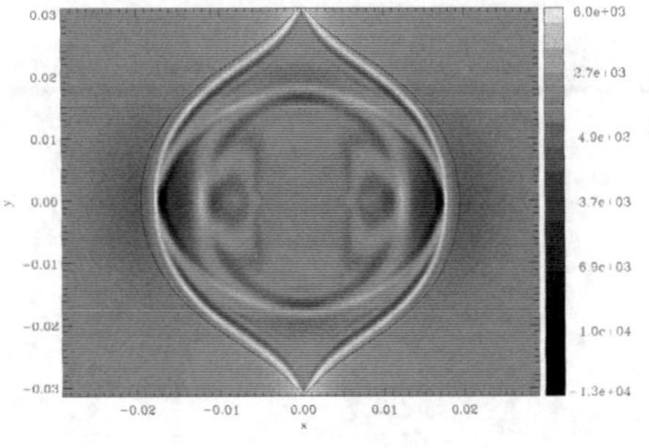

Figure 3. Contour plot of the parallel electric field for $\rho_i = 0.004$ m. The figure is taken at a step of the nonlinear phase of the reconnection process characterized by a width of the magnetic island close to the island width in fig. 2.

Journal of Physics: Conference Series **260** (2010) 012015 doi:10.1088/1742-6596/260/1/012015

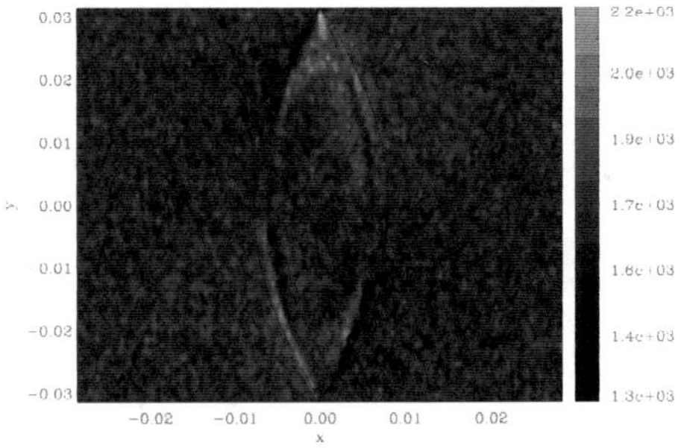

Figure 4. Contour plot of the parallel electron temperature during the nonlinear phase of the reconnection process ($\rho_i = 0.004$ m).

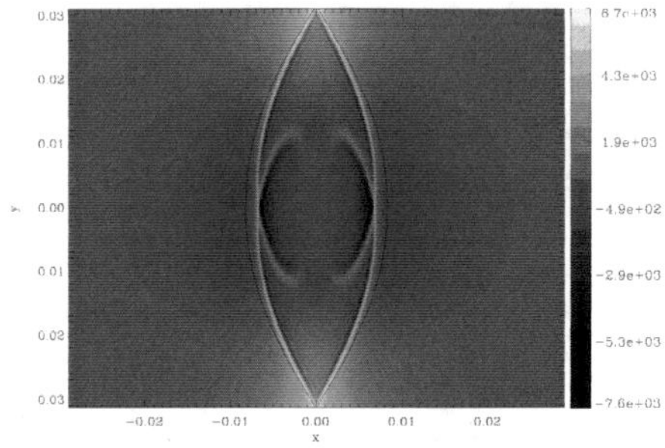

Figure 5. Contour plot of the parallel electric field taken at the same time step of the nonlinear phase of the reconnection process as fig. 4 ($\rho_i = 0.004$ m).

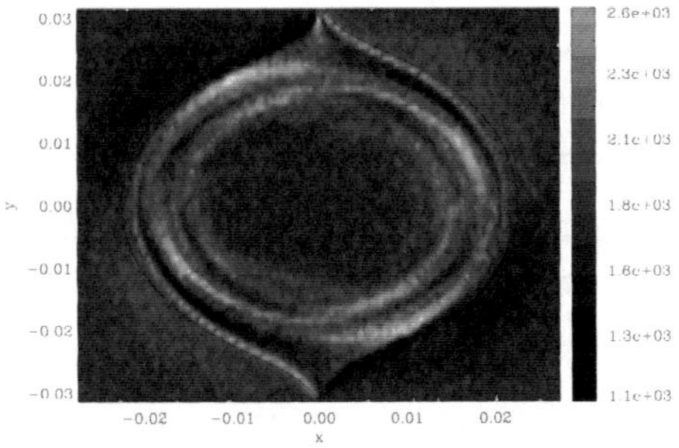

Figure 6. Contour plot of the parallel electron temperature at the saturated stage of reconnection ($\rho_i = 0.004$ m).

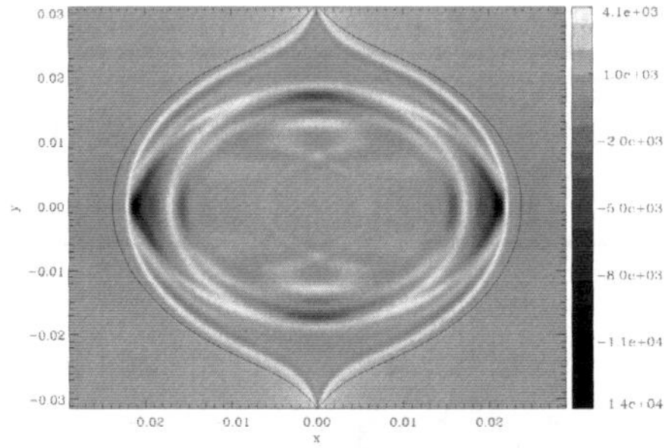

Figure 7. Contour plot of the parallel electric field taken at the same time of the saturated stage of reconnection as fig. 6 ($\rho_i = 0.004$ m).

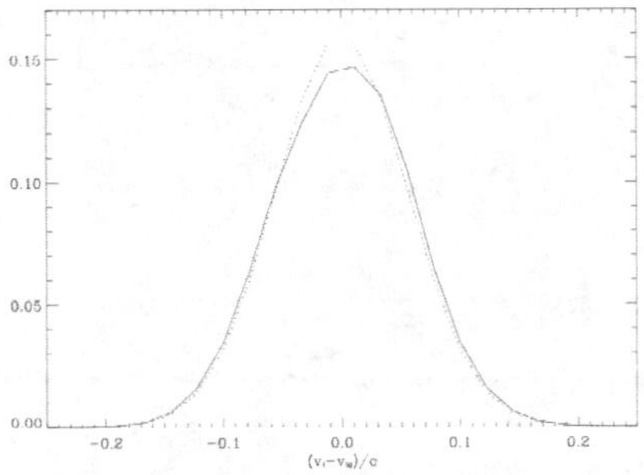

Figure 8. Profile of the distribution function of electrons as function of their parallel velocity normalized to the speed of light (continuos line). The dashed line represents the distribution function at $t = 0$ m. The distribution function refers to the case $\rho_i = 0.002$ m.

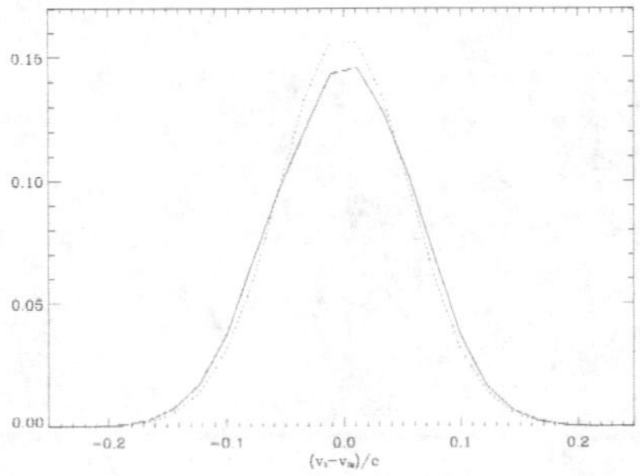

Figure 9. Profile of the distribution function of electrons as function of their parallel velocity normalized to the speed of light (continuos line) The dashed line represents the distribution function at $t = 0$ m. The figure refers to the case $\rho_i = 0.004$ m. As in fig. 8, the distribution function has been normalized to the total number of particles.

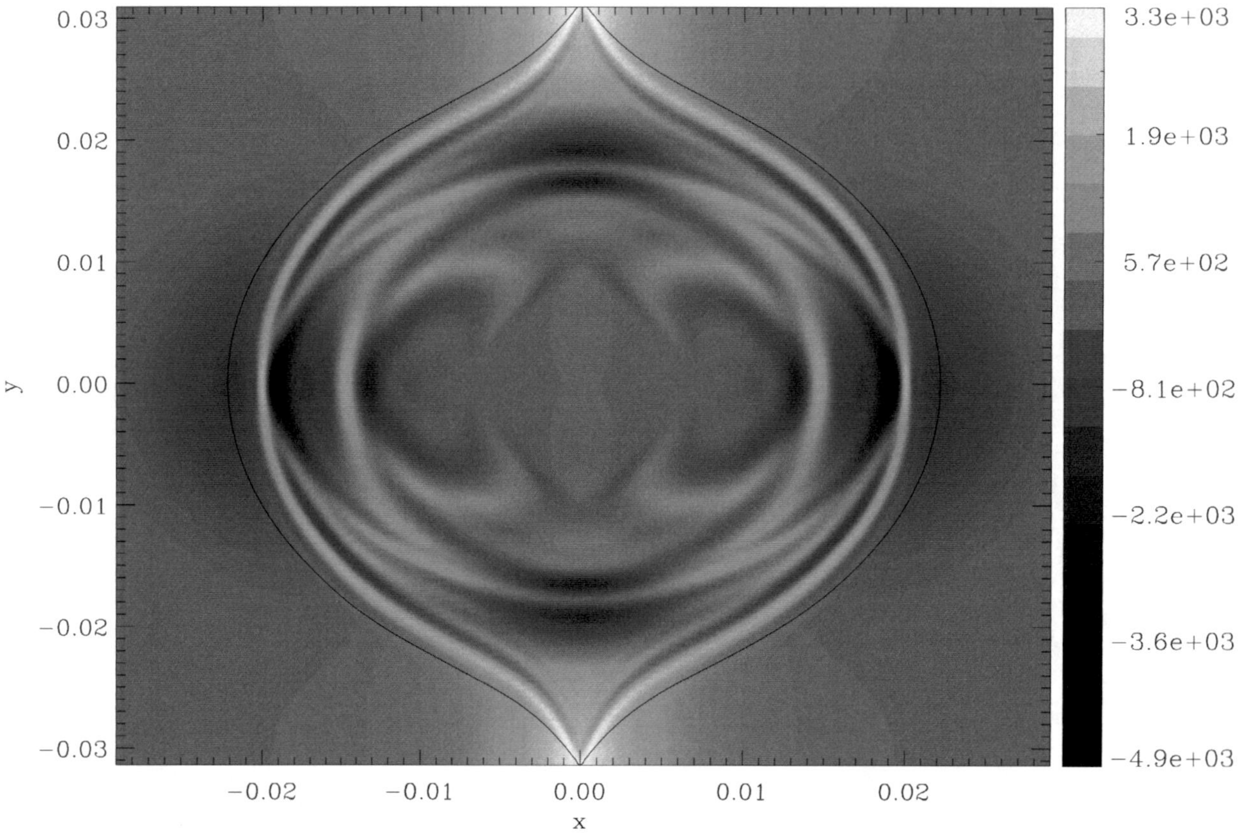

Figure 10. Contour plot of the parallel electric field taken at a step of the reconnection process characterized by an island width close to the island width in fig. 7 ($\rho_i = 0.002$ m).

Acknowledgments
The authors would like to thank dr L-G Eriksson for helpful discussions. This work was partly supported by the Euratom Communities under the contract of Association between EURATOM/ENEA. The views and opinions expressed herein do not necessarily reflect those of the European Commission.

References
[1] Speiser T W 1968 *Earth's Particles and Fields* ed B M McCormac (New York: Reinhold Book Corp.) p. 393.
[2] Savrukhin P V 2001 *Phys. Rev. Lett.* **86** 3036.
[3] Klimanov I, Fasoli A, Goodman T P, and the TCV team 2007 *Plasma Phys. Control. Fusion* **49** L1
[4] Lin R P, Krucker S, Hurford G J, Smith D M, Hudson H S, Holman G D, Schwartz R A, Dennis B R, Share G H, Murphy R J, Emslie A G, Johns-Krull C, and Vilmer N 2003 *The Ap. J.* **595** L69.
[5] Oieroset M, Pin R P, Phan T D, Larson D E, and Bale S D 2002 *Phys. Rev. Lett.* **89** 195001.
[6] Drake J F, Swisdak M, Che H, and Shay M A 2006 *Nature* **443** 553.
[7] Pritchett P L 2006 *J. Geophys. Res.* **111** A10212.
[8] Pritchett P L 2008 *Phys. Plasmas* **15** 102105.
[9] Fu X R, Lu Q M, and Wang S 2006 *Phys. Plasmas* **13** 012309.
[10] Wan W, Lapenta G, Delzanno G L, and Egedal J 2008 *Phys. Plasmas* **15** 032903.
[11] Porcelli F, Borgogno D, Califano F, Grasso D, Ottaviani M, and Pegoraro F 2002 *Plasma Phys. Control. Fusion* **44** B389.
[12] Grasso D, Califano F, Pegoraro F, and Porcelli F 2001 *Phys. Rev. Lett.* **86** 5051.
[13] Grasso D, Pegoraro F, Porcelli F, and Califano F 1999 *Plasma Phys. and Control. Fusion* **41** 1497.
[14] Perona A, Eriksson L-G, and Grasso D 2010 *Phys. Plasmas* **17** 042104.
[15] Waelbroeck F L, Hazeltine R D, and Morrison P J 2009 *Phys. Plasmas* **16** 032109.
[16] Grasso D, Tassi E, and Waelbroeck F L 2010 *Phys. Plasmas* **17**, 082312.
[17] Grasso D, Califano F, Pegoraro F, and Porcelli F 2000 *Plasma Phys. Rep.* **26** 548.
[18] Loureiro N F, and Hammett G W 2008 *J. Comput. Phys.* **86227** 4518.
[19] Brizard A J, and Chan A A 1999 *Phys. Plasmas* **6** 4548.
[20] Candy J 1996 *J. Comput. Phys.* **129** 160.
[21] Porcelli F 1991 *Phys Rev. Lett.* **66** 425.

Journal of Physics: Conference Series **260** (2010) 012016 doi:10.1088/1742-6596/260/1/012016

Diffusion coefficient and convective velocity profiles of Ni-impurity in JET plasmas with ICRH

N I Pometescu

Association EURATOM-MECTS, University of Craiova, Department of Physics, 13 A I Cuza, 200585 Craiova, Romania

E-mail: npomet@yahoo.com

Abstract. The influence of the RF heating on the Ni-impurity transport is studied from comparative analysis of two discharges in JET: the reference discharge #69808 without RF power and discharge #68383 with the maximum ICRF power of 8.3 MW which is applied to electrons in Hydrogen Minority Heating scheme. With this goal we use multi-fluid Weiland model with trace impurity approximation and input profiles of density and temperature from the two shots. For the most instable determined eigen modes are calculated diffusivity, convective velocity and peaking factor for the two shots, respectively.

1. Introduction

Experiments have shown that auxiliary heating can influence impurity accumulation and transport and, by consequence, the plasma confinement. In JET, various type of discharges at ITER relevant collisionality had shown that the density profile of Ni (Z=28) impurity is flattening when ion cyclotron resonance heating (ICRH) is applied [1]. Influence of ICRH on the impurity density profile was intensively studied but the present understanding of impurity transport and accumulation is still insufficient to predict the behaviour of elements over wide Z range. Also, some dicrepancies between calculated peaking factor and experimental measurements rest to be explained and clarified. The remainder of the paper is organized as follows. In section 2 are given the essential characteristics of the theoretical model as the impurity density perturbation and impurity particle flux due to ITG/TE modes in plasmas (obtained in [2] by using multi-fluid Weiland model). In section 3 analitycal expressions of the profiles are given to fit the experimental data for impurity and electron density and temperature profiles from shots 69808 and 68383 in JET ([3], [4]). The reference discharge 69808 is without RF power and the discharge 68383 is with the maximum ICRF power of 8.3 MW in H Minority Heating. The influence of the heating on the density perturbation profile is taken into account by the specific profiles of electron density and electron/impurity temperature profiles for the two shots. Eigenvalues for the ITG mode in the inner region of the plasma are determined in section 4. The final section is dedicated to the conclusions and evaluation of the diffusivity, convective velocity and normalized impurity density peaking factor for Ni in the most unstable eigen-mode determined previously.

2. Characteristics of the model

The Weiland multi-fluid model with trace impurity approximation is used to describe ITG/TE mode turbulence and the impurity species.The equation for the impurity/ion density perturbation due to ITG modes in plasmas with radio-frequency heating, obtained in [2] read as

$$\widetilde{n}_z = \left\{\widetilde{\omega}\left(\frac{R}{2L_{nz}}-1\right)-\tau_z^*\left(\frac{R}{2L_{Tz}}-\frac{7}{3}\frac{R}{2L_{nz}}+\frac{5}{3}\right)+\frac{Z}{A_z q_*^2}\left(\frac{\widetilde{\omega}+5\tau_z^*/3}{\widetilde{\omega}-2\tau_z^*}\right)\right\}\frac{\widetilde{\phi}}{N} \quad (1)$$

where $N=\widetilde{\omega}^2+10\tau_z^*\widetilde{\omega}/3+5\tau_z^{*2}/3$. Here $\widetilde{\phi}=e\phi/T_e$ is the normalized potential, $\widetilde{n}_z=\delta n_z/n_z$ is the normalized impurity perturbation, $\widetilde{\omega}=\omega/\omega_{De}$ the normalized frequency (where ω_{De} is the electron magnetic drift frequency). The other notations are $\tau_z^*=T_z/ZT_e$, $1/L_{nj}=-d\ln n_j/dr$, $1/L_{Tj}=-d\ln T_j/dr$, $A_z=m_z/m_i$, $q_*=2qk_\theta\rho_s$, $\rho_s=c_s/\Omega_{ci}$, $c_s=\sqrt{T_e/m_i}$.

The quasilinear impurity particle flux

$$\Gamma_{nz}=-D_z\nabla n_z+n_zV_z \quad (2)$$

is given by - see eq.(8) in [2],

$$\begin{aligned}\frac{\Gamma_{nz}}{n_zc_s} &= \frac{k_\theta\rho_s\overline{\gamma}\left|\widetilde{\phi}_k\right|^2}{|N|^2}\left\{\frac{R}{2L_{nz}}\left(|\widetilde{\omega}|^2+\frac{14\tau_z^*}{3}\widetilde{\omega}_r+\frac{55\tau_z^{*2}}{9}\right)-\frac{R}{2L_{Tz}}\left(2\tau_z^*\widetilde{\omega}_r+\frac{5\tau_z^{*2}}{3}\right)\right. \\ &\left.-\left(|\widetilde{\omega}|^2+\frac{10\tau_z^*}{3}\widetilde{\omega}_r+\frac{35\tau_z^{*2}}{9}\right)+\frac{Z}{A_zq_*^2|N_1|^2}\left[\tau_z^*\left(\frac{19}{3}\widetilde{\omega}_r^2-\frac{\overline{\gamma}^2}{3}+\frac{100\tau_z^*}{9}\widetilde{\omega}_r-2\widetilde{\omega}_r|\widetilde{\omega}|^2\right)\right]\right\}\end{aligned} \quad (3)$$

where $N_1=\widetilde{\omega}-2\tau_z^*$. In (2) D_z and V_z are the impurity diffusivity and convective velocity, respectively.

3. Input profiles

In the following are given the profiles which will be used as input data. They fit the profiles given in [3] for the reference discharge #69808 without RF power and discharge #68383 with the maximum ICRF power of 8.3 MW applied to electrons in Hydrogen Minority Heating scheme in JET. A subscript h will distinguish the quantities in the presence of ICRH from those without RF heating. The impurity considered here is Ni (Z=28). With $x=r/a$, the impurity density profiles n_z , n_{z-h}, electron density n_e, n_{e-h}, impurity temperature T_z, T_{z-h}, electron temperature T_e, T_{e-h} and safety factor q and q_h profiles are given, respectively, in (4)- (12) and plotted in figures 1-5. The values for n_{0e}, T_{0z} and T_{0e} are given as $n_{0e}=3.5\times10^{19}$ m^{-3}, $T_{0z}=9.4$ keV, $T_{0e}=5.66$ keV.

$$\begin{aligned}n_z &= n_{0z}\left\{0.54-\frac{x}{2}+0.4\exp\left[-30(x-0.01)^2\right]+0.14\exp\left[-15(x-0.34)^2\right]\right. \\ &\left.+0.13\exp\left[-20(x-0.8)^2\right]+0.12\exp\left[-200(x-0.95)^2\right]\right\}\end{aligned} \quad (4)$$

$$\begin{aligned}n_{z-h} &= n_{0z}\left\{0.38\exp\left[-10x^2\right]+0.43\exp\left[-30(x-0.39)^2\right]+0.28\exp\left[-12(x-0.6)^2\right]\right. \\ &+0.33\exp\left[-8(x-0.7)^2\right]+0.4\exp\left[-10(x-0.8)^2\right] \\ &\left.+0.28\exp\left[-50(x-0.93)^2\right]-0.9\exp\left[-10^3(x-1.01)^2\right]\right\}\end{aligned} \quad (5)$$

$$n_e = n_{0e}\left\{0.91-0.91\,x^2+0.1\exp\left[-15(x+0.1)^2\right]+0.07\exp\left[-10(x-0.85)^2\right]\right\} \quad (6)$$

$$\begin{aligned}n_{e-h} &= n_{0e}\left\{0.8-0.78\,x^3+0.07\exp\left[-30x^2\right]+0.08\exp\left[-20(x-0.3)^2\right]\right. \\ &\left.+0.03\exp\left[-80(x-0.7)^2\right]+0.1\exp\left[-100(x-0.9)^2\right]\right\}\end{aligned} \quad (7)$$

$$
\begin{aligned}
T_z &= T_{0z}\left\{0.37 - 0.4x + 0.64\exp[-9x^2] + 0.06\exp\left[-16\,(x-0.6)^2\right]\right. \\
&\quad \left. + 0.12\exp\left[-20\,(x-0.9)^2\right]\right\} && (8)\\
T_{z-h} &= T_{0z}\left\{1.04\exp[-4\,(x+0.14)^2] + 0.21\exp\left[-5\,(x-0.7)^2\right]\right\} && (9)\\
T_e &= T_{0e}\left\{0.6 - 0.7x + 0.4\exp[-14(x-0.08)^2] + 0.2\exp\left[-30\,(x-0.35)^2\right]\right. && (10)\\
&\quad \left. + 0.18\exp\left[-20\,(x-0.65)^2\right] + 0.2\exp\left[-90\,(x-0.92)^2\right]\right\} \\
T_{e-h} &= T_{0e}\left\{0.93 - 1.04x + 0.97\exp[-10(x+0.1)^2] + 0.45\exp\left[-30\,(x-0.35)^2\right]\right. \\
&\quad \left. + 0.24\exp\left[-20\,(x-0.65)^2\right] + 0.28\exp\left[-90\,(x-0.92)^2\right]\right\} && (11)\\
q &= 1.5 + x^2 + 4x^3 + 2x^4, \qquad q_h = 1 + x^2 + 4x^3 && (12)
\end{aligned}
$$

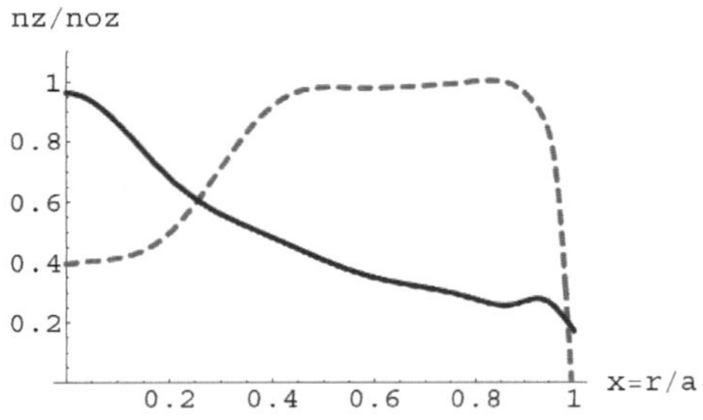

Figure 1. Ni-impurity density profiles in #69808 (full line) without RF power and discharge #68383 with ICRF power applied to electrons in Hydrogen Minority Heating (dashed line).

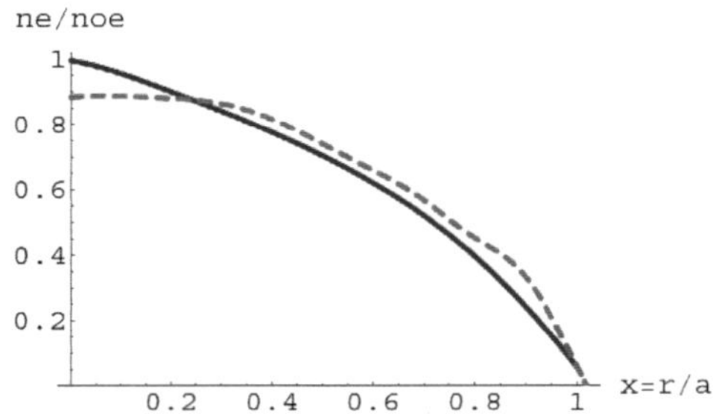

Figure 2. Electron density profiles in #69808 (full line) without RF power and discharge #68383 with ICRF power applied to electrons in Hydrogen Minority Heating (dashed line).

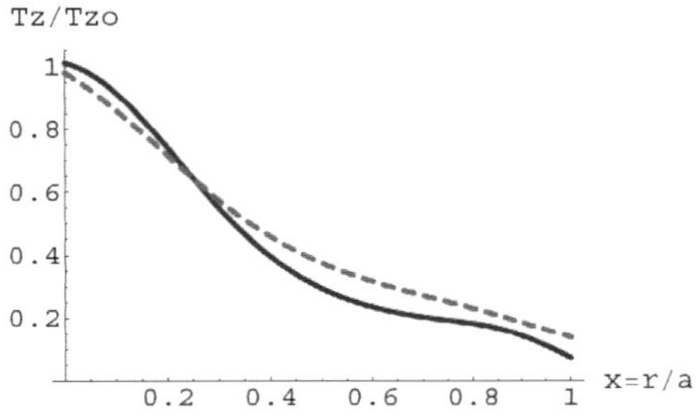

Figure 3. Ni-impurity temperature profiles in discharge #69808 (full line) and discharge #68383 (dashed line).

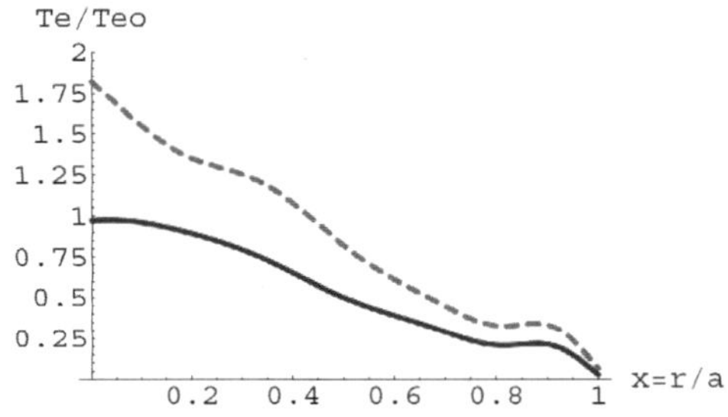

Figure 4. Electron temperature profiles in discharge #69808 (full line) and discharge #68383 (dashed line).

4. Quasineutrality condition and ITG eigenmodes

In order to represent the radial variation of the impurity density perturbation for the two shots we need to find the eigenfrequency modes for ITG instability. This imply to solve equation

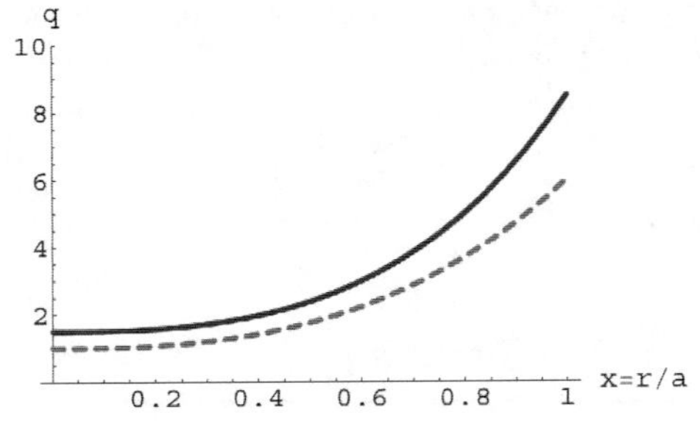

Figure 5. Safety factor profiles in discharge #69808 (full line) and discharge #68383 (dashed line).

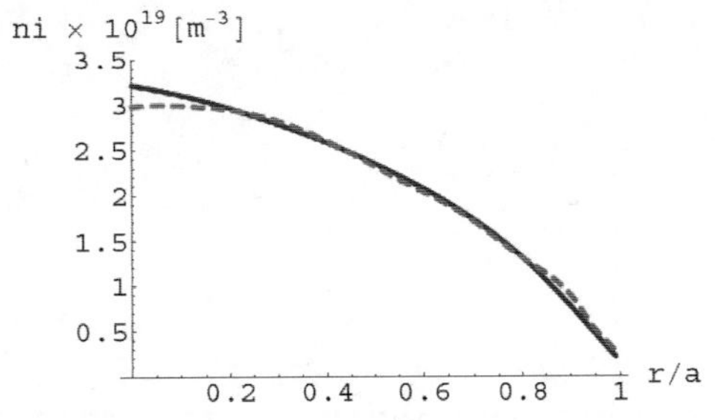

Figure 6. Ion density profiles in discharge #69808 (full line) and discharge #68383 (dashed line).

resulting from the quasineutrality condition

$$-e\ \delta n_e + \sum_i e_i\ \delta n_i = 0 \tag{13}$$

where summation over i correspond to summation over all ion/impurity plasma species. The resulting dispersion equation, which is of great complexity, was solved in different approximations; see for example [5]. In the following we assume one ion species with charge number Z_i and one impurity species (Ni) with charge number $Z = 28$. In this case quasineutrality condition (13) read as

$$\frac{\delta n_e}{n_e} = Z_i\ f_i \frac{\delta n_i}{n_i} + Z f_z\ \frac{\delta n_z}{n_z} \tag{14}$$

where

$$f_z = \frac{n_z}{n_e}\ \ ,\ \ f_i = \frac{n_i}{n_e} \tag{15}$$

From neutrality condition

$$Z_i n_i + Z n_z = n_e \tag{16}$$

and (15) we obtaine the density ion profiles - see figure 6,

$$n_i = \frac{n_e - Z n_z}{Z_i} \tag{17}$$

In the following the disscution will focuss on the inner region of plasma with $0 < r/a < 0.6$, where the contribution from trapped electrons will be neglected and the quasineutrality condition is approximated as

$$(1 - f_t)\frac{\delta n_{ef}}{n_{ef}} = (1 - Z\ f_z)\frac{\delta n_i}{n_i} + Z f_z\ \frac{\delta n_z}{n_z} \tag{18}$$

where δn_z is the impurity density perturbation, δn_i - ion density perturbation, δn_{ef} - circulating electrons density perturbation and f_t - fraction of trapped particles. The dispersion equation will be obtained from (18) by introducing $\widetilde{n_z} = \delta n_z / n_z$ and $\widetilde{n}_i = \delta n_i / n_i$ from (1) and , $\widetilde{n}_{ef} = \delta n_{ef} / n_{ef}$ for free electrons, with assuption of quasi-adiabatic behavior,

$$\frac{\delta n_{ef}}{n_{ef}} = \frac{e\ \delta\phi}{T_e} \tag{19}$$

Also, the ion (Hydrogen) temperature profile is assumed the same with impurity temperature profile. In this case, from the dispersion equation result four eigenvalues

$$\widetilde{\omega} = \widetilde{\omega}_{re} + i\widetilde{\gamma} \tag{20}$$

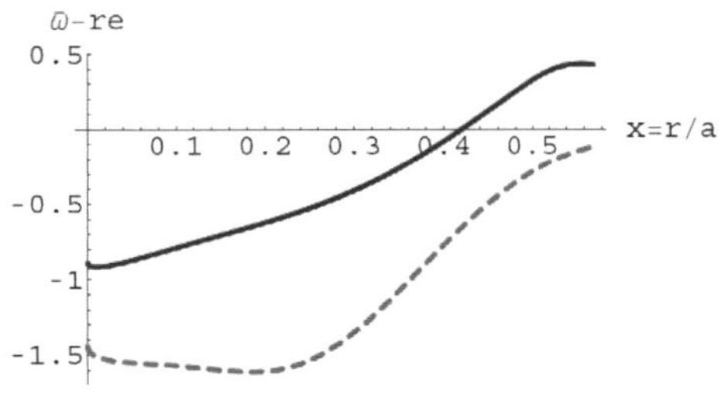

Figure 7. Re[$\widetilde{\omega}$] in discharge #69808 (full line) and discharge #68383 (dashed line).

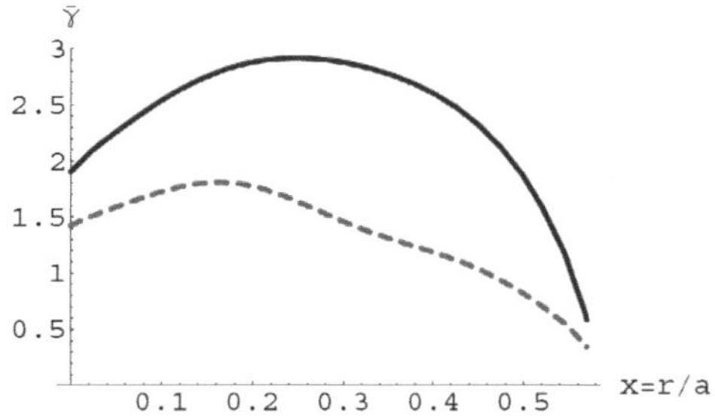

Figure 8. Growth rate of ITG mode in discharge #69808 (full line) and discharge #68383 (dashed line).

Between the four eigenmodes we choose the unstable mode with the real and imaginary part represented in figure 7 and figure 8, respectively. As can be seen, in the presence of RF heating the growth rate $\widetilde{\gamma}$ becomes smaller than in the case without RF heating which means a more 'quiet' ITG instability.

5. Discussions and conclusions

The Ni-impurity diffusivity D_z and convective velocity V_z are calculated from (3) for ITG mode eigenvalue $\widetilde{\omega}$ with previously determined characteristics plotted in figure 7 and 8 for a fixed turbulence scale $k_\theta \rho_s = 0.2$ and with potential fluctuation level $|\phi_k| = (\gamma/\omega_{*e})(1/k_\theta L_{ne})$ - see for example [5].

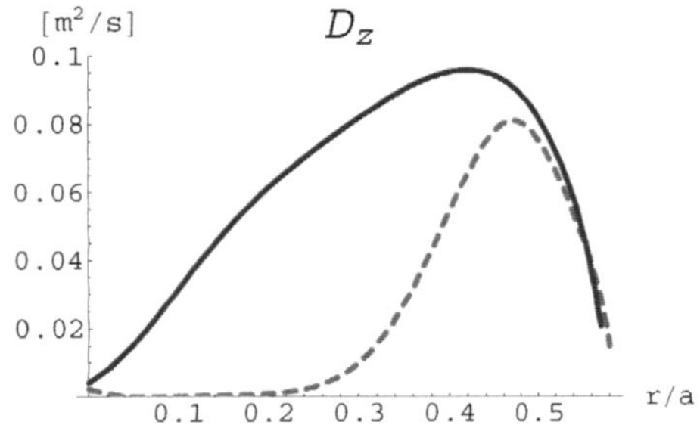

Figure 9. Diffusivity in discharge #69808 (full line) and discharge #68383 (dashed line).

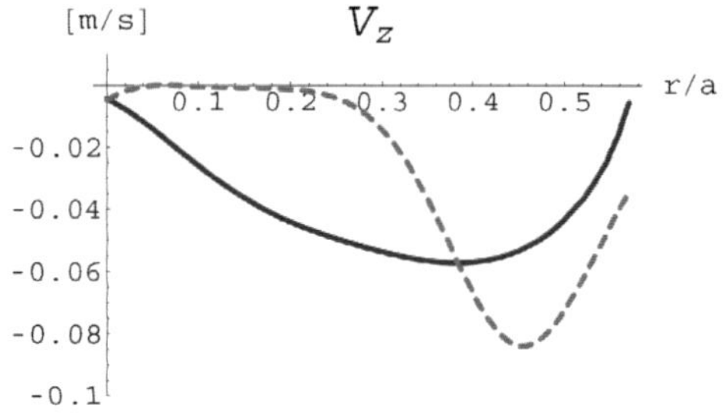

Figure 10. Convective velocity in discharge #69808 (full line) and discharge #68383 (dashed line).

In figures 9 and 10 are shown the radial profiles of D_z and V_z in the two shots. We note that in the region with $0 < r/a < 0.5$ the diffusivity in the presence of RF heating is smaller than diffusivity in absence of RF heating, with a very small value of D_{z-h} in the interval $0 < r/a < 0.3$ and a maximum value at $r/a = 0.47$ that correspond to the position where start the plateau for n_{z-h}. In the case without RF heating the diffusivity D_z rich its maximum at $r/a = 0.43$ and is much greater than D_{z-h} for $0 < r/a < 0.4$. As for the diffusivity, the presence of RF heating reduced nearly to zero the magnitude of the convective velocity in the region with $0 < r/a < 0.3$. Hence, in the central region of the plasma the transport due to ITG instability is practically suppressed as was confirmed by experimentally observations.

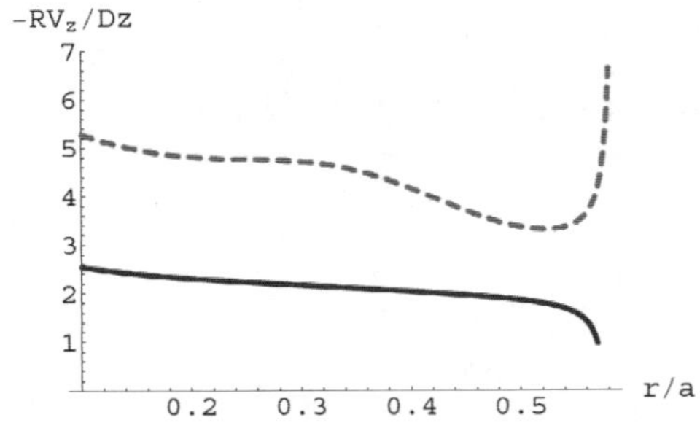

Figure 11. Normalized impurity density peaking factor in discharge #69808 (full line) and discharge #68383 (dashed line).

The normalized impurity density peaking factor $-RV_z/D_z$ (assuming a surce-free plasma) of Ni-impurities as function of radial coordinate is shown in figure 11.The density peaking factor in the case without RF heating decrease linearly with r/a in the region with $0 < r/a < 0.5$. In the case with RF heating the density peaking factor decrease with r/a also for $r/a < 0.5$ but with a plateau for $0.2 < r/a < 0.3$. The results obtained in the present approximation are relevant for ITG mode in the region with $r/a < 0.5$ and can be used as a starting point for more precise calculations in iterative procedure.

Acknowledgments

This work has been carried out within Contract of Association EURATOM-MECTS Romania. Acknowledgements to Professor Tünde Fülöp and Hans Nordman for discussions.

References

[1] Puiatti M E, Valisa M, Angioni C, Garzotti L, Mantica P, Mattioli M, Carraro L, Coffey I and Sozzi C 2006 *Phys Plasmas* **13** 042501

[2] Nordman H, Singh R, Fulop T, Eriksson L-G, Dumont R, Anderson J, Kaw P, Strand P, Tokar M and Weiland J 2008 *Phys Plasmas* **15** 042316

[3] Carraro L, Angioni C, Giroud C, Puiatti M E, Valisa M, Buratti P, Buttery R, Coffey I, Garzotti L, Van Eester D, Lauro Taroni L, Lawson K, Lerche E, Pantica P, Mattioli M and Naulin V 2007 *Proceedings of the 34th EPS Conference on Controlled Fusion and Plasma Physics* Warsaw, July 2-July 6, 2007 ECA Vol **31F**, O-4.028

[4] Valisa M, Angioni C, Carraro L, Coffey I, Predebon I, Puiatti m E, Lauro Taroni L, Alper B, Belo P, Giroud C, Van Eester D, Lerche E, Naulin V, Tala T, Tsala M and JET-EFDA contributors 2009 *Proceedings of the 36th EPS Conference on Plasma Physics*, Sofia, June 29-July 3, 2009 ECA Vol **33E**, P-2.169

[5] Weiland J 2000 *Collective Modes in Inhomogeneous Plasma*, IOP, Bristol.

Journal of Physics: Conference Series **260** (2010) 012017 doi:10.1088/1742-6596/260/1/012017

Flux-driven gyrokinetic simulations of ion turbulent transport at low magnetic shear

Y Sarazin[1], A Strugarek[1,2], G Dif-Pradalier[3], J Abiteboul[1], S Allfrey[1], X Garbet[1], Ph Ghendrih[1], V Grandgirard[1], G Latu[1]

[1] CEA, IRFM, F-13108 Saint-Paul-lez-Durance, France.
[2] CEA, IRFU, F-91191 Gif-sur-Yvette Cedex, France.
[3] Center for Astrophysics and Space Science, U.C.S.D., La Jolla, CA 92093, USA.

E-mail: `yanick.sarazin@cea.fr`

Abstract.
Ion Temperature Gradient driven turbulence is investigated with the global full-f gyrokinetic code GYSELA for different magnetic equilibria. Reversed shear and monotonous q profile cases do not exhibit dramatic changes nor in the dynamics nor in the level of turbulence, leading to similar mean profiles. Especially, no transport barrier is observed in the vicinity of $s = 0$ in the general case, although the radial extent of the gap without resonant modes is larger than the typical turbulence correlation length. Conversely, a transport barrier is found to develop in the gap region if non resonant modes are artificially suppressed from the simulation. Such simulations tend to reconcile previously published contradictory results, while extending the analysis to more realistic flux-driven gyrokinetic regimes.

1. Introduction

Turbulence and transport studies in tokamak plasmas primarily aim at investigating routes towards improved confinement regimes. In the core of fusion plasmas, internal transport barriers (ITBs) are commonly reported in magnetically confined fusion plasmas, provided the heating power exceeds some critical value (see [1] for a review on experimental results). Once the barrier is triggered, the following positive feed back loop is supposed to play a key role in its sustainment: the strong pressure gradient (and possibly its curvature as well) at the barrier position drives strongly sheared poloidal flows via the radial force balance; this shear has now long proven efficient in stabilizing the turbulence through decorrelation processes of convective cells [2]. Two main ingredients are invoked to explain the empirical threshold for the onset of ITBs (see e.g. [3] for a review): velocity shear and magnetic topology. The former can be either intrinsic, i.e. self-generated by the turbulence itself, or extrinsic in the presence of sources of momentum. This issue has been extensively studied in the literature and will not be addressed here (see [4] for an overview). The later can affect plasma turbulence at least in two ways: either through large (negative) magnetic shear $s = \mathrm{d}\log q/\,\mathrm{d}\log r$, with q the safety factor profile, which is linearly stabilizing, or due to the rarefaction of resonant surfaces close to low order rational q surfaces and/or low shear regions $s \approx 0$ [5]. As a matter of fact, the formation of ITBs has been experimentally associated with the presence of rational q surfaces in both JET and ASDEX Upgrade plasmas [8, 9].

From the non-linear theoretical side, the possibility to trigger ITBs in the vicinity of low magnetic shear and/or low order rational q values remains an open issue. Indeed, contradictory results have been obtained with first principle numerical simulations of core plasma turbulence. Fluid simulations retaining resonant modes only have reported impressive agreements with experimental observations, especially regarding the splitting of the ITB when two $q = 2$ magnetic surfaces were present in the plasma [5]. However, such results were not recovered with both local and global gyrokinetic simulations [6]. Non-resonant modes were found to play a critical role in driving finite turbulent heat flux in the vicinity of $s \approx 0$. Later on, similar simulations with fully kinetic electrons, including trapped electron modes, reported fine scale corrugations on the electron temperature and density profiles close to low order q rational values, resulting from the local increase of self generated zonal flows [7].

The role of a large gap region without resonant modes in the case of reversed shear profiles is re-investigated in this paper and extended to flux-driven global gyrokinetic simulations of ion temperature gradient driven turbulence. Section 2 highlights the main equations which are solved. Two cases are then compared in section 3, with (hollow q profile) and without (monotonous q profile) any large radial gap in resonant modes. Finally, the case with an even larger gap is considered in section 4, when Fourier filtering only retains those modes which are resonant within the simulation domain.

2. The model

The numerical investigation is performed with the global gyro-kinetic code GYSELA, which models the self consistent time evolution of the full ion distribution function f in the electrostatic limit. The electron response is taken adiabatic. The semi-Lagrangian numerical scheme is detailed in reference [10]. Most of the simulations reported in this paper are performed with the set of equations detailed in reference [11]. Such a system neglects small order terms (of order $\beta \ll 1$, the ratio of kinetic over magnetic pressure) which are proportional to the parallel current. The simulations reported in section 4 make use of the recently upgraded set of equations, which accounts for the conservative formulation consistent with modern gyro-kinetic theory [12]. They read:

$$B_{\parallel}^{*}\frac{\partial f}{\partial t} + \boldsymbol{\nabla}\cdot\left(B_{\parallel}^{*}\frac{\mathrm{d}\mathbf{x}_G}{\mathrm{d}t}\,f\right) + \frac{\partial}{\partial v_{G\parallel}}\left(B_{\parallel}^{*}\frac{\mathrm{d}v_{G\parallel}}{\mathrm{d}t}\,f\right) = \mathcal{C}(f) + S \tag{1}$$

where the equations of motion of the guiding centers are given below:

$$\begin{aligned} B_{\parallel}^{*}\frac{\mathrm{d}\mathbf{x}_G}{\mathrm{d}t} &= v_{G\parallel}\mathbf{B}^{*} + \frac{\mathbf{b}}{e_s}\times\boldsymbol{\nabla}\Xi && (2)\\ B_{\parallel}^{*}\frac{\mathrm{d}v_{G\parallel}}{\mathrm{d}t} &= -\frac{\mathbf{B}^{*}}{m_s}\cdot\boldsymbol{\nabla}\Xi && (3)\\ \text{with}\quad \boldsymbol{\nabla}\Xi &= \mu_s\boldsymbol{\nabla}B + e_s\boldsymbol{\nabla}\bar{\phi} \\ \mathbf{B}^{*} &= \mathbf{B} + (m_s/e_s)\,v_{G\parallel}\boldsymbol{\nabla}\times\mathbf{b} \end{aligned}$$

These modified equations only differ from the old ones by small terms, proportional to $(m_s v_{G\parallel}/e_s B^2)\mu_0 j_{\parallel}$. This ratio is indeed very small, of the order of $\rho_*(a/qR) \ll 1$ for thermal particles ($\rho_* = \rho_{c0}/a$, with ρ_{c0} the thermal gyro-radius and a the minor radius of the torus). Therefore, no significant impact on the results presented here is to be expected. Here, $\bar{\phi}$ denotes the gyro-averaged electric potential, where the Bessel function is replaced by a Padé approximation. The scalar $B_{\parallel}^{*}$ is simply $B_{\parallel}^{*} = \mathbf{B}^{*}\cdot\mathbf{b}$, with $\mathbf{b} = \mathbf{B}/B$ the unit vector along the magnetic field line at the guiding-center position. The magnetic equilibrium is made of tori with concentric and circular cross-sections. The Fokker-Planck type collision operator $\mathcal{C}(f)$ acts on $v_{G\parallel}$ only [13], and has been modified to preserve mass, momentum and energy. The source term

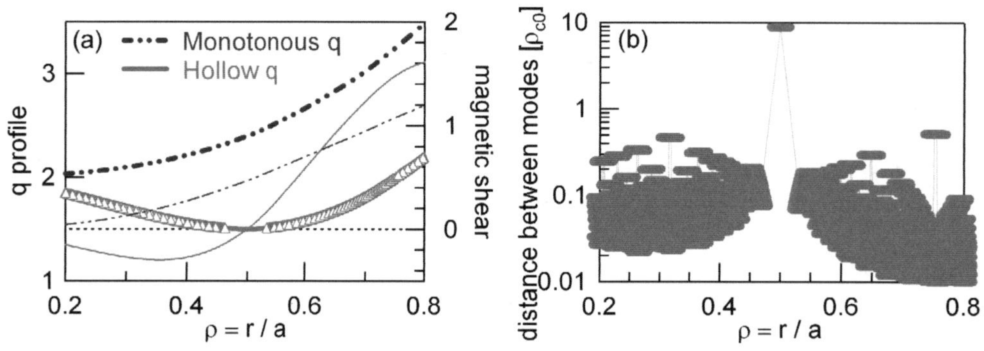

Figure 1. (a) Profiles of q (bold lines) and magnetic shear s (thin lines) used in the simulation runs. The triangles on the hollow q profile refer to the locations of the resonant modes present in the simulation; (b) distance between resonant modes for the hollow q case, normalized by the thermal gyro-radius ρ_{c0}.

S provides constant heat in the system. It is given a double *tanh* shape, localized close to the inner radial boundary of the simulation domain (see Fig.2a). Such flux-driven simulations are particularly time-consuming since the equilibrium state typically develops on time scales of the order of the energy confinement time τ_E. Since $\omega_c\tau_E$ scales like ρ_*^{-3} in the gyro-Bohm regime, small ρ_* relevant simulations remain extremely challenging. Still, as evident in the following, turbulence reaches statistical steady state on much smaller time scales, which then becomes accessible on nowadays supercomputers.

3. Comparative dynamics with monotonous and hollow q profiles

Two main simulations are reported here. They are characterized by the following set of parameters. The relatively small value of $\rho_* = 1/128$ already requires about $8.6\,10^9$ grid points in the 5D phase space: $N_r \times N_\theta \times N_\varphi \times N_{v_\parallel} \times N_\mu = 256 \times 256 \times 128 \times 128 \times 8$. At the considered collisionality $\nu_* = 0.5$, neoclassical transport is mainly governed by trapped particles (banana regime). Finally, the source magnitude is $S_0 = 10^{-2}$, and vanishes for $\rho > 0.3$, while the simulation domain is $0.2 < \rho < 0.8$. Similar results were obtained with smaller (factor 1/2) and larger (factor 2) forcings as well. As detailed in reference [11], the corresponding additional power P_{add} that is injected in the system reads, in physical units and for a purely Deuterium plasma:

$$P_{add} \approx 2.89 \frac{R_0}{a} \frac{S_0}{\rho_*} \frac{n_{0[10^{20}m^{-3}]} T_{0[keV]}}{B_{0[T]}} \quad [MW] \tag{4}$$

where n_0, T_0 and B_0 are the arbitrary normalizing density, temperature and magnetic field, respectively. Considering a Deuterium plasma with $n_0 = 2.10^{19}m^{-3}$, $T_0 = 4keV$ and $B_0 = 3T$, and given the parameters of the simulations, namely $A = R_0/a = 3.2$ and $\rho_* = 1/128$, one finds that $S_0 = 10^{-2}$ would correspond to $P_{add} \approx 3.2\,MW$ of injected heating power.

These two simulations only differ by their prescribed q profile: monotonous and hollow q profiles. They are depicted on Fig.1a. The absolute magnitude of the magnetic shear is comparable in each case. The hollow q profile case, which we will hereafter refer to as "RS case" (for "reversed shear"), is characterized by a large gap of resonant modes in the vicinity of the low order rational q_{min} 1.5, at $\rho = 0.5$. The resonant (m, n) (poloidal and toroidal wave numbers) modes are such that there exists a radial location ρ_{mn} in the simulation domain where

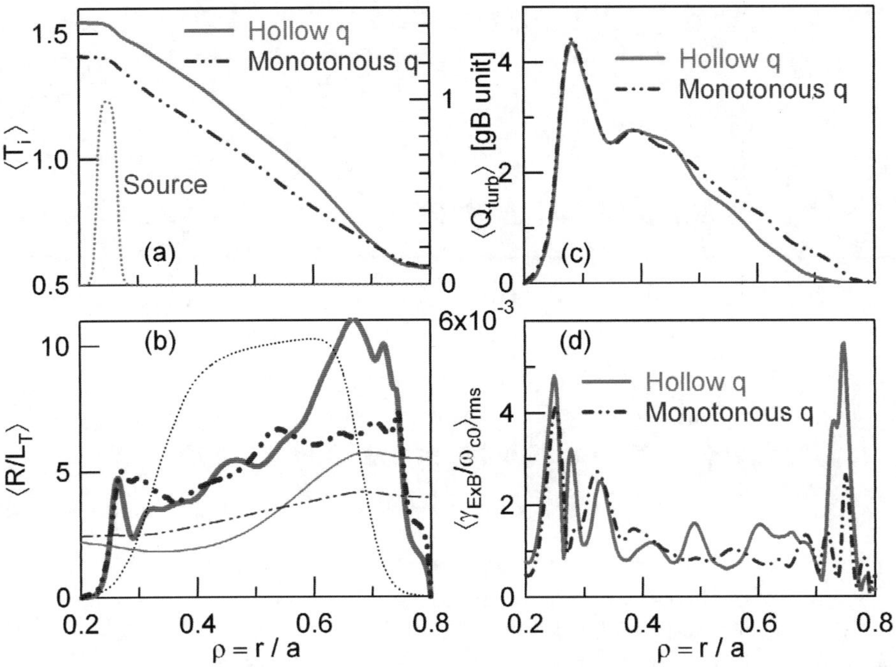

Figure 2. Time averaged profiles of (a) temperature (left axis) and source (normalized to its maximum, right axis); (b) normalized temperature gradient (bold lines) and critical one (thin lines; from [14]) (black dotted line = initial profile); (c) turbulent heat flux, and (d) root mean square value of the $E \times B$ shearing rate in the two compared simulations.

$q(\rho_{mn}) = m/n$. Indeed, as visible on Fig.1b, while most of the resonant modes are distant from fractions of one thermal Larmor radius ρ_{c0}, this normalized distance reaches about $\Delta_{gap}/\rho_{c0} \approx 9$ at $\rho = 0.5$. We shall refer to this region as the "gap" in the following. Simulations with even larger gaps have been performed (up to $\Delta_{gap}/\rho_{c0} \approx 17$, cf. section 4) at this ρ_* value, leaving unchanged the main conclusions.

These two cases exhibit roughly the same linear growth rate. As inferred from the early time evolution of those (m, n) modes which exhibit the fastest exponential growth at $\rho = 0.5$, they are of the order of $\gamma_{lin} \approx 1.5\,10^{-3}\omega_{c0}$ for the monotonous q profile, and about $\gamma_{lin} \approx 1.17\,10^{-3}\omega_{c0}$ for the RS case.

The time averaged profiles, averaged over the saturated non linear phase of turbulence (cf. Fig.3), are displayed on Fig.2 for both of these simulations. Neither the temperature gradient length nor the turbulent heat flux Q or the $E \times B$ shearing rate exhibit significant differences in the gap region, comprised within $0.464 \leq \rho \leq 0.534$. Conversely, the temperature is larger inside $\rho \approx 0.7$ for the RS case, as a consequence of the larger value of R/L_T in this case at the edge, from $\rho \approx 0.6$ to $\rho \approx 0.75$. As evident on Fig.2b, the radial shape of $\langle R/L_T \rangle$ mimics the one of the critical gradient, as derived from [14] (we do not have included the $\epsilon = r/R$ correction to the calculation of $R/L_{T,crit}$ to get a better fit of the non linear results). Especially, the RS case exhibits a significant increase of $R/L_{T,crit}$ at the edge, due to the larger values of the ratio s/q in this region. The larger $E \times B$ shearing rate in this region (cf. Fig.2d) should contribute to further increase the effective critical gradient. Consistently, the turbulent flux is smaller at

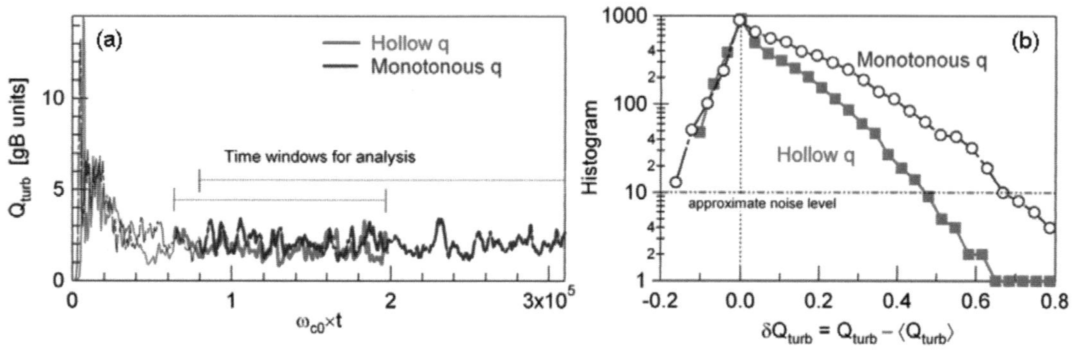

Figure 3. (a) Time histories of the turbulent heat flux at $\rho = 0.5$ in both simulations. The time windows considered for the various analyses are highlighted; (b) Histogram of the flux in each case. The "noise level" refers to the threshold below which the statistics becomes too poor.

the edge for the RS case, from about 50% at $\rho = 0.6$ to a factor 2 at $\rho = 0.75$.

The time evolution of the turbulent heat flux at $\rho = 0.5$ is shown on Fig.3a. From this graph, it is evident that both simulations have reached statistical steady state on turbulent time scales. Although they share approximately the same magnitude at this location, the fluxes exhibit significant differences with respect to their dynamics. Indeed, they both exhibit skewed probability density functions, characteristic of avalanche-like dynamics and already many times reported in fluid (see e.g. [15]) and more recently gyrokinetic [16, 17, 18, 11] flux-driven simulations of turbulence. However, the RS case is characterized by much less large scale transport events, as evident from the less asymmetric histogram of the flux. Such a property possibly further reinforces the reduction of the transport level in the outer region, as noticed in Fig.2c.

Still, avalanches are observed to cross the gap region. In previous gyrokinetic local simulations [6], it was reported that non resonant modes can fill in the gap region. Indeed, modes of the electric potential are clearly visible in this region on Fig.4b. One possible mechanism for their excitation is when the propagation time of an avalanche event into this depleted region, namely $\tau_{aval} \approx \Delta_{gap}/v_{aval}$, becomes comparable or even smaller than the time τ_{damp} for such modes to be locally damped. On the one hand, the avalanche velocity is of the order of the diamagnetic velocity $v_{aval} \approx \rho_* v_T$ [11]. On the other hand, the damping time is likely to be dominated by Landau damping for non resonant modes. Landau damping is all the larger since the modes are far from their resonant surface, i.e. basically since $k_\parallel v_T$ is large (γ_L scales approximately like $\gamma_L \sim -k_\parallel v_T \exp\{-\frac{1}{2}(\omega/k_\parallel v_T)^2\}$, with ω the real frequency of the mode, of the order of the diamagnetic frequency $\omega \approx (k_\theta \rho_i) v_T/L_n$ with $k_\theta = m/r$). In the gap region, $k_\parallel v_T/\omega_{c0} = \rho_*(n + m/q)/A$ remains finite, whatever the considered couple of poloidal and toroidal mode numbers (m, n). Therefore, in small ρ_* simulations, $k_\parallel v_T/\omega_{c0}$ remains small even in the gap region for two reasons. First because it scales like ρ_*. Second because the maximum m and n mode numbers of the simulation *must* increase when decreasing ρ_* (linear modes peak at around $k_{\theta,peak}\rho_i \approx 0.3$, which translates into $m_{peak} \approx 0.3(r/a)\rho_*^{-1}$). In this case, high order rational q_{min} values only are likely to satisfy $(n + m/q) \neq 0$ whatever $-m_{max} \leq m \leq m_{max}$ and $-n_{max} \leq n \leq n_{max}$. In other words, for a given hollow q profile, the gap width decreases with ρ_*, like $\Delta_{gap} \approx \rho_*^{1/2}(2q_{min}/n\rho_{min}q''_{min})^{1/2}$ [5]. The fact that avalanches do cross the gap then suggests that, in this case, the following hierarchy of times prevails: $\tau_{aval} < \tau_{damp}$.

The turbulent convection cells exhibit similar k_θ spectrum in both simulations, peaking

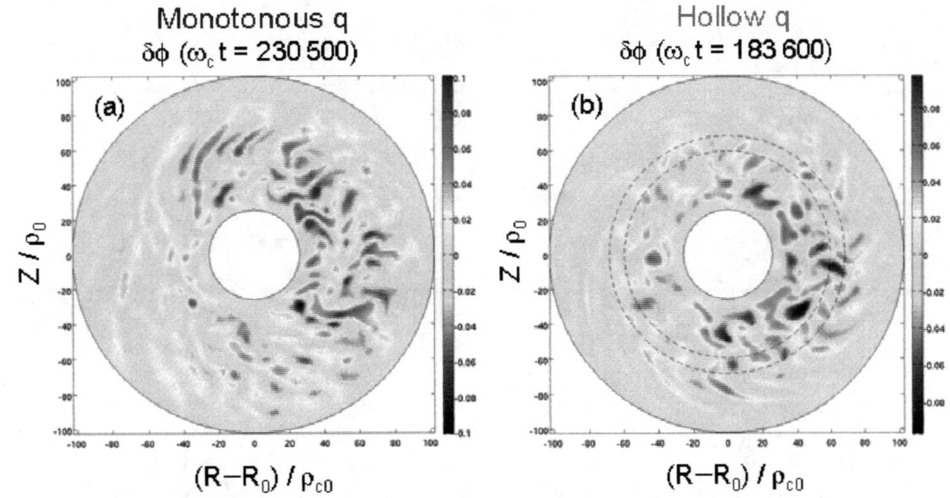

Figure 4. Snapshots of the poloidal cross-section of the electric potential fluctuations during one large outburst in the (a) monotonous and (b) hollow q profile cases.

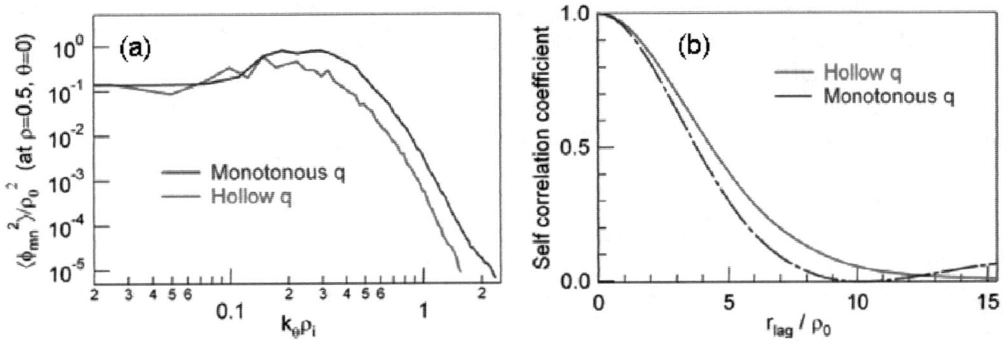

Figure 5. (a) Time averaged Fourier spectra of the $k_\theta = nq/r$ modes at $\rho = 0.5$ and $\theta = 0$ in both cases; (b) Two point radial auto-correlation function of the electric potential. The radial window slightly exceeds the gap region $0.4 < \rho < 0.6$.

around $k_\theta \rho_{c0} \sim 0.2 - 0.3$, Fig.5a. As far as their radial property is concerned, Figure 4b shows that they can even cross the gap (cf. for instance at $\theta \approx 0$). This striking observation is reinforced by the structure of the two-point correlation function of the electric potential fluctuations, Fig.5b. Indeed, the correlation function is even wider in the RS case than for the monotonous q profile, with a long tail extending approximately to $r_{lag}/\rho_{c0} \approx 15$. Candy *et al.* have already shown that, in the absence of global effects such as profile variations, the radial mode structure of those modes resonant at the edge of the gap region can significantly overlap [6]. Unfortunately, at such a still moderate ρ_* value, it has not been possible with the global GYSELA code to further investigate this point, and decouple global effects from linear properties.

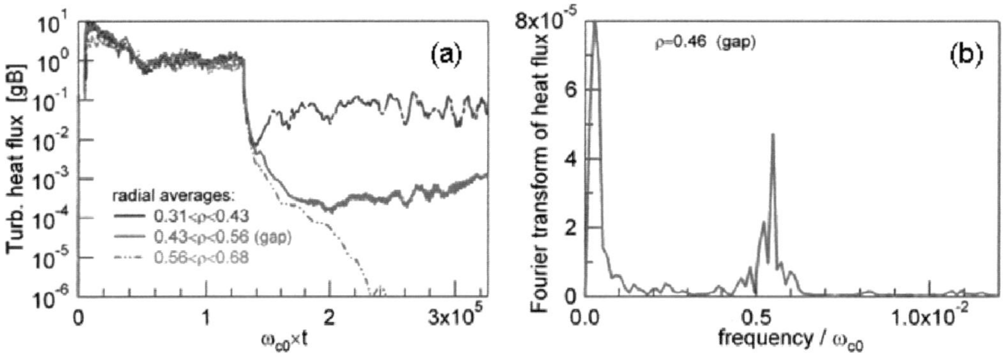

Figure 6. (a) Time evolution of the heat flux (logarithmic scale) when the conical filter is applied (from $\omega_{c0}t_0 = 128\,000$). Each curve corresponds to the average over radial domains either left (core) or right (edge) from the gap, and within the gap; (b) Fourier frequency spectrum of the heat flux fluctuations at $\rho = 0.46$, within the gap region.

4. Case without non-resonant modes

In order to sort out the role played by non resonant modes in the triggering of transport barriers in low shear regions, another simulation has been run, making use of the conservative expression of the gyrokinetic equation, eq.1. Also, for the simple reason that these simulations are part of another set of runs, they also differ from the ones of section 3 by two other parameters: the number of μ values is larger by a factor 2 ($N_\mu = 16$), and the collisionality is smaller ($\nu_* = 5.10^{-2}$). As far as the hollow q profile is concerned, it has essentially the same shape as the one reported on Fig.1, but with $q_{min} = 2.008$ (it is slightly above 2 to prevent the existence of any resonant mode in the vanishing shear region), such that the gap region is even larger $\Delta_{gap}/\rho_{c0} \approx 17$.

The simulation run is divided in two time windows. Up to $\omega_{c0}t_0 = 128\,000$, all modes are retained like in standard simulations. From t_0 onwards, only those modes resonant within the simulation domain are kept, the other being artificially set to zero. For each toroidal n mode, the retained poloidal m modes satisfy the relation $nq_{min} \le m \le nq_{max}$, with q_{min} and q_{max} the minimum and maximum values of the q profile, such that a conic filter is applied in Fourier space. Notice that non-resonant modes can still exist locally (such that $(n + m/q) \neq 0$, with m and n still satisfying the previous relation). However, their number is reduced. It is even more reduced for the extremal q values, and particularly at q_{min} for hollow q profiles.

The beginning of the simulation does not depart from what we have learned in the previous section. Especially, the turbulent heat flux does not exhibit any drop in the gap region, although it is significantly larger than in the previous cases, Fig.6a. From t_0 however, the dynamics changes dramatically. First of all, the turbulent heat flux rapidly decreases everywhere, as a result of the energy sink introduced by the conic filter. While the flux partly recovers left from the gap, i.e. close to the heat source, it remains at an extremely low value within the gap, about 2 orders of magnitude below. More surprisingly, the heat flux does completely vanish right from the gap. It is worth noticing that such a behavior is specific of this RS case. Indeed, when applying the same filter to a case with monotonous q profile, the turbulent heat flux was marginally and homogeneously reduced in the whole radial domain, with no drastic change on the simulation results. Incidently, we note in these new simulations the impact on the resonance condition of the Doppler shift $\omega_E = k_\theta v_E$ induced by any mean $E \times B$ flow $v_E = \phi'_{00}/B$. Indeed, the most prominent (m, n) modes are then such that $k_\parallel v_T + \omega_E \approx 0$, or alternatively

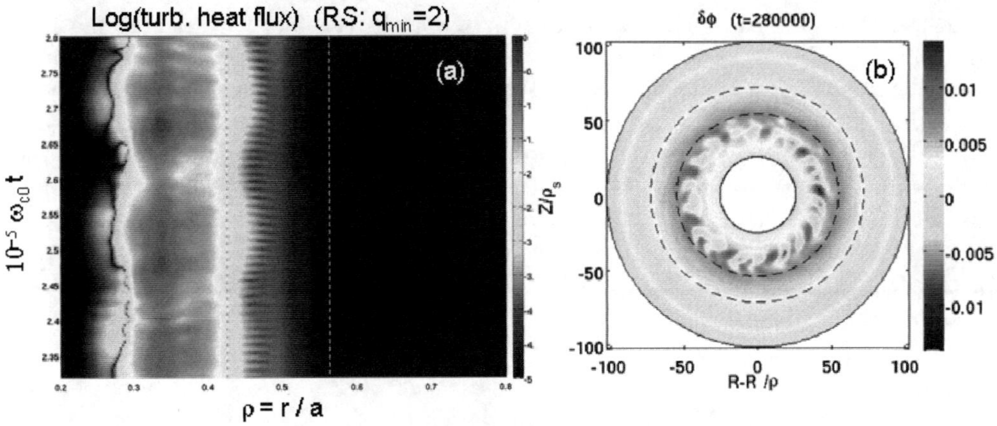

Figure 7. (a) Space-time evolution of the turbulent heat flux when the conical filter is active; (b) Snapshot of the poloidal cross-section of the electric potential fluctuations when the filter is active. In each graph, thin dashed lines indicate the boundaries of the gap region.

$n = -(m/q)\{1 \pm (qR_0/r)v_E/v_T\}$. Such a time dependent (v_E evolves in time) relationship is well fulfilled in the course of the simulation.

The impact of the conic filter on the RS simulation can be understood as follows. It introduces an asymmetry between, on the one hand, the radial regions where q is close to the extremum q_{min} and q_{max}, and, on the other hand, the rest of the radial domain. Neglecting the Doppler shift discussed above, those (m,n) modes satisfying $m = n(q \pm \delta q)$ (i.e. from one side to another of the resonant condition $m = nq$) are excited at each position, either linearly or via non linear coupling. Typical values of the parallel wave vector $k_{\|}qR \approx 1$ suggest that δq remains small, of the order of $\delta q \approx 1/n$. Then, focussing on the region close to q_{min}, only half of these non resonant modes are allowed when the filter is present, namely those such that $m = n(q_{min} + \delta q)$. In the case of monotonous q profile, it is not expected to significantly impact the simulation results, since it occurs at the radial boundary of the domain, where fluctuations are usually forced to zero in global codes. Conversely, when q_{min} stands in the middle of the considered radial domain, as is the case for hollow q profiles, such a drastic reduction of the number of accessible modes is likely to become visible.

As a matter of fact, turbulence is almost completely quenched in the gap, as evident in Fig.7a: the turbulent flux stops at the entrance of the gap region. No avalanche succeeds in crossing the gap in this case. Such a behavior characterizes transport barriers. Yet, no strong temperature gradient could be observed in this too short simulation run. Indeed, the temperature profile builds up on energy confinement time scale, which is much larger than the explored time window. Also, the root mean square value of the $E \times B$ shearing rate is lower in the gap region when the Fourier filter is active. Finally, consistently with the radial structure of the turbulent flux, turbulence eddies do not appear in the gap region. They develop characteristic structures of toroidal ITG modes in the inner region, but vanish from the inner boundary of the gap outwards, Fig.7b.

Two additional observations can be made with respect to the transport barrier. First of all, its inner boundary seems to breeze, in the sense that low frequency oscillations (the period is about $\omega_{c0}\tau_{slow} \approx 2.10^4$) are visible on the color plot of the flux, around e.g. $\rho \approx 0.45$. This slow dynamics was also reported in fluid flux-driven simulations of interchange turbulence in the scrape-off layer of tokamak plasmas, for barriers triggered by strong velocity shear [19]. In this case, they were understood as the result of turbulence spreading into otherwise the linearly stable sheared region. Secondly, fast oscillating outburts are also visible, with decreasing magnitude from the inner gap boundary toward almost $\rho \approx 0.5$. Surprisingly, the high characteristic frequency is very close to the one expected for Geodesic Acoustic modes, which is equal to $\omega_{GAM} = [(2 + 1/q^2)T_e/T_i + 7/2]^{1/2}c_s/R \approx 6.10^{-3}\omega_{c0}$ in the present simulation, as evident on Fig.6b. Such a result is surprising since those specific modes characteristic of the GAM dynamics, namely $(m, n) = (\pm 1, 0)$, are absent from the simulation when the filter is applied. As a matter of fact, a detailed analysis (not shown here) reveals that the dominant (m, n) modes with $5 \leq m \leq 20$ do exhibit large frequency oscillations. This frequency increases with m, approximately linearly like $\omega/\omega_{c0} \sim 3.10^{-4}(m + 9)$. We do not have any explanation for this observation so far.

5. Conclusions

Flux-driven simulations of ITG turbulence in the cases of hollow and monotonous q profiles have been performed with the global and full-f gyrokinetic GYSELA code. In the vicinity of q_{min} where $s \approx 0$, the reversed shear cases exhibit large radial gap regions without any resonant mode. Still, no transport barrier has been observed so far in these cases, although different heat source magnitudes and increasing gap widths have been explored. Conversely, turbulent transport dramatically drops (by 2 orders of magnitude) in the gap region when a conical Fourier filter is applied, which only retains those modes which are resonant in the simulation domain. Such gyrokinetic simulations tend to reconcile – and extend to the turbulence flux-driven regime – previously published contradictory results on the topic. Still, understanding and reproducing the experimental triggering of internal transport barriers in JET remain an open issue.

Acknowledgments

It is our pleasure to acknowledge C. Passeron for her constant and invaluable help in the course of running GYSELA, and T. Görler for useful comments. This work was granted access to the HPC resources of CCRT and CINES under the allocation 2009052224 made by GENCI (Grand Equipement National de Calcul Intensif), as well as on HPC-FF in Jülich. This work, supported by the European Communities under the contract of Association between EURATOM and CEA, was carried out within the framework of the European Fusion Development Agreement. The views and opinions expressed herein do not necessarily reflect those of the European Commission.

References

[1] Wolf R C 2003 *Plasma Phys. Control. Fusion* **45** R1-R91
[2] Biglari H, Diamond P H and Terry P 1990 *Phys. Fluids* B **2** 1
[3] Garbet X, Baranov Y, Bateman G *et al* 2003 *Nucl. Fusion* **43** 975
[4] Terry P W 2000 *Rev. Mod. Phys.* **72** 109
[5] Garbet X, Bourdelle C, Hoang G T *et al* 2001 *Phys. Plasmas* **8** 2793
[6] Candy J, Waltz R E and Rosenbluth M N 2004 *Phys. Plasmas* **11** 1879
[7] Waltz R E, Austin M E, Burrell K H and Candy J 2006 *Phys. Plasmas* **13** 052301
[8] Challis C D *et al* 2002 *Plasma Phys. Control. Fusion* **44** 1031
[9] Joffrin E, Challis C D, Conway G D *et al* 2003 *Nucl. Fusion* **43** 1167
[10] Grandgirard V, Brunetti M, Bertrand P *et al* 2006 *J. Comput. Phys.* **217** 395
[11] Sarazin Y, Grandgirard V, Abiteboul A *et al* 2010 *Nucl. Fusion* **50** 054004
[12] Brizard A J and Hahm T S 2007 *Rev. Mod. Phys.* **79** 421
[13] Dif-Pradalier G, Grandgirard V, Sarazin Y, Garbet X and Ghendrih Ph 2009 *Phys. Rev. Lett.* **103** 065002
[14] Jenko F, Dorland W and Hammett G W 2001 *Phys. Plasmas* **8** 4096
[15] Garbet X, Sarazin Y *et al* 1999 *Nucl. Fusion* **39**, 2063
[16] Idomura Y, Urano H, Aiba N and Tokuda S 2009 *Nucl. Fusion* **49** 065029
[17] Ku S, Chang C S, Adams M *et al* 2006 *J. Phys.: Conf. Ser.* **46** 87
[18] McMillan B F, Jolliet S, Tran T M, Villard L, Bottino A and Angelino P 2009 *Phys. Plasmas* **16** 022310
[19] Ghendrih Ph, Sarazin Y, Ciraolo G *et al* 2007 *J. Nucl. Mater.* **363-365** 581

Journal of Physics: Conference Series **260** (2010) 012018 doi:10.1088/1742-6596/260/1/012018

Microturbulence studies in RFX-mod

F Sattin[1], S C Guo[1], I Predebon[1], S F Liu[2], X Garbet[3] and M Veranda[1]

[1] Consorzio RFX, Associazione Euratom-ENEA per la fusione, Padova, Italy
[2] Department of Physics, Nankai University, Tianjin, Peoples Republic of China
[3] CEA, IRFM, F-13108 Saint Paul Lez Durance, France

E-mail: `fabio.sattin@igi.cnr.it`

Abstract. Present-days Reversed Field Pinches (RFPs) are characterized by quasi-laminar magnetic configurations in their core, whose boundaries feature sharp internal transport barriers, in analogy with tokamaks and stellarators. The abatement of magnetic chaos leads to the reduction of associated particle and heat transport along wandering field lines. At the same time, the growth of steep temperature gradients may trigger drift microinstabilities. In this work we summarize the work recently done in the RFP RFX-mod in order to assess the existence and the impact upon transport of such electrostatic and electromagnetic microinstabilities as Ion Temperature Gradient (ITG), Trapped Electron Modes (TEM) and microtearing modes.

1. Introduction

The current paradigm for high-performance discharges in Reversed Field Pinches (RFPs) is provided by SHAx states: magnetic equilibria characterized by field lines twisting around one helically wound axis. These states feature well conserved magnetic surfaces in the core; the possibility of their existence was theoretically proposed several years ago [1], experimentally identified more recently [2, 3, 4], and represent nowadays the regular working scenario, at least in the RFX-mod device, to which we will refer throughout this paper.

The border of the magnetically ordered volume is characterized thermally by steep internal energy transport barriers (ITBs). The region of high magnetic chaos and consequently degraded transport is pushed to the edge, downhill to the ITB[1]. Remarkable progresses have been done with respect to the earlier chaotic RFPs as far as energy losses are concerned, but power-balance estimates of the heat conductivity suggest that its value is still above the collisional level [4]: either some residual overlapping of MHD tearing modes provides non-negligible transport of magnetic nature, or other instabilities are at work. An extensive coverage of the actual status of knowledge about the physics of present-days RFPs is provided by Cappello*et al* [6]. Recent theoretical research has focused on investigating both possibilities: as far as the topological characterization of the core from the MHD point of view is concerned, see the work by Bonfiglio *et al* presented at this conference [7]. In this work, we will investigate the possible existence and role of microinstabilities in SHAx states, how and whether these instabilities could be triggered, and to what extent they could affect overall transport. We aim at providing a brief survey of the results of this activity obtained at RFX-mod along the past two years. We supplement existing

[1] Note that–just like tokamaks–RFPs feature edge transport barriers as well [5]. These barriers are not subject of the present study

Journal of Physics: Conference Series **260** (2010) 012018 doi:10.1088/1742-6596/260/1/012018

results with novel unpublished material; furthermore, hints to new lines of research as well as to issues not yet completely settled, will be given. The outline of the work is as follows: In section 2 Ion Temperature Gradient (ITG) modes will be addressed. Their stability properties will be examined in connection with pure electron-ion (pure) plasmas (Sec. 2.1), and then possible modifications due to plasma pollution by impurities will be considered for (Sec. 2.2). Section 3 is devoted to studies of Trapped Electron Modes. Finally, an extensive coverage is provided in Sec. 4 of Microtearing Modes. On the basis of current evidence, these modes appear likely candidates as possible sources of residual turbulent transport at the location of ITBs.

2. Ion Temperature Gradient Modes

2.1. ITGs in pure plasmas

ITGs are regarded as one of the main sources of turbulent transport in tokamaks; accordingly, they were the first to be addressed by investigations. It was early noticed by Guo [8] (see also [9]) that RFPs are more resilient than Tokamaks to this kind of instability. The rationale for this behaviour stays in the different level of Landau damping due to the different connection length L_c of magnetic field lines between the two devices. Waves with helicity defined by the angular numbers (m, n) are damped through energy exchange to those particles that are close to the resonance condition

$$v_{||} = \omega k_{||}^{-1} \quad k_{||} = 2\pi \times (m - nq) \times L_c^{-1} \tag{1}$$

where $v_{||}$ is the parallel (to the magnetic field) component of the particle velocity, $k_{||}$ the parallel wavenumber of the mode, ω its angular frequency, and q the safety factor. In RFPs, by virtue of their geometry, L_c is shorter than in Tokamaks by about a factor (minor radius)/(major radius), accordingly, $k_{||}$ is larger by the same amount. Landau damping of the wave is the most effective the more resonant particles are present. If the velocity distribution function $f(v_{||})$ is close to a Maxwellian, the smaller $v_{||}$ the more resonant particles are available. As a consequence, the critical temperature scale length needed for triggering the instability turns out to be substantially shorter than in Tokamaks. Earlier semianalytical calculations where later reinforced by full numerical linear electrostatic gyrokinetic calculations using GS2 code [10], suitably adapted to account for RFP magnetic topology [11]: under common plasma conditions, it may be stated that ITG instability at mid-radius arises for $R/L_{Ti} > 20 \div 25$ (R being the major radius): at least a fourfold factor larger than for similar plasma conditions in tokamak geometry (Fig. 1). An independent investigation using linearly the code TRB (adapted to RFP geometry) led to the same quantitative result [12][2]: see Fig. (1), where TRB result (the large circle) lies quite accurately on top of GS2 ones.

In Fig. (2) an instance of a nonlinear time-dependent TRB simulation is presented, out of a yet preliminary set of simulations aiming to investigate a possible sinergy between the ITG-driven and other sources of turbulence in SHAx discharges. An initially ITG-unstable profile (red curve) is let to evolve on top of a fixed background diffusivity, χ_{back} , that attempts to modelize the residual MHD tearing-mode driven diffusivity, and is chosen small in the inner half radius and large in the outer half. The total heat conductivity is thus $\chi = \chi_{back} + \chi_{ITG}$, where the former term is known and fixed, and the latter is self-consistently calculated by TRB. In principle, the existence of a irreducible level of conductivity might enforce temperature gradients, and thus provide a constant free energy source for triggering ITGs. However, within the numerical range of diffusivites scanned–still rather preliminary–we did not find evidence for such mechanism: all curves collapsed towards practically the same profile, basically driven just by ITGs.

[2] Despite being a fluid code, TRB may account for such a kinetic effect as the Landau damping by implementing the Hammett-Perkins (H-P) closure scheme [13]. Thus, agreement between TRBs and GS2 fully kinetic treatment, provided as a side effect further confidence about the reliability of the H-P approximation.

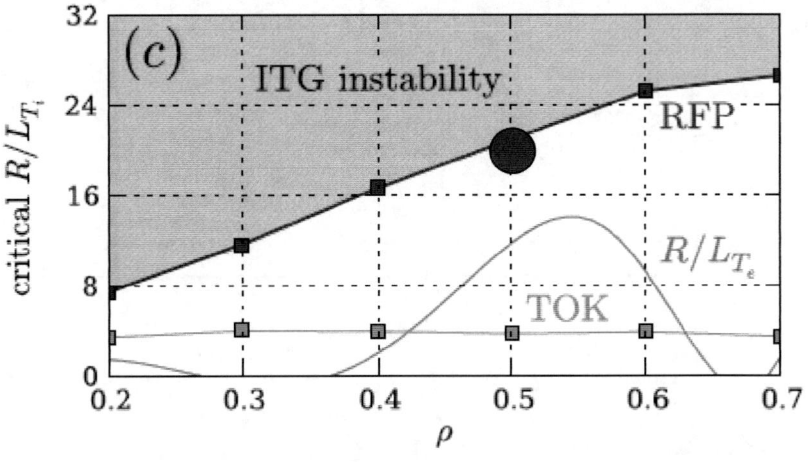

Figure 1. ITG stability threshold versus normalized minor radius $\rho = r/a$ for typical RFX-mod plasma conditions (RFP) and for a tokamak profile (TOK). The threshold is the R/L_{Ti} value at which the growth rate become positive. The red curve is the typical experimental value of R/L_{Te}. The blue curve with symbols are GS2 results for the RFP. The violet dot is the equivalent quantity computed by TRB. The green curve with symbols is the stability threshold for a tokamak with the same plasma parameters.

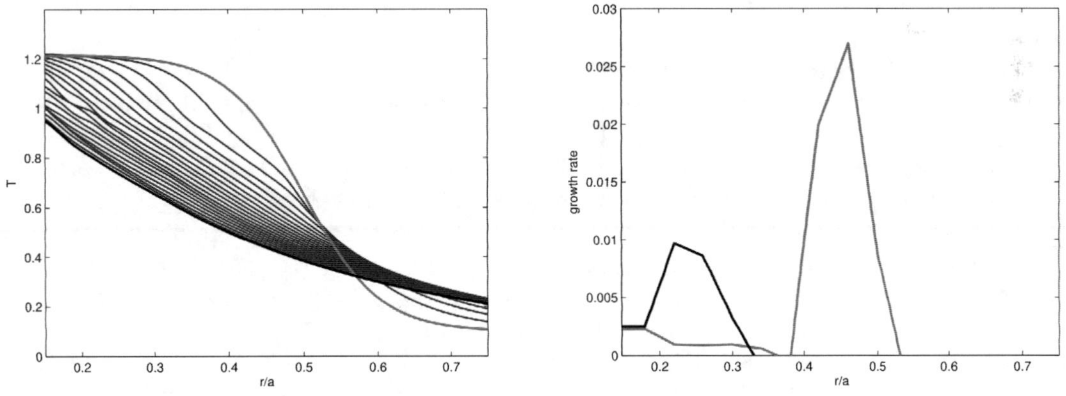

Figure 2. Left panel, an initially ITG-unstable ion temperature profile (red curve) is let to evolve (blue curves) under self-generated turbulent transport. The black curve is the final profile computed in the numerical simulation. It is quite close to the asymptotical stable profile, as it can be guessed from the right panel: a linear study of both the initial and final temperature profiles shows that in the former case the maximum growth rate is positive, whereas the final profile is close to marginal stability (red and black curves respectively). In these simulations the ratio T_i/T_e is held fixed. The heat source is kept constant throughout the whole radius: it is a simplifying but not really satisfactory assumption in RFPs, where the heating is purely ohmic. Finally, the q profile is held fixed, too.

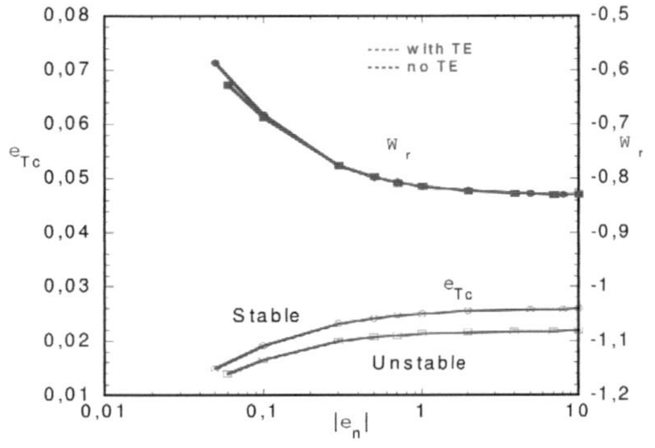

Figure 3. Comparison of the ITG threshold $\varepsilon_{Tc} = L_{T(critical)}/R$ and corresponding real frequency Ω_r (taken negative) between the cases with and without TE. The x axis is $\varepsilon_n = L_n/R$. Curves have been computed at fixed wavenumber: $k_y\rho_i \approx 0.45$.

An accurate comparison between codes predictions and experimental profiles is not easy to be done, in part because some of the information needed by codes is still not available with accuracy from measurements in RFX-mod: the safety factor q, which is reconstructed by an equilibrium code based upon measurements by external probes, and the ion temperature profile, which is inferred from spectroscopic measurements of impurities [4], but not actually measured. Furthermore, if ITG modes are due to be the main drive of transport, one expects to encounter only marginally stable situations, thus growth rates close to zero. With allowance for the uncertainty in some plasma conditions, the present evidence in RFX-mod favours the view that ITGs are probably sub-critical in most discharges, although it cannot be ruled out the possibility thatat the location of particularly steep ITBsmarginal instability conditions may be attained (see, e.g., [8]). Accordingly, we may argue that ITGs might become of future concern for RFPs, should the present trend towards steeper ITBs continue.

2.2. ITGs and impurities: preliminary results

RFX-mod plasmas are ordinarily polluted by intrinsic carbon and oxygen impurities, yielding $Z_{eff} \geq 2$ throughout most of the radius. This raises the question whether impurities may play a role in (de)stabilizing ITGs. The conclusion from the analysis of the linear stability properties of the mode by the gyrokinetic integral eigenmode equation, is that impurities are destabilizing for ITG modes[3] at the locations where $L_{eZ} = L_n/L_Z < 0$; that is, the impurity profile must be hollow where plasma density profile is concave, and viceversa. The rationale for this result relies yet on Eq. (1): the mode frequency is increased as $-L_{eZ}$ increases towards larger positive values. For a given wavenumber $k_{||}$ this change acts to shift the resonance condition (1) to a position in velocity space lesser populated, and thus diminish the impact of the Landau damping. This study is obviosly highly simplified, being carried out for just one impurity. Actually, each impurity in its several charge states should be accounted for. It is known that, in RFX-mod, impurities do not accumulate in the core, and peak quite outwards [14]. Therefore, it is quite likely that a region where the criterion $L_{eZ} < 0$ be fulfilled does exist.

3. Trapped Electron Modes

The fraction of trapped electrons (TE) in a RFP is almost equal to its Tokamak counterpart throughout most of the volume, and somewhat smaller only in the outermost region [15], thus

[3] This is partially at odds with the Tokamak case, where impurities are generally stabilizing for ITGs, and destabilizing for TEMs (see, e.g., [17]).

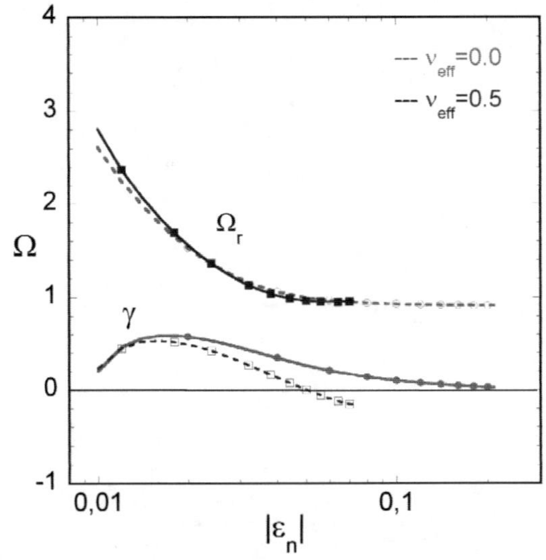

Figure 4. The frequency Ω_r and the growth rate γ of TEM are plotted as functions of $|\varepsilon_n|$ for different normalized collision frequencies ($\nu_{eff} = \nu Rq/c_s$). In this plot, $R/L_{Te} = 0.025$ and $R/L_{Ti} \approx 0$.

TE could potentially provide as large a contribution as in tokamaks in destabilizing turbulence. Two possible effects of TE were investigated. (I) First of all, there is the possible coupling with ITGs, with destabilizing effects, since TE-driven modes resonate at the same wavelengths as ITGs. However, studies both with TRB (by switching off the TE fraction) and GS2 [16] showed that threshold of the mode is not quantitatively altered by their presence (Fig. 3). (II) Further analytical results show that, instead, Trapped Electron Mode (TEM) instability can arise in a RFP, essentially driven by strong density gradients (Fig. 4): actually, under the limit $R/L_n >> 1$ (where L_n is the density scale length), GS2 simulations show that TEMs are the dominant electrostatic instability. However, it is important to recall that present RFPs do not feature ordinarily core peaked density profiles, since there is not appreciable central particle source. Therefore, the present results do not apply to ordinary plasma conditions, rather to non-standard conditions attained, e.g., transiently by means of pellet injections.

4. Microtearing Modes

4.1. Linear growth rates

The Microtearing Mode (MT) is a high-wavenumber drift-tearing mode driven linearly unstable by electron temperature gradients, unlike its long-wavelength counterpart, which is essentially current-driven. The growth of MT may lead to chains of overlapping magnetic islands and subsequent local stochastization of magnetic field lines near mode rational surfaces. Hence, it was speculated in the 70s that MTs might provide an effective contribution to the thermal diffusivity through electron parallel motion along stochastic field lines. However, since it was also argued that, when electron collision frequency decreases substantially below the electron diamagnetic frequency these modes become stabilized [18], and most present-days tokamaks are effectively collisionless in the core, the study of MTs was relegated to the edge region. More recently, interest in these modes was revived in connection with ITBs, where large temperature gradients are sustained (see, e.g., [19]). RFX-mod features presently a moderate peak temperature ($1 \div 1.5$ keV) which, coupled to the strong internal temperature gradients, should represent an optimal environment for these modes to grow. Simulations were carried out using GS2: linear runs, with

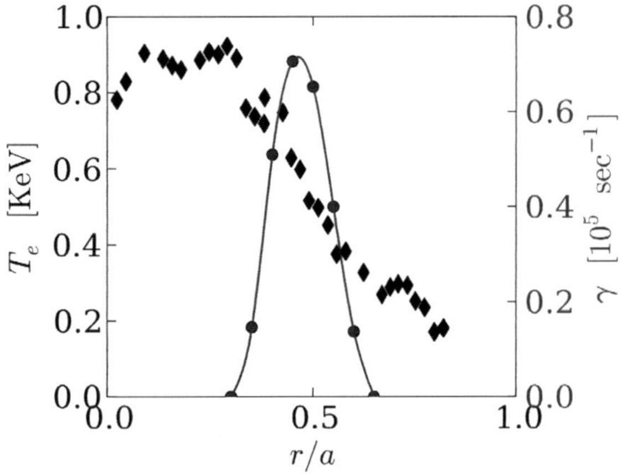

Figure 5. Electron temperature profile (diamonds), and the growth rate of the most unstable MT mode (solid curve with circles).

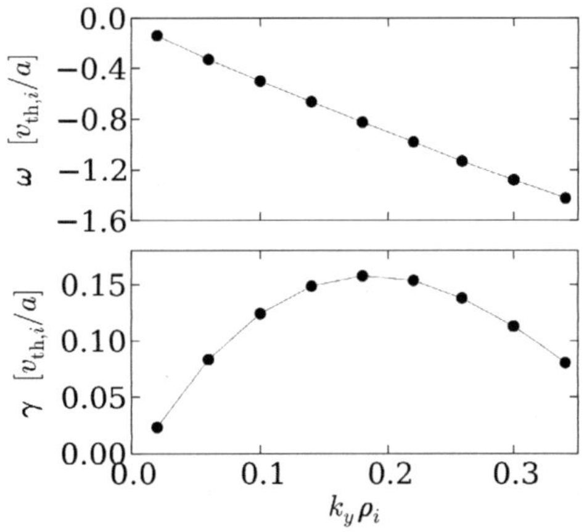

Figure 6. Mode real frequency (upper plot) and growth rate (lower plot) of the most unstable mode as a function of the wavenumber, at mid-radius. Frequencies are normalized to (ion thermal speed)/(minor radius). Note that here the sign convention opposite to that of figs. 3,4 has been used: the ion diamagnetic drift has the positive sign.

fluctuations of the magnetic vector potential in both parallel and perpendicular components included. The analysis presented is based on the profiles of one reference experimental SHAx case (shot 23977). For such discharge the collision frequency and the electron diamagnetic frequency stay roughly in the range $\nu \approx 10 \times \omega^*$. The dominant core instability turns out to be of MT type (Fig. 5), essentially driven by the electron temperature gradient. The tearing nature is revealed by the parity of the mode, odd for the electrostatic potential and even for the parallel magnetic vector potential. In Fig. (6) mode frequency and growth rate are plotted versus the wavenumber, for some radial positions.

MTs thus appear to provide a sensible contribution to turbulent transport, at the location of the ITB, whereR/L_T is maximum. The question naturally arises: quantify the MT-related transport in terms of an effective heat conductivity in present plasma conditions.

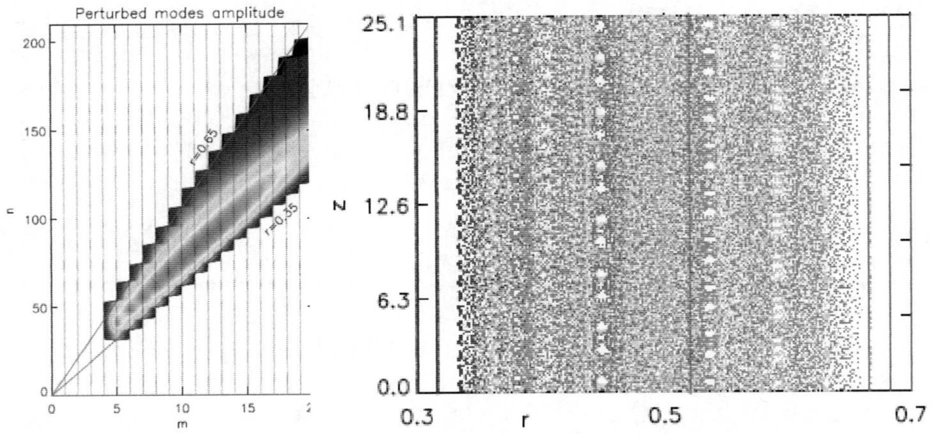

Figure 7. Left plot, Perturbed modes amplitude contour plot. Right plot, Poincaré plot (in the radial-toroidal plane) built from tracking a large number of field lines. Different colors label different starting positions.

4.2. Heat diffusivity due to MTs

A self-consistent calculation of the MT-driven heat flux using GS2 turned out unfeasible due to the heavy computational requirements for the nonlinear run. We adopted therefore the quasi-linear estimate

$$\chi \approx \tilde{b}^2 u_{th,e} L_{corr} \tag{2}$$

Inside this expression, the only quantity known accurately is the electron thermal speed $u_{th,e}$. The normalized amplitude of magnetic field perturbations is estimated according to the criterion suggested by Drake *et al* [18]: $\tilde{b} \approx \rho_e/L_T, \rho_e$ being the electron Larmor radius. The longitudinal correlation length L_{corr} is usually taken in tokamaks equal to the connection length, however–to the best of our knowledge–no checks have been done to ensure that this expression makes sense in RFPs as well. Thus, we resorted to direct numerical evaluation of L_{corr}: a sinthetic magnetic field was built by superposing a large population of modes with different wave-numbers (m, n) and with amplitude such that their cumulated effect yields a field amplitude of order $\tilde{b}$. A plot showing the pattern of modes taken into account is given in fig. (7a). A large number of field lines was tracked using the field-line-tracing code NEMATO [20]. The resulting field topology turns out to be fairly stochastic (Fig. 7b). Finally, L_{corr} is estimated as

$$L_{corr} = \int_0^\infty \frac{< b(l)b(0) >}{< b^2(0) >} dl \tag{3}$$

In this expression, b is the amplitude of the magnetic field, l the distance travelled along the trajectory of the field line, and the average $<>$ is taken over the ensemble of field lines. We recovered that $L_{corr} \approx 2\pi a$. This figure is expected on the basis of the näive identification (correlation length = connection length); on the other hand, at first sight it is a bit puzzling that it does not depend upon the typical wavelength of the magnetic field perturbations involved, as one can argue by a comparison with the related work by DAngelo and Paccagnella [21].

We could eventually estimate Eq. 2; fairly interestingly, it turned out $\chi \approx 5 \div 20$ m^2/s: the precise figure depends upon plasma properties as well as upon the detailed way as the perturbed field is assembled[4]. This figure is fairly compatible with independent power-balance estimates of

[4] It is interesting to notice that DAngelo and Paccagnella [20] performed an exercise similar to the present one by using a magnetic field perturbation rather different from the present one, and recovered similar figures for the conductivity.

χ at the ITB, that yield $5 \div 50$ m^2/s [4]. This strongly supports the view that a relevant fraction of the turbulent transport across the ITB is driven by MTs. Note however that it is not the only transport source: this is suggested by the fact that we found definitely positive growth rates in correspondence of real plasma profiles. If transport was mainly determined by MTs, profiles would accommodate along marginal stability curves. In particular, pressure-driven resistive g-modes have been experimentally identified in the outer plasma region, and might contribute even more internally [22].

5. Conclusions

The RFP, even in the new quasi-laminar SHAx state, confirms to be a device essentially prone to electromagnetic tearing instabilities, although now localized on the microscopic ion Larmor radius scale. Conversely, it appears to be resilient to purely electrostatic modes (ITGs and TEMs): these modes are found potentially unstable, but they are so in plasma conditions characterized by strong gradients (either temperature or density gradients) which are not common in present operating conditions, although intrinsic impurities may affect the extent of this assertion.

Acknowledgments

S. Cappello, D. Escande, E. Fable and R. Paccagnella helped improving the manuscript. This work was supported by the European Communities under the contract of Association between EURATOM/ENEA. The views and opinions expressed herein do not necessarily reflect those of the European Commission.

References

[1] D F Escande, *et al, Phys. Rev. Lett.* **85**, 1662 (2000)
D F Escande, *et al, Phys. Rev. Lett.* **85**, 3169 (2000)
D F Escande *et al, Plasma Phys. Control. Fusion* **42**, B243 (2000)
[2] R Lorenzini, *et al, Phys. Rev. Lett.* **101**, 025005 (2008)
[3] R Lorenzini, *et al, Nature Physics* **5**, 570 (2009)
[4] P Martin, *et al, Nucl. Fusion* **49**, 104019 (2009)
[5] G Spizzo, *et al, Phys. Rev. Lett.* **96**, 025001 (2006)
N Vianello, *et al, Nucl. Fusion* **49**, 045008 (2009)
[6] S Cappello, *Theory of Fusion Plasmas* (AIP Publishing 2008)
[7] D Bonfiglio, *et al, Magnetic chaos healing in the helical reversed field pinch: indications from the volume preserving field line tracing code NEMATO*, this conference
[8] S C Guo, *Phys. Plasmas* **15**, 122510 (2008)
[9] S Liu, S C Guo and J Q Dong, *Phys. Plasmas* **17**, 052505 (2010)
[10] M Kotschenreuther, G Rewoldt, and W M Tang, *Comp. Phys. Comm.* **88**, 128 (1995)
W Dorland, F Jenko, M Kotschenreuther and B N Rogers, *Phys. Rev. Lett.* **85**, 5579 (2000)
[11] I Predebon, C Angioni and S C Guo, *Phys. Plasmas* **17**, 012304 (2010)
[12] F Sattin, X Garbet and S C Guo, *Plasma Phys. Control. Fusion* **52**, 105002 (2010)
[13] G W Hammett and F W Perkins, *Phys. Rev. Lett.* **64**, 3019 (1990)
[14] S Menmuir, *et al, Plasma Phys. Control. Fusion* **52**, 095001 (2010)
[15] M Gobbin, *et al., J. Plasma Fusion Res. Series* **8**, 1147 (2009).
[16] S C Guo, *et al*, poster P1.1102, *37th EPS Conference on Plasma Physics* (Dublin, 2010)
[17] R Paccagnella *et al, Nucl. Fusion* **30**, 546 (1990)
[18] J F Drake, N T Gladd, C S Liu and C L Chang, *Phys. Rev. Lett.* **44**, 994 (1980)
[19] W Horton, *Microtearing and ETG Modes*, talk held at the Workshop *Multiscale Interaction between magnetic islands and micro-turbulence in magnetic fusion devices* (Aix-en Provence, June 21, 2010). *Website: http://sites.univ-provence.fr/ lia-fjmfl/workshop2010.html*
[20] J M Finn and L Chacon, *Phys. Plasmas* **12**, 054503 (2005)
[21] F DAngelo and R Paccagnella, *Phys. Plasmas* **3**, 2353 (1996)
[22] M Zuin, *et al, Nucl. Fusion* **50**, 052001 (2010)

Journal of Physics: Conference Series **260** (2010) 012019 doi:10.1088/1742-6596/260/1/012019

Gyrokinetic investigation of magnetic islands in tokamaks

M. Siccinio, E. Poli, W. A. Hornsby*, A. G. Peeters**

Max Planck Institut für Plasmaphysik, Boltzmannstrasse 2, 85748 Garching bei München, Germany

* Centre for Fusion, Space and Astrophysics, Physics Department, University of Warwick, CV4 7AL Coventry, United Kingdom

** University of Bayreuth, 95440 Bayreuth, Germany

E-mail: `mattia.siccinio@ipp.mpg.de`

Abstract. A powerful new numerical tool to carry out investigations into magnetic islands by means of the gyrokinetic equations is described in the present paper. An island structure, with imposed width and rotation, has been implemented in the flux-tube code GKW. Numerical simulations retaining finite Larmor radius effects (FLRs), toroidicity, kinetic electrons and self-consistent electrostatic potential are presented, concerning, among others, the incomplete density profile flattening for small magnetic islands, the dependency of the island potential on the island rotation frequency and the emission of drift waves connected to the island rotation. Turbulent fluctuations are not fully resolved by filtering out short wavelengths.

The tearing mode is a non-ideal magnetohydrodynamic instability leading to a degradation of the confinement capability of a tokamak, and acting at times as a trigger for disruptive events. Therefore, a precise comprehension of its dynamics represents a primary objective in the present fusion research. In this paper, a numerical investigation has been worked out by means of the flux-tube gyrokinetic code GKW (GyroKinetics@Warwick [1]), which allows, in principle, to account for most of the physical effects of possible interest, e.g. finite Larmor radius effects, collisions, toroidicity and self-consistent calculation of electromagnetic disturbances. In this preliminary work, however, we restrict ourself to some particular aspects of this otherwise extremely broad topic.

1. Implementation of a magnetic island structure

The code GKW is a gyrokinetic flux-tube code, which employs Hamada coordinates [2] and exploits the periodicity constraint of the flux-tube approximation to solve the equations in the Fourier space on the plane perpendicular to the magnetic field lines (such codes are sometimes referred to as *spectral codes*). The gyrokinetic equation is solved for both ions and electrons, and the perturbed electric and magnetic fields can be calculated self-consistently from the particles response by means of Poisson and Ampére equations, respectively. In the simulations presented in this paper, however, the Ampére's law has been switched off, as, for the moment, we are not interested in the island evolution (the island is completely described by a perturbation on the parallel vector potential). Subsequently, an island of fixed width and angular rotation frequency is imposed at every timestep, entering the gyrokinetic equation for both species. It is worth in any case to stress that the problem of the self-consistent island evolution is in principle treatable with GKW.

The coordinate system of the code is constituted by a coordinate *along* the magnetic field lines (indicated by s, varying from -1/2 to 1/2 in the poloidal plane and such that $s = 0$ corresponds to the outer midplane of the tokamak), a poloidal flux ψ, defined to be zero at the centre of the flux-tube domain and playing the role of a radial-like coordinate, and a helical coordinate $\zeta = qs - \gamma$, where q is the safety factor (which is in general a function of ψ) and γ is a toroidal angle normalized to 2π. The flux-tube approximation constrains every perturbed quantity in the code to be periodic in the directions $\nabla\zeta$ and $\nabla\psi$, calculated at the centre of the computational domain. In general, the perturbed vector potential $A_\parallel$ associated to the magnetic island is identified by a mode of the form

$$A_\parallel = C \exp\left[2\pi i \left(ms - n\gamma\right) - i\omega t\right], \tag{1}$$

where C is the amplitude, m, n are the poloidal and toroidal mode number, respectively, and ω is the island rotation frequency. Supposing the island to be located at the centre of the computational box (where by consequence one has to fix $q = m/n$), and expanding the safety factor as $q \approx m/n + (\partial q/\partial\psi)\psi$, Eq.(1) can be rewritten as

$$A_\parallel = C \exp\left[2\pi i n \left(\zeta - \frac{\partial q}{\partial \psi}\psi s\right) - i\omega t\right]. \tag{2}$$

It is evident that this mode is not in general periodic in ψ on the computational domain. Thus, an approximation of the mode is required. The solution we adopted consists in projecting Eq.(2) on the radial Fourier harmonics, adding moreover an artificial Gaussian smoothing, depending on s, in order to avoid the appearance of abrupt "jumps" at the boundary of the computational domain. The interested reader is referred to [3] for further details. The amplitude of the mode C, as well as the island rotation frequency ω and the magnetic shear $\partial q/\partial\psi$ are given as input values (C is calculated from the island width, which is the actual input data).

All the calculations performed in this paper are collisionless, and in addition the number of poloidal harmonics considered has been kept low in order to cut off turbulent effects, which are left for future work.

2. Preliminary results

In this section, numerical simulations confirming some well-known physical effects connected to the presence of a magnetic island in a tokamak plasma are presented. This is helpful both to validate our implementation and to give a further support to the existing theories. Two aspects of the island dynamics are discussed in this section: the island electrostatic potential and the nonlinear density flattening in the island region.

A rotating magnetic island can be thought of as a time-varying magnetic perturbation, and therefore it possesses an associated electrostatic potential, according to Faraday's law. In the existing literature (see for example [4]), an analytical expression for such potential is derived under the assumption of electrons immediately shorting out every parallel electric field, i.e.

$$E_\parallel = 0 \Rightarrow \nabla_\parallel \phi + \frac{1}{c}\frac{\partial A_\parallel}{\partial t} = 0. \tag{3}$$

Recalling Eq.(2), it is clear that $\partial A_\parallel/\partial t = -\omega \partial A_\parallel/\partial\xi$, where $\xi = m\theta - n\varphi - \omega t$ is a helical coordinate (with θ and φ being the poloidal and toroidal angle, respectively). After some straightforward algebra, one obtains

$$\phi = \frac{\omega B}{k_\theta c}\left[x - h\left(\Omega\right)\right], \tag{4}$$

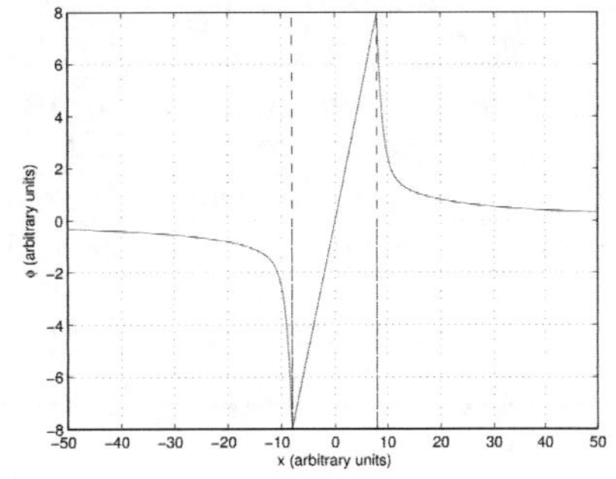

Figure 1. Qualitative behaviour of the island electrostatic potential through the island O-point according to Eq.(4) for a positive rotation frequency ω. Vertical dashed lines denote the location of the island separatrix.

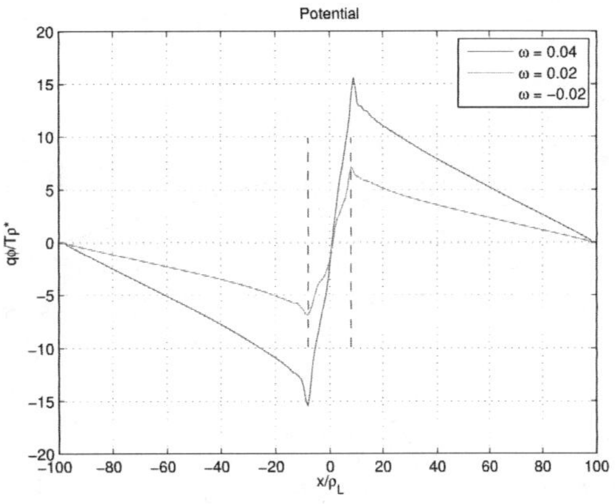

Figure 2. Self-consistent electrostatic potential as a function of the radial coordinate through the island O-point for various values of the island rotation frequency ω calculated with GKW. The potential is normalized to $q_i\phi/T\rho_*$, where q_i and T are the ion charge and temperature, respectively, while $\rho_* = \rho_L/R$ is the ratio between the thermal ion Larmor radius and the tokamak major radius. For the presented simulations, $\rho_* = 2 * 10^{-3}$. The radial coordinate is normalized to ρ_L, while ω is normalized to v_{th}/R, where v_{th} is the ion thermal velocity.

where B is the magnetic field, x is a radial coordinate defined to be zero on the rational surface, $k_\theta = m/r$, $\Omega = 2x^2/w^2 - \cos\xi$ (with w indicating the island half-width) and $h(\Omega)$ a function, playing the role of an integration constant (as $\nabla_\parallel\Omega = 0$), which has to be determined from boundary conditions. Fig.1 shows a qualitative plot of Eq.(4) for a positive value of ω, under the assumption of vanishing electric field at some distance from the island.

This behaviour of the potential has been confirmed by GKW simulations. As presented in Fig.2, the potential is found to scale linearly with the frequency, presenting in addition a behaviour which closely resembles the analytical results. It is important to mention that Fig.2 has been derived in absence of equilibrium pressure gradients. In fact, gradients allow some particular finite orbit effect to take place, as discussed in the following sections.

The density profile inside a magnetic island is known to flatten, as the appearing perturbed component of the magnetic field leads to a fast parallel radial transport. Indeed, such flattening is typically the first check of a proper implementation of the magnetic island. For sufficiently large islands, we see that the flattening is indeed present in our simulations, except for a relatively narrow region around the separatrix where finite banana-width effects prevent the pressure profile to completely flatten (see next section). It is known moreover (see [5]), that in the separatrix region a complete flattening of the pressure profile is inhibited, because of the competition between perpendicular and parallel transport. However, in these simulations, where no collisions are present and the turbulence is not fully resolved, we expect this latter mechanism to be of minor importance. Fig.3 highlights what discussed.

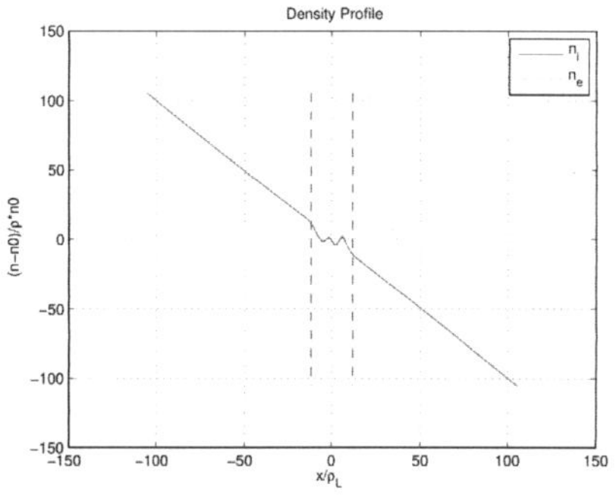

Figure 3. Ion and electron density profiles for a non-rotating island of width $w = 12\rho_L$. The equilibrium density gradient has been fixed to $R/L_n = 1$.

3. Incomplete flattening for small islands

Once the fundamental aspects of the island dynamics have been benchmarked, some more recent result of the literature can be investigated. As a matter of fact, GKW is able to take into account a large number of physical mechanisms, allowing to model a wide variety of phenomena with a very high degree of self-consistency.

When the island width starts to be close to or below the ion banana width, an important finite orbit effect occurs. Because of their large banana width, in fact, trapped ions are able to spend part of their bounce time inside the separatrix and part outside. It has been observed [6] that for islands of the order of the banana width, ions exhibit an "adiabatic" response to the banana-averaged island potential ϕ_b, defined as

$$\phi_b(x) = \frac{\phi(x+\rho_b) + \phi(x-\rho_b)}{2}. \tag{5}$$

Actually, the comparison between the perturbed ion density calculated with GKW and the quantity n_b, defined as

$$n_b = -\sqrt{\epsilon} n_0 \frac{q_i \phi_b}{T_i} \tag{6}$$

(where the factor $\sqrt{\epsilon}$ appears as only trapped particles contribute), exhibits a remarkable agreement in the island region, as shown in Fig.4. On the contrary, the difference between the two curves outside the island depends on the fact that the adiabatic response, according to the existing theoretical models [7], takes place only if the scale of variation of the potential is small (or comparable) with respect to the ion orbit. This circumstance is fulfilled inside the island, but not outside.

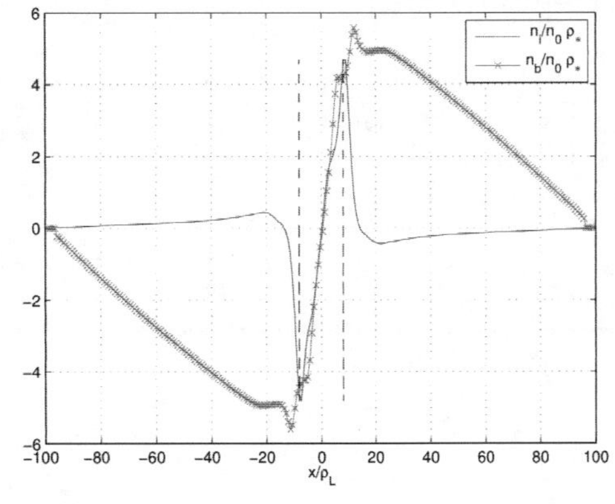

Figure 4. Comparison of the perturbed ion density and n_b for an island of width $w = 8\rho_L$ and $\omega = -0.04$. With the parameters chosen for the simulations, the inverse aspect ratio amounts to $\epsilon = 0.19$, while the ion banana width ρ_b amounts to $\rho_b/\rho_L = 3.44$.

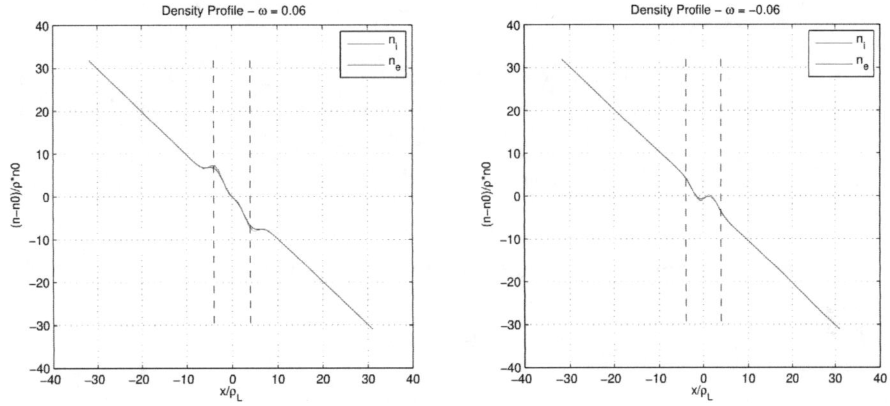

Figure 5. Ion and electron density profiles for an island of width $w = 4\rho_L$ for $\omega = 0.06$ and $\omega = -0.06$ (this means $|\omega|$ slightly larger than the diamagnetic frequency).

This effect on the perturbed density profile implies that for islands rotating in the electron diamagnetic direction (i.e. for positive ω) the ion response *opposes* the flattening, while on the contrary the rotation in the opposite direction leads to a complete flattening, or even to a reversed pressure profile. The density profiles in a small island calculated by GKW for different frequencies are represented in Fig.5, confirming the drift-kinetic results of Ref.[6]. This phenomenon has potentially a major impact on the bootstrap current suppression inside the island region [8]. The numerical investigation of this effect, which requires a full benchmark of the collision operator, is left however for future work.

4. Emission of drift-waves

One of the most important effects which can be observed *only if* FLRs *are retained* is the emission of drift-waves by a rotating magnetic island. This phenomenon, which is supposed to have a significant impact on the background turbulence, has been first pointed out by Waelbroeck and others [9]. In that paper, a difference between islands rotating in the electron diamagnetic

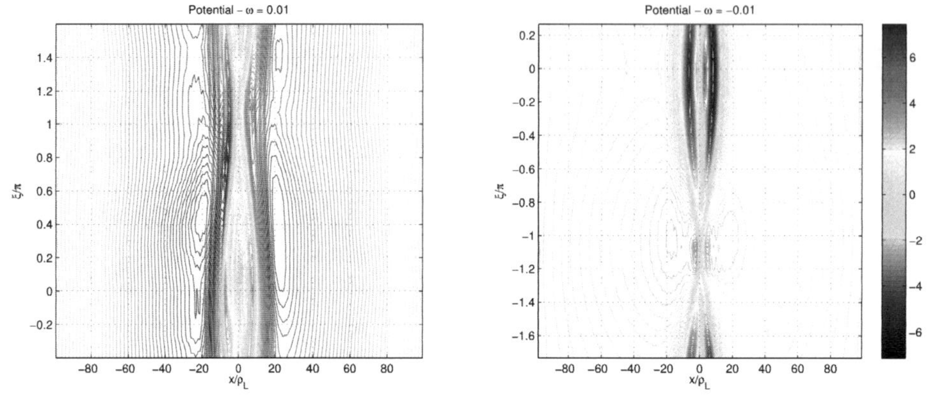

Figure 6. Contour plot of the potential for an island of width $w = 8\rho_L$ and rotating in the electron- (left) and ion- (right) diamagnetic directions. With the chosen parameters, we have $\omega_{*,e} = -\omega_{*,i} = 0.015$. The O-point is identified by $\xi = 0$, and the two figures have been realized with the same number of contours.

direction and those rotating in the ion diamagnetic direction has been pointed out. In fact, the emission of drift-waves has been foreseen to take place only in the former case, as the expression for the potential in the Fourier space exhibits a pole for $0 < \omega < \omega_{*,e}$ [9]. This is clearly visible in the GKW numerical simulations, see Fig.6. For an island rotating in the ion-diamagnetic direction (right) the potential is well localized in the island region, quickly going to zero outside the island separatrix. On the contrary, for an island rotating in the electron-diamagnetic direction, the potential is much more spread on the computational domain. The

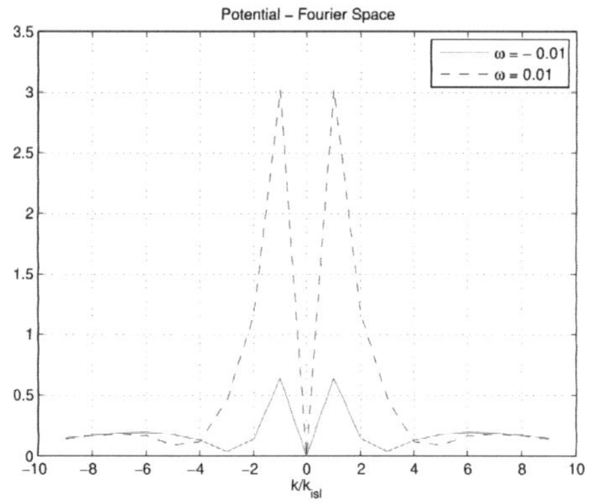

Figure 7. Potential in the Fourier space (radial wavenumbers) for $k_\zeta = k_{\zeta,isl}$, for the same island and the same frequencies considered in Fig.6. k_{isl} is the wavenumber corresponding to the radial box extent.

appearence of the drift-waves is moreover confirmed in Fig.7, where the "radial" Fourier spectra of the electrostatic potentials of Fig.6 (for $k_\zeta = k_{\zeta,isl}$ corresponding to the island periodicity) are depicted. It is clear that, for islands rotating in the electron-diamagnetic direction, the Fourier components are much larger than in the opposite case, and this can be related to the occurrence of drift waves.

Journal of Physics: Conference Series **260** (2010) 012019 doi:10.1088/1742-6596/260/1/012019

5. Summary and Outlook

We have employed the flux-tube gyrokinetic spectral code GKW to carry out an investigation of some relevant physical phenomena caused by the presence of a magnetic island in the plasma. This code allows to take into account many relevant physical aspects (such as FLRs, toroidicity and self-consistent electrostatic potential), yielding therefore a more complete description of the island dynamics in comparison to most of the numerical investigations available in literature. This work represents the first step in the direction of achieving a self-consistent numerical description of the island dynamics retaining the large majority of physical effects potentially playing a role.

[1] Peeters A G et al., 2009 *Comp. Phys. Comm.* **180**, 2649
[2] Hamada S, 1962 *Nucl. Fusion* **2**, 23
[3] Hornsby W A et al., 2010 *Phys. Plasmas* **17**, accepted for publication
[4] Wilson H R et al., 1996 *Phys. Plasmas* **3**, 248
[5] Fitzpatrick R, 1995 *Phys. Plasmas* **2**, 825
[6] Bergmann A, Poli E and Peeters A G, 2009 *Phys. Plasmas* **16**, 092507
[7] Connor J W and Wilson H R, 1995 *Phys. Plasmas* **2**, 4575
[8] Carrera R, Hazeltine R D and Kotschenreuther M, 1986 *Phys. Fluids* **29** 899
[9] Waelbroeck F, Connor J W and Wilson H R, 2001 *Phys. Rev. Letters* **87**, 215003

Gyrofluid simulations of collisionless reconnection in the presence of diamagnetic effects

E Tassi[1], **F L Waelbroeck**[2] **and D Grasso**[3,4]

[1]Centre de Physique Théorique, CNRS – Aix-Marseille Universités, Campus de Luminy, case 907, F-13288 Marseille cedex 09, France

[2]Institute for Fusion Studies, The University of Texas at Austin, Austin, TX 78712-1060, USA

[3] Istituto dei Sistemi Complessi - CNR, Roma, Italy

[4] Dipartimento di Energetica, Politecnico di Torino, Italy

E-mail: `tassi@cpt.univ-mrs.fr`

Abstract.
The effects of the ion Larmor radius on magnetic reconnection are investigated by means of numerical simulations, with a Hamiltonian gyrofluid model. In the linear regime, it is found that ion diamagnetic effects decrease the growth rate of the dominant mode. Increasing ion temperature tends to make the magnetic islands propagate in the ion diamagnetic drift direction. In the nonlinear regime, diamagnetic effects reduce the final width of the island. Unlike the electron density, the guiding center density does not tend to distribute along separatrices and at high ion temperature, the electrostatic potential exhibits the superposition of a small scale structure, related to the electron density, and a large scale structure, related to the ion guiding-center density.

1. Introduction

Magnetic reconnection is responsible for many phenomena occurring in laboratory and astrophysical plasmas. [1, 2] In this paper, we restrict consideration to reconnection in low-beta configurations (i.e. $m_e/M_i < \beta \ll 1$, with m_e and M_i indicating electron and ion mass, respectively) characterized by the presence of a magnetic "guide" field perpendicular to the reconnection plane. The guide field ensures that the ions undergo rapid Larmor gyration as they cross the reconnection region. Guide field reconnection is thought to be responsible, in particular, for the sawtooth crash in tokamaks. Kinetic simulations of guide-field reconnection are generally more costly and difficult than in the alternative case of "component reconnection," due to the separation of scales imposed by the guide field. [3, 4, 5] In the linear [6] as well as in the nonlinear regime, [7, 8, 9] magnetic reconnection is characterized by processes occurring in very narrow resonant layers. For high-temperature plasmas, such as those existing in the present generation of tokamaks, the resonant layer is narrower than the ion Larmor radius, so that the ions may traverse the entire layer during a cyclotron period.[10] Their response is then a nonlocal functional of the fields they sample along their orbit. [11, 12]

Conventional fluid models treat finite Larmor radius (FLR) effects by expanding the ion response to first order in $(k_\perp \rho_i)^2$. [13, 14, 15, 16] In the linear regime, the predictions of fluid models that include FLR effects [17] are qualitatively similar to the results of kinetic calculations [11, 12] that account for the nonlocal response of the ions. In particular, both fluid and kinetic models predict the stabilization of ideal instabilities when the growth rate γ_0 computed in the absence of diamagnetic drifts is less than half of ω_{*i}, the ion diamagnetic drift angular frequency. [11, 17, 18] It is unclear *a priori*, however, whether fluid FLR calculations such as the investigations of the nonlinear saturation of reconnecting modes in Refs. [13, 15, 19], offer an adequate description of ion dynamics in the nonlinear regime. In particular, it is unclear whether they are adequate for the task of extrapolating the experimental observations to different values of ρ_i/L, where L describes the plasma size. Some indications are provided by nonlinear simulations using gyrokinetic codes, [20] but numerical limitations make scaling studies with such codes impractical.

The need for a simpler nonlinear model for the nonlocal ion response has motivated the derivation and investigation of gyrofluid models that account for these physical effects in a fluid context. Grasso *et al.* [21] introduced the first electromagnetic gyrofluid model and applied it to the study of magnetic reconnection. Their gyrofluid model, although unable to describe kinetic phenomena such as wave particle resonance (Landau damping), [12] nevertheless illuminated the qualitative properties of the reconnection dynamics. More elaborate models that do account for Landau damping were subsequently developed by Snyder and Hammett, but the applications of their models have been limited to the study of electromagnetic effects on turbulence driven by pressure gradients. [22]

In the present contribution, we investigate a recently derived 3-field gyrofluid model [23] by means of numerical simulations. This gyrofluid model is part of a family of Hamiltonian fluid models [24, 21, 25, 26] that have been investigated in past years, but is distinguished from these other Hamiltonian models by its nonlocal treatment of the ion density evolution. In the linear regime, the nonlocal description of the ion density evolution improves on FLR models by describing the evanescence of drift waves excited at frequencies in the ion drift direction but below the diamagnetic frequency. [23] The propagation and evanescence properties of drift waves can be important for determining the role of the polarization current in island evolution. [27]

In recent work, [28] we have used the 3-field model of Ref. [23] to revisit earlier investigations [21] of gyrofluid magnetic reconnection in a *homogeneous* plasma, where ion diamagnetic effects vanish. The present article extends these studies to the case of *inhomogeneous* plasma, where ion diamagnetic drifts affect the reconnection. Such ion drift effects can be important for sawtooth dynamics in tokamaks [9, 13, 15], where it accounts for the saturation of the resistive kink instability which can prevent complete reconnection during sawtooth crashes. [29, 30] It is also important for reconnection at the dayside magnetopause [31, 32] where the interface between the incoming solar wind and the Earth's magnetosphere is characterized by a strong density gradient. Unfortunately, our assumption of a strong guide field makes our model inapplicable to the magnetopause.

The paper is organized as follows. In Sec. 2 we review the model equations. Sec. 3 is devoted to the simulation results in the linear phase and to the comparison with analytical predictions. In Sec. 4 we consider the evolution of the system in the nonlinear phase. We conclude in Sec. 5.

2. The model

We consider the two-dimensional version of the Hamiltonian gyrofluid model described in Ref.[23]. The model equations are

$$\frac{\partial n_i}{\partial t} + [\Phi, n_i] = 0, \tag{1}$$

$$\frac{\partial n_e}{\partial t} + [\phi, n_e] - [\psi, \nabla^2\psi] = 0, \tag{2}$$

$$\frac{\partial}{\partial t}(\psi - d_e^2\nabla^2\psi) + [\phi, \psi - d_e^2\nabla^2\psi] + \rho_s^2[\psi, n_e] = 0, \tag{3}$$

$$n_e = \Gamma_0^{1/2} n_i + (\Gamma_0 - 1)\phi/\rho_i^2. \tag{4}$$

These represent the continuity equation for the ion guiding centers, the electron continuity equation, the Ohm's law and the Poisson's equation, respectively. In Eqs.(1)-(4) n_i is the ion guiding center density, n_e the electron density, ϕ the electrostatic potential, ψ the poloidal magnetic flux function, d_e the electron skin depth, ρ_s the sonic Larmor radius and $\Phi = \Gamma_0^{1/2}\phi$ is the gyro-averaged electrostatic potential. For our analysis the Padé approximant version $\Gamma_0^{1/2} = (1 - \rho_i^2\nabla^2/2)^{-1}$ of the gyro-average operator will be adopted, where ρ_i is the ion Larmor radius. Given a Cartesian coordinate system (x, y, z), we assume all the fields be translationally invariant along z and we define the canonical bracket between two generic fields f and g by $[f, g] = \hat{z} \cdot \nabla f \times \nabla g$. We also recall that we normalize the time with respect to the Alfvèn time and distances with respect to a magnetic equilibrium scale length L. Dependent variables are normalized in the following way

$$n_i = \frac{L}{\hat{d}_i}\frac{\hat{n}_i}{n_0}, \qquad n_e = \frac{L}{\hat{d}_i}\frac{\hat{n}_e}{n_0}, \qquad \psi = \frac{\hat{A}_z}{BL}, \qquad \phi = \frac{\hat{\rho}_s^2}{L^2}\frac{L}{\hat{d}_i}\frac{e\hat{\phi}}{T_e},$$

where carets denote dimensional variables, d_i is the ion skin depth, n_0 a background density amplitude, A_z the magnetic potential, e the unit charge, B a characteristic magnetic field amplitude and T_e the electron temperature, which is assumed to be constant.

3. Numerical simulations in the linear regime

We devote this section to the analysis of the numerical solutions of the model (1)-(4) in the phase in which nonlinear terms are not dominant.
We solve the model equations over the domain $\{(x, y) : -\pi \le x < \pi, -a\pi \le y < a\pi\}$, where we prescribe the value of the number a at each simulation. The grid is composed by 1024×128 points and double periodic boundary conditions are imposed on the field perturbations, which are the quantities that the code advances in time. Unlike Ref. [28] we consider now *inhomogeneous* density equilibria. The simulations are then started by perturbing the equilibrium

$$n_{i_{eq}}(x) = n_0' x, \qquad n_{e_{eq}}(x) = n_0' x, \qquad \psi_{eq}(x) = \sum_{n=-11}^{11} a_n \exp(inx), \tag{5}$$

where n_0' is a constant and the a_n are the Fourier coefficients of the function $f(x) = 1/\cosh^2 x$. With respect to the simulations of Ref. [28] the presence of equilibrium density gradients makes the evolution of n_i no longer negligible and introduces the diamagnetic effects. A global dispersion relation for the system in the presence of diamagnetic effects was derived in Ref. [23]. This showed also how the Padé approximant version of the gyrofluid model was able to reproduce adequately dispersive properties of the system, such as a spectral gap in the frequencies, which are not caught by Finite Larmor Radius (FLR) models.
The equilibrium (5) is perturbed in n_i with a four-cell pattern disturbance of the form $\tilde{n}_i \propto \cos(x + y/a) - \cos(x - y/a)$. The field ϕ is also perturbed according to (4), in such a way that the initial perturbation on n_e is zero.

If we indicate with $\chi(x, y, t)$ any of the three dynamical variables of our system, i.e. n_i, n_e and ψ, then we can decompose the fields as $\chi = \chi_{eq}(x) + \tilde{\chi}(x, y, t)$. We write then the perturbation $\tilde{\chi}$ as a Fourier series truncated at N modes, so that $\tilde{\chi} = \sum_{-N}^{N} \hat{\tilde{\chi}}_{\mathbf{k}} \exp(i\mathbf{k} \cdot \mathbf{x} + \lambda t)$, where the hat indicates the Fourier transform, $\mathbf{k}$ is the wave vector and $\lambda = \gamma - i\omega$ is a complex number having the growth rate γ as real part and the rotation frequency $-\omega$ as imaginary part. Simulations with a relatively small amplitude of the initial perturbation, corresponding to 10^{-6}, have been carried out, in order to obtain a long linear phase ($\sim$ 25 Alfvèn times for

Table 1. Table displaying values of linear growth rate γ and rotation frequency ω, for the dominant mode. A comparison is made between values obtained from numerical simulations $(\omega_{num},\gamma_{num})$ and values $(\omega_{theor},\gamma_{theor})$ predicted by the asymptotic theory of Ref. [11]. The agreement between numerical and analytical results is better for $\Delta' = 59.9$, which is a value closer to the asymptotic regime of validity of the theory. For all cases $d_e = 0.2$.

Δ'	ρ_i	ρ_s	v_*	γ_{num}	γ_{theor}	ω_{num}	ω_{theor}	Error
34.2	0.4	0.4	-0.16	0.33	0.44	0	0	24 %
34.2	0.4	0.4	-0.4	0.23	0.37	-0.02	0	38 %
34.2	0.35	0.2	-0.04	0.28	0.36	0.03	0.11	22 %
59.9	0.4	0.4	-0.16	0.27	0.33	0	0	19 %
59.9	0.4	0.4	-0.4	0.21	0.28	-0.01	0	27 %
59.9	0.35	0.2	-0.04	0.23	0.27	0.03	0.08	15 %

most of the simulations) that can be effectively investigated. In particular, we are interested in extracting from the simulations the values of γ and ω, which can permit us to observe the effects of equilibrium density gradients and finite gyro-radius, on the linear mode stabilization and island rotation frequency. We find it also of interest to compare the obtained numerical results with those predicted by the theory developed in Ref. [11]. The linear analysis by Porcelli refers namely to the collisionless case in the presence of finite ion temperature. Here we limit ourselves to recall the results of the theory that are relevant for our analysis. By adopting a matching asymptotic technique, in Ref. [11], the following dispersion relation, for the main mode, is obtained

$$\lambda^2 - \omega_{*e}\omega_{*i} + i\lambda(\omega_{*e} + \omega_{*i}) \approx \gamma_0^2. \tag{6}$$

In (6) the electron diamagnetic frequency ω_{*e} is defined by $\omega_{*e} = -k_y cT_e n_0'/(eBn_0 v_A)$, where v_A is the Alfvèn speed based on B. The ion diamagnetic frequency is $\omega_{*i} = -\tau\omega_{*e}$, with $\tau = \rho_i^2/\rho_s^2$, and γ_0, for our case, reads $\gamma_0 = 2k_y(2d_e(\rho_s^2+\rho_i^2)/\pi)^{1/3}$. Splitting the dispersion relation into its real and imaginary part yields

$$\omega \approx \frac{\omega_{*e}}{2}(1-\tau), \tag{7}$$

$$\gamma^2 \approx \gamma_0^2 - \left[\frac{\omega_{*e}}{2}(1+\tau)\right]^2, \tag{8}$$

which give us a prediction for the growth rate and the rotation frequency. Such results, however, are valid only in the limit $\rho_i \gg d_e$ and $\Delta' \gg \rho_i^{1/3} d_e^{2/3}$, where Δ' is the standard stability parameter for tearing modes.

Examples of theoretical and numerical values are provided in Table 1. Two main cases are considered, corresponding to two different values of Δ' (which correspond to the values $a = 1.5$ and $a = 2$ of the parameter scaling the length of the domain along y). For each of these two cases the same sets of parameters ρ_i, ρ_s and $v_* = \omega_{*e}/k_y$ are considered. From these representative examples we can see that the numerical results follow the trends predicted by the theory, for the dominant mode (the latter is $k_y = 0.66$ for $\Delta' = 34.2$ and $k_y = 0.5$ for $\Delta' = 59.9$). In particular, for fixed ρ_i and ρ_s, the cases with $v_* = -0.4$ exhibit smaller growth rate than those with $v_* = -0.16$. This indicates the stabilizing effect operated by diamagnetic flows and corresponding to the term proportional to ω_{*e}^2 in (8). We report also that another simulation, again with $\rho_i = \rho_s = 0.4$, but with $v_* = -0.64$, showed no growth of the magnetic island, indicating the complete linear stabilization due to diamagnetic effects. For these two sets of parameters, the numerical solutions show almost no rotation frequency. On the other hand, for the case $\rho_i = 0.35$, $\rho_s = 0.2$ and $v_* = -0.04$, the numerical solutions show a positive value of ω, which corresponds to a propagation along the direction of the ion diamagnetic drift (which in our case corresponds to the positive y direction). The propagation along the ion diamagnetic drift direction is indeed also what is predicted by the theory, according to (7), when ions are hotter than electrons. With regard to the quantitative agreement with the theory, Table 1 shows that this improves when increasing Δ'. In the case with larger Δ' the distances between the numerical and predicted values get shorter and we can quantify that the relative error drops from an average of 28% to 18.7%. The reduction of the error is of course a consequence of the fact that, by increasing Δ', we get closer to the asymptotic regime in which the theory is valid.

4. Numerical simulations in the nonlinear regime

In this section we focus on the evolution of the system in the nonlinear phase.

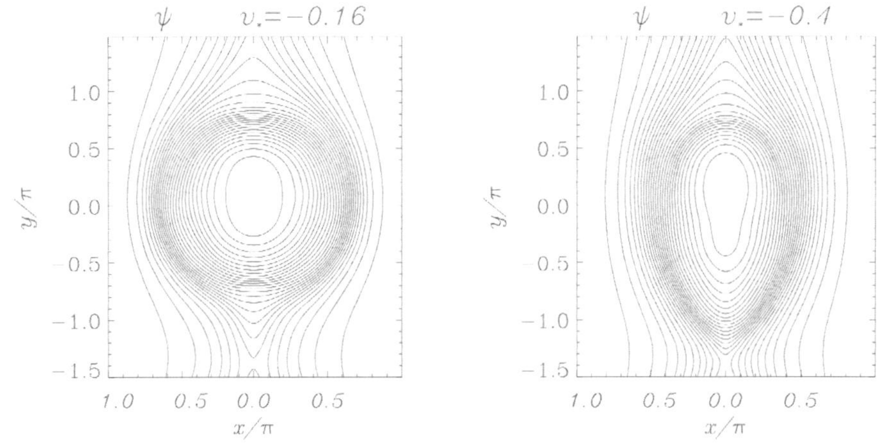

Figure 1. Contour plots of ψ for $v_* = -0.16$, at $t = 39$ (left) and $v_* = -0.4$ at $t = 50$ (right). The values of the other parameters are: $\rho_s = \rho_i = 0.4$, $d_e = 0.2$. The plots indicate that stronger diamagnetic effects lead to a smaller saturated island.

Fig. 1 shows a comparison between contour plots of ψ for $v_* = -0.16$ and $v_* = -0.4$, at times at which the two simulations reached the same phase of the dynamical evolution. From the figure we can see that the presence of stronger diamagnetic effects, apart from reducing the linear growth rate, also leads to a nonlinear island of smaller width. We recall that a nonlinear stabilizing effect of n_0' was observed in a cold-ion, three-field model in cylindrical geometry [15]. In that context diamagnetic flows were indeed suggested to be responsible for the saturation of islands at a small width. We also note that, in spite of the fact that, in the linear phase, the islands exhibited very little rotation for these parameters, in the nonlinear regime, upward shifts of the O-points are present, in particular in the case $v_* = -0.4$. Island propagation in the ion diamagnetic drift direction, caused either by viscosity [33, 16] or by nonlinear zonal flows, [34, 35] has been reported in FLR simulations of visco-resistive reconnection.

An effective way to look at the nonlinear dynamics of our system is to express the model in terms of its "normal fields", suggested by the Hamiltonian structure. We recall that the model equations (1)-(4) can be rewritten as [23]

$$\frac{\partial n_i}{\partial t} + [\Phi, n_i] = 0, \tag{9}$$

$$\frac{\partial G_+}{\partial t} + [\phi_+, G_+] = 0, \tag{10}$$
$$\frac{\partial G_-}{\partial t} + [\phi_-, G_-] = 0, \tag{11}$$

where $G_\pm = \psi - d_e^2\nabla^2\psi \pm d_e\rho_s n_e$ and $\phi_\pm = \phi \pm (\rho_s/d_e)\psi$. The fields n_i and $G_\pm$, which we denote as normal fields, represent natural variables for the system because, in terms of them, the model equations take the form of simple advection equations. As pointed out in a number of previous papers on reconnection based on Hamiltonian fluid models [36, 37, 38, 39, 26, 40], the stream functions $\phi_\pm$ correspond to velocity fields that rotate in opposite directions. The fields $G_\pm$, advected by such velocity fields, get then stirred in opposite directions. The stirring process causes $G_\pm$ to form finer and finer structures, leading to an energy cascade toward small scales. Because n_e is proportional to the difference between G_+ and G_-, such cascade inevitably reflects on the electron density structures.

In order to investigate the nonlinear structures, we double the number of grid points in the y direction, which enables us to better resolve small scale structures that form nonlinearly. We also bring up to 10^{-3} the amplitude of the initial perturbation. From Fig. 2 indeed we see that electron density, as is typical of Hamiltonian reconnection, develops lobes around the separatrices. These emerge as the difference between the spiral arms present in G_+ and G_- (the latter not shown here but such that $G_-(x,y,t) = G_+(-x,y,t)$). Unlike the above mentioned studies of Hamiltonian reconnection, however, the structures of Fig. 2 are of course not symmetric with respect to the $y = 0$ axis, because of diamagnetic effects. The pattern of ϕ_+ (and, analogously of $\phi_-(x,y,t) = -\phi_+(-x,y,t)$) are such to create thinner and shorter layers in the lower part of the magnetic island whereas wider and longer lobes are present in the upper part of the magnetic island. Inside the island, stirring of $G_\pm$ takes place and the initial linear density profile gets flattened. The ion guiding center density gradient also gets partly flattened in the center of the island, as a consequence of the advection operated by Φ. Such stream function, however, has a circulation pattern quite different from that obtained by the combination of ϕ_+ and ϕ_-. Therefore n_i does not tend to accumulate around the separatrices, unlike n_e. The four convective cells of the gyroaveraged electrostatic potential, on the other hand, tend to lie along the magnetic separatrices. Such behavior is also evident in the electrostatic potential ϕ. By virtue of the relation $\phi = (1 - \rho_i^2\nabla^2/2)\Phi$, we can also infer that ϕ exhibits, in general, finer structures, compared to Φ. We can indeed observe regions with steep gradients of electrostatic potential, around the separatrices, similarly to what happens to the electron density. Nevertheless, in this respect, it is important to recall that, for $n_0' = 0$, the growth of n_i was negligible, and one had $\tilde{\phi}_{\mathbf{k}} \sim -\rho_i^2\tilde{n}_{e\mathbf{k}}$, as $\rho_i \to +\infty$. This led to conclude [28] that the filamentation in n_e implied an analogous phenomenon in ϕ, with consequent formation of strong electric fields. In the presence of diamagnetic effects, on the other hand, Eq. (4), in the large ρ_i limit, yields

$$\tilde{\phi}_{\mathbf{k}} \sim -\rho_i^2\tilde{n}_{e\mathbf{k}} + \frac{2}{k^2}\tilde{n}_{i\mathbf{k}}. \tag{12}$$

Therefore, ϕ is no longer proportional to $\tilde{n}_e$, but is given by the superposition of two contributions: one, related to $\tilde{n}_e$, which is responsible for the formation of steep gradients, and a second one, due to $-2\nabla^{-2}\tilde{n}_i$ which, on the contrary, is relevant at large scales.

5. Conclusions

Numerical simulations of a Hamiltonian electromagnetic gyrofluid model have been carried out and investigated. The present analysis represents a step forward, with respect to a recent related work [28], because it accounts for inhomogeneous density equilibria, which lead to diamagnetic effects. In the linear phase, diamagnetic effects are shown to stabilize the growth rate of the

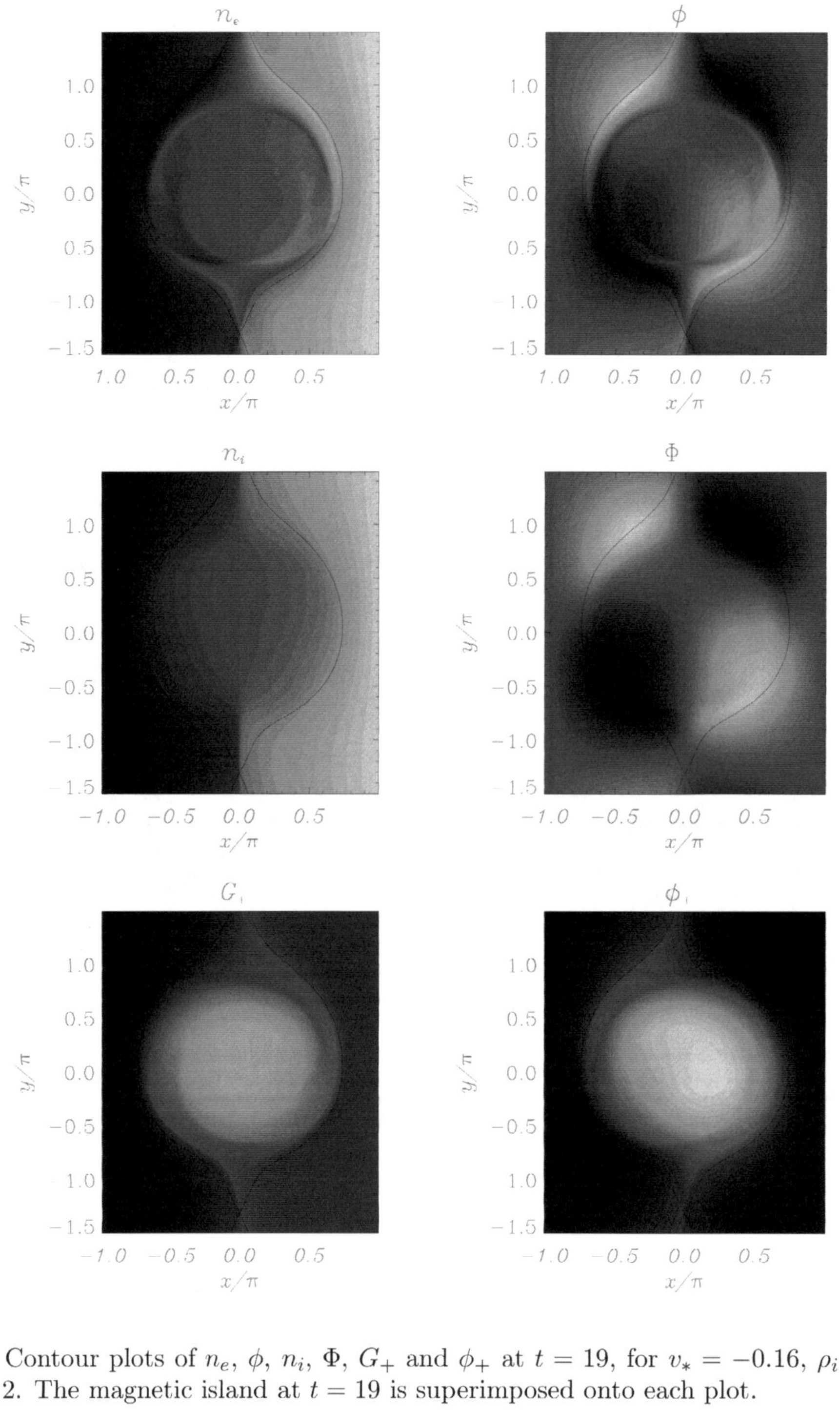

Figure 2. Contour plots of n_e, ϕ, n_i, Φ, G_+ and ϕ_+ at $t = 19$, for $v_* = -0.16$, $\rho_i = \rho_s = 0.4$ and $d_e = 0.2$. The magnetic island at $t = 19$ is superimposed onto each plot.

dominant mode and induce a propagation of the magnetic island. Simulations also indicate that the direction of propagation tends to move from that of the electron to that of the ion diamagnetic drift, as the ion temperature is increased. Such features are in agreement with what is predicted by the linear theory of Ref.[11] and the quantitative agreement improves as the values of parameters approach the asymptotic regime of validity of the theory. In the nonlinear phase, stronger equilibrium gradients tend to yield smaller "saturated" islands, as already observed in Ref.[15]. We note also a different distribution of the ion guiding center density with respect to the electron density, with the former retaining a four-cell pattern, whereas the latter concentrates around separatrices, although in an asymmetric way, also with respect to the $y = 0$ axis, as a consequence of the diamagnetic terms. Compared to the homogeneous equilibrium case [28], the potential possesses now a smoother contribution, due to the finite n_i fluctuations. We finally indicate the derivation of a refined version of the linear theory, and the extension of the model to include parallel dynamics, as subjects of work in progress.

Acknowledgments

This work was supported by the European Community under the contracts of Association between EURATOM and ENEA and between EURATOM, CEA, and the French Research Federation for fusion studies. FW was supported in part by a fellowship from the Katholieke Universiteit Leuven and by the US Department of Energy under contract DE-FG02-04ER-54742. The views and opinions expressed herein do not necessarily reflect those of the European Commission. Financial support was also received from the Agence Nationale de la Recherche (ANR EGYPT).

References

[1] Priest E R and Forbes T G 2000 *Magnetic Reconnection* (Cambridge, Cambridge University Press)
[2] Biskamp D 2000 *Magnetic Reconnection in Plasmas* (Cambridge, Cambridge University Press)
[3] Wesson J A 1986 *Plasma Phys. Control. Fusion* **28** 243–248
[4] Udintsev V S, Ottaviani M, Maget P *et al.* 2005 *Plasma Physics and Controlled Fusion* **47** 1111–1136
[5] Lazarus E A, Waelbroeck F L, Luce T C *et al.* 2006 *Plasma Phys. Control. Fusion* **48** L65–L72
[6] Coppi B, Galvao R, Pellat R, Rosenbluth M and Rutherford P 1976 *Sov. J. Plasma Phys.* **2**
[7] Waelbroeck F L 1989 *Phys. Fluids B* **1** 2372–2380
[8] Zakharov L, Rogers B and Migliuolo S 1993 *Physics of Fluids B: Plasma Physics* **5** 2498–2505
[9] Rogers B and Zakharov L 1996 *Physics of Plasmas* **3** 2411–2422
[10] Drake J F 1978 *Physics of Fluids* **21** 1777–1789
[11] Porcelli F 1991 *Phys. Rev. Lett.* **66** 425–428
[12] Berk H L, Mahajan S M and Zhang Y Z 1991 *Physics of Fluids B: Plasma Physics* **3** 351–362
[13] Biskamp D 1981 *Phys. Rev. Lett.* **46** 1522–1525
[14] Hazeltine R D, Kotschenreuther M and Morrison P J 1985 *Phys. Fluids* **28** 2466–2477
[15] Biskamp D and Sato T 1997 *Phys. Plasmas* **4** 1326–1329
[16] Fitzpatrick R, Watson P and Waelbroeck F L 2005 *Phys. Plasmas* **12** 082510
[17] Zakharov L and Rogers B 1992 *Physics of Fluids B: Plasma Physics* **4** 3285–3301
[18] Huysmans G T A, Sharapov S E, Mikhailovskii A B and Kerner W 2001 *Physics of Plasmas* **8** 4292–4305
[19] Rogers B and Zakharov L 1995 *Phys. Plasmas* **2** 3420
[20] Matsumoto T, Tokuda S, Kishimoto Y, Takizuka T and Naitou H 1999 *Journal of Plasma and Fusion Research* **75** 1188–1194
[21] Grasso D, Califano F, Pegoraro F and Porcelli F 2000 *Plasma Physics Reports* **26** 512–518
[22] Snyder P B and Hammett G W 2001 *Phys. Plasmas* **8** 3199–3216
[23] Waelbroeck F L, Hazeltine R D and Morrison P J 2009 *Physics of Plasmas* **16** 032109
[24] Cafaro E, Grasso D, Pegoraro F, Porcelli F and Saluzzi A 1998 *Phys. Rev. Lett.* **80** 4430–4433
[25] Tassi E, Morrison P J, Waelbroeck F L and Grasso D 2008 *Plasma Physics and Controlled Fusion* **50** 085014
[26] Tassi E, Morrison P J, Grasso D and Pegoraro F 2010 *Nuclear Fusion* **50** 034007
[27] Fitzpatrick R and Waelbroeck F 2005 *Phys. Plasmas* **12** 122511
[28] Grasso D, Tassi E and Waelbroeck F L 2010 *Phys. Plasmas* accepted for publication
[29] O'Rourke J 1991 *Plasma Physics and Controlled Fusion* **33** 289
[30] Nagayama Y *et al.* 1991 *Phys. Rev. Lett.* **67** 3527
[31] Swisdak M, Rogers B, Drake J and Shay M 2003 *J. of Geophys. Research-Space Phys.* **108**
[32] Cassak P A and Shay M A 2007 *Physics of Plasmas* **14** 102114
[33] Fitzpatrick R and Waelbroeck F L 2004 *Phys. Fluids* **12** 22307
[34] Uzawa K, Ishizawa A and Nakajima N 2010 *Plasma and Fusion Research* **5** S1016

[35] Uzawa K, Ishizawa A and Nakajima N 2010 *Phys. Plasmas* **17** 042508
[36] Grasso D, Califano F, Pegoraro F and Porcelli F 2001 *Phys. Rev. Lett.* **86** 5051–5054
[37] Porcelli F, Borgogno D, Califano F, Grasso D, Ottaviani M and Pegoraro F 2002 *Plasma Physics and Controlled Fusion* **44** B389
[38] Del Sarto D, Califano F and Pegoraro F 2005 *Phys. Plasmas* **12** 012317
[39] Del Sarto D, Califano F and Pegoraro F 2006 *Modern Phys. Lett. B* **20** 931–961
[40] Tassi E, Grasso D and Pegoraro F 2010 *Commun. Nonlin. Sci. Numer. Simulat.* **15** 2

Journal of Physics: Conference Series **260** (2010) 012021 doi:10.1088/1742-6596/260/1/012021

Global collisional gyrokinetic simulations of ITG microturbulence starting from a neoclassical equilibrium

T Vernay[1], S Brunner[1], L Villard[1], B F McMillan[1], O Sauter[1], S Jolliet[2], T M Tran[1] and A Bottino[3]

[1] Centre de Recherches en Physique des Plasmas, Ecole Polytechnique Fédérale de Lausanne, Association EURATOM – Confédération Suisse, CH-1015 Lausanne, Switzerland
[2] Japan Atomic Energy Agency, Higashi-Ueno 6-9-3, Taitou, Tokyo, 110-0015, Japan
[3] Max-Planck-Institut für Plasmaphysik, Boltzmannstrasse 2, EURATOM Association, D-85748 Garching, Germany

E-mail: `thibaut.vernay@epfl.ch`

Abstract. Linearized operators describing inter-species and like-species collisions have been discretized and implemented in the gyrokinetic Particle-In-Cell (PIC) code ORB5 [S. Jolliet, Comp. Phys. Comm. **177**, 409 (2007)] based on the delta-f approach. Simulation results for neoclassical transport are compared with both analytical predictions as well as results from other codes. This new version of ORB5 including collisional dynamics thus makes it possible to carry out simulations of microturbulence starting from a global neoclassical equilibrium including self-consistent electric fields. First results of ITG microturbulence simulations carried out in this way are presented. The issue of numerical noise, inherent to the PIC approach and further accentuated by the implementation of collisions in the delta-f scheme, is addressed. It is shown how a noise reduction scheme based on a coarse graining procedure [Y. Chen and S. E. Parker, Phys. Plasmas **14**, 082301 (2007)] ensures the physical relevance of such simulations. Furthermore, a novel delta-f algorithm is presented, which switches between a canonical and a local Maxwellian background for carrying out the collisonless and collisional dynamics respectively, and its advantages are discussed.

1. Introduction

Gyrokinetic codes are devoted to the study of microturbulence and resulting anomalous transport in magnetic fusion plasmas. In these hot, low density plasmas, collisional effects are a relatively slow process compared to the characteristic time scale (i.e. frequencies and growth rates) of the microinstabilities underlying the small scale turbulence. Collisions have thus often been neglected in gyrokinetic simulations under the rationale of both reduced complexity of the required algorithm and reduced numerical cost. There is indeed a computational price for implementing collisions into gyrokinetic codes, especially in the frame of the δf Particle-In-Cell (PIC) approach in the form of increased numerical noise [1]. Nonetheless, in order for such simulations to account for full realistic interactions between particles, collisions have to be considered, as they may lead to various important phenomena regarding turbulent transport (e.g. modified growth rates of instabilities, damping of zonal flows).

Journal of Physics: Conference Series **260** (2010) 012021 doi:10.1088/1742-6596/260/1/012021

This paper shows how collision operators, combined with a noise reduction scheme [2], make it possible to reach a neoclassical equilibrium, with a low noise level, which is a sound basis for starting global gyrokinetic simulations of microturbulence. A suitable collisional δf algorithm for carrying out runs at low collisionality is proposed [3], which takes advantage of the collisionless stationary solutions of the gyrokinetic equation. It results in some reduction in the computational cost of the simulation and thus provides a practical way to get relevant results related to collisional microturbulence.

The paper is organized as follows: Sec. 2 briefly presents the simulation model of the global gyrokinetic code ORB5 [4] including collisions. Sec. 3 describes the numerical methods used to solve the model. Sec. 4 shows some neoclassical benchmarks validating the collisional model. Sec. 5 proposes a novel algorithm for collisional gyrokinetic simulations switching between a local and a canonical Maxwellian background, for respectively carrying out the collisional and collisionless dynamics. This algorithm, combined with a noise reduction procedure, is then applied to global collisional gyrokinetic simulations of ITG microturbulence in Sec. 6. Conclusions are drawn in Sec. 7.

2. Simulation model

For the results presented in this paper, adiabatic electrons are assumed and the only kinetic species which is considered is Deuterium. In the frame of the gyrokinetic theory, f is the gyro-averaged particle distribution function. The distribution is usually expressed in terms of the gyrocenter variables $(\vec{R}, v_{||}, \mu)$, where $\vec{R}$ is the guiding centre position, $v_{||} = \vec{v}\cdot\vec{B}/B$ is the parallel velocity of the particle, $\vec{B}$ is the magnetic field and μ the magnetic moment. The gyrokinetic equation with collisions, governing the evolution of the ion distribution f, reads:

$$\frac{D}{Dt}f = -C(f), \tag{1}$$

where C is the ion self-collision operator and D/Dt is the collisionless gyrokinetic operator:

$$\frac{D}{Dt} = \frac{\partial}{\partial t} + \frac{\mathrm{d}\vec{R}}{\mathrm{d}t}\cdot\frac{\partial}{\partial\vec{R}} + \frac{\mathrm{d}v_{||}}{\mathrm{d}t}\frac{\partial}{\partial v_{||}}. \tag{2}$$

The time-derivatives of the gyrocenter variables $(\mathrm{d}\vec{R}/\mathrm{d}t, \mathrm{d}v_{||}/\mathrm{d}t, \mathrm{d}\mu/\mathrm{d}t = 0)$, implemented as particle trajectories in ORB5 [4], are given by the equations derived by Hahm [5]. The code uses a δf algorithm, where f is decomposed into a background f_0 and a perturbation δf: $f = f_0 + \delta f$. The background f_0 may be choosen as a local Maxwellian f_{LM} or a canonical Maxwellian f_{CM}. The self-collision operator appearing on the right hand side of Eq.(1) is linearized with respect to a local Maxwellian background: $C(f) \approx \hat{C}(\delta f_{LM})$ where $f = f_{LM} + \delta f_{LM}$, according to the procedure described in [1]. Moreover, the finite Larmor radius (FLR) effects are neglected in the collision operator, while they are retained in the collisionless dynamics for turbulence studies. The gyrokinetic equation for δf thus reads:

$$\frac{D}{Dt}\delta f = -\left[\frac{D}{Dt}f_0 + \hat{C}(\delta f_{LM})\right]. \tag{3}$$

The gyrokinetic equation is coupled with the quasi-neutrality relation, providing an equation for the self-consistent electrostatic potential ϕ:

$$\frac{en_{i0}(\Psi)}{T_e(\Psi)}[\phi(\vec{x},t) - \langle\phi\rangle(\Psi,t)] - \nabla_\perp\cdot\left(\frac{n_{i0}(\Psi)}{B\Omega_i}\nabla_\perp\phi\right) = \bar{\delta n}_i(\vec{x}), \tag{4}$$

where Ω_i is the ion cyclotron frequency, T_e is the electron temperature, e is the elementary charge and $\langle\rangle$ stands for the flux surface average operator. n_{i0} is the ion background density and $\bar{\delta n}_i$ is the perturbed gyrodensity contribution.

3. Numerical methods

Equation (3) is solved using the low-noise δf PIC method. N numerical particles, called markers, are introduced. They follow trajectories according to the gyrocenter equations [5]. In addition to phase space coordinates, each marker r has two so-called weights w_r and p_r [6], which are also functions of time and represent the fraction of the physical distributions δf and f_0, respectively, carried by each marker. The distribution functions f_0 and δf in gyrocenter variables are thus written as follows:

$$\delta f(\vec{R}, v_{||}, \mu, t) = \sum_r w_r(t)\delta[\vec{R} - \vec{R}_r(t)]\delta[v_{||} - v_{||,r}(t)]\delta[\mu - \mu_r(t)], \quad (5)$$

$$f_0(\vec{R}, v_{||}, \mu, t) = \sum_r p_r(t)\delta[\vec{R} - \vec{R}_r(t)]\delta[v_{||} - v_{||,r}(t)]\delta[\mu - \mu_r(t)]. \quad (6)$$

Evolving the gyrokinetic equation is thus performed by integrating numerically in time the marker trajectories together with weight equations. The toroidal angle φ and the square root of the normalized poloidal magnetic flux $s = \sqrt{\Psi/\Psi_{\mathrm{edge}}}$ are used respectively as the toroidal and the radial coordinate. The poloidal coordinate is the straight-field-line angle θ^*, such that $\mathrm{d}\varphi/\mathrm{d}\theta^* = q_s =$ constant along a field line, with $q_s(\Psi)$ the safety factor on the magnetic surface $\Psi =$ constant. Collision operators, representing a diffusion process, add to the collisionless scheme random kicks in velocity space as well as modifications to the weight equations in order for the collisional algorithm to conserve particle number, parallel momentum and kinetic energy [7], [1]. The quasi-neutrality equation is solved in toroidal and poloidal Fourier spaces using a finite-element method [8]. The energy conservation of the code has been extensively tested [4].

4. Neoclassical benchmark

In order to test the implementation of the self-collision operator, the results presented in this section benchmark ion neoclassical transport obtained with ORB5 against analytical results, as well as previous numerical results from the drift-kinetic PIC code FORTEC-3D and the gyrokinetic Eulerian code GT5D [9]. The benchmark parameters are described in Tab. 1, with density and temperature profiles peaked at $r = 0.5a$. R_0 is the major radius, a the minor radius, T_e and T_i the electron and ion temperature respectively, $L_{Ti} = T_i/\nabla T_i$ and $L_n = n/\nabla n$ the characteristic lengths of the temperature and density gradient respectively at $r = 0.5a$. One defines $\rho^* = \rho_s/a$, where $\rho_s = c_s/\Omega_i$ is the sound Larmor radius for ions. The collisionality parameter ν^* is defined as the ratio between the detrapping collision frequency and the bounce frequency:

$$\nu^* = \frac{R_0 q_s}{\tau_{ii} v_{thi} \epsilon^{3/2}}, \text{ with } \tau_{ii} = \frac{6\sqrt{\pi}}{\nu_{ii}}, \quad (7)$$

where ν_{ii} is the ion-ion collision frequency, q_s is the safety factor and $\epsilon = r/R_0$ the local inverse aspect ratio. Figure 1 shows a collisionality scan for both the ion heat diffusivity χ_{Hi} and the coefficient k appearing in the neoclassical force balance relation (Eq.(20) in Ref. [9]). The heat diffusivity is given in units of $\chi_{GB} a/L_n$, where $\chi_{GB} = \rho_s^2 c_s/a$ is the Gyro-Bohm diffusivity. Analytical values of the heat diffusivity derived in the limit of large aspect ratio by Chang and Hinton (C-H) [10] are plotted as well. Predictions from the moment equation approach by Hirshman and Sigmar (H-S) [11] are shown in addition to results derived by Hinton and

$a/R_0 = 0.2$	$T_e/T_i = 1$	$R_0/L_{Ti} = 6$	$R_0/L_n = 6$	$1/\rho^* = 150$

Table 1. Run parameters, ion neoclassical transport benchmark

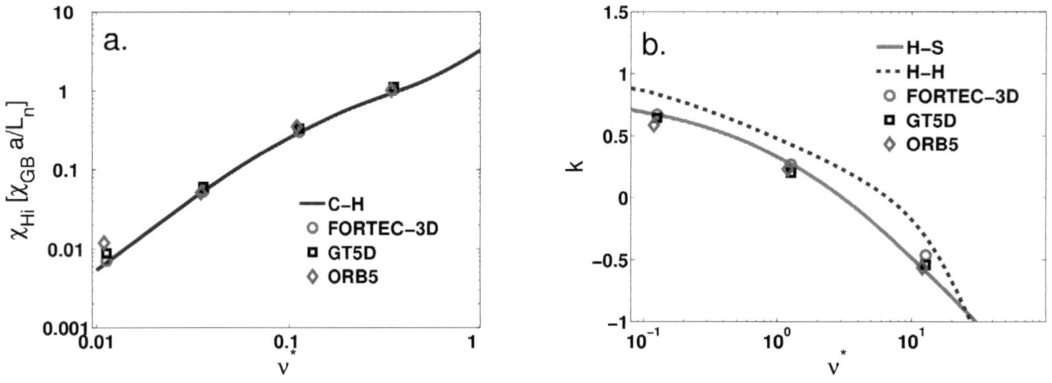

Figure 1. Considering parameters of Table 1. a) Ion heat diffusivity χ_{Hi} and b) neoclassical force balance coefficient k as a function of the effective collisionality. ORB5 results, shown for position $r/a = 0.5$ and at time $t \sim \tau_{ii}(r/a = 0.5)$, are compared to simulations from GT5D and FORTEC-3D, as well as analytical results from Chang and Hinton (C-H), Hinton and Hazeltine (H-H), Hirshman and Sigmar (H-S).

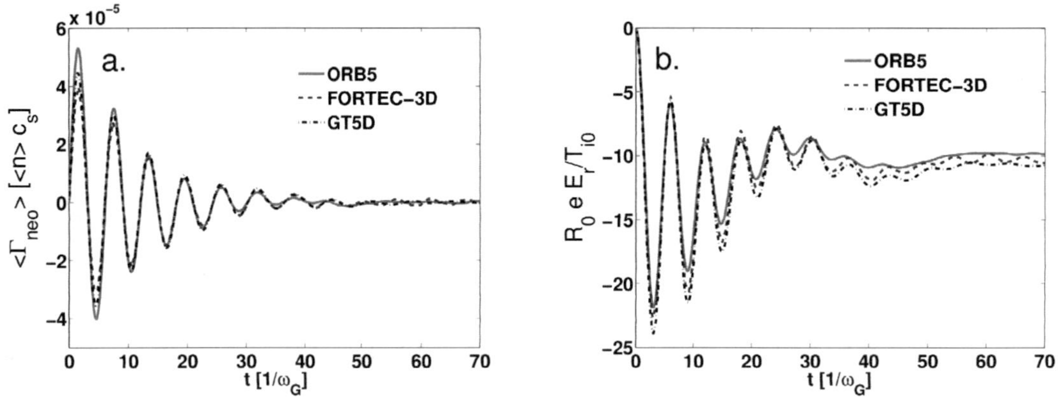

Figure 2. Time evolutions of a) ion gyrocenter flux Γ and b) radial electric field $E_r = -\partial\langle\phi\rangle/\partial r$ at $r = 0.5a$. Comparisons between results from ORB5, GT5D and FORTEC-3D. The ambipolarity is satisfied and an equilibrium neoclassical electric field is established. $\nu^*(r/a = 0.5) = 0.12$. ω_G is the GAM frequency.

Hazeltine (H-H) [12] for the coefficient k. The ambipolarity condition, i.e. a vanishing ion flux since the electrons are adiabatic, is verified in Figure 2.a, showing the time evolution of the ion gyrocenter flux at $r = 0.5a$. The setting up of the neoclassical radial electric field at $r = 0.5a$ is shown in Figure 2.b. The agreement between ORB5 results and those described in [9] is clearly very satisfactory.

Journal of Physics: Conference Series **260** (2010) 012021 doi:10.1088/1742-6596/260/1/012021

5. Collisional δf-scheme with local/canonical Maxwellian background switching

The collisionless version of ORB5 usually makes use of a canonical Maxwellian (CM) background, which is a stationary state of the gyrokinetic equation:

$$f_{CM} = \frac{\mathcal{N}(\Psi_0)}{(2\pi\mathcal{T}(\Psi_0)/m)^{3/2}} \exp\left[-\frac{mv_{\parallel}^2}{2\mathcal{T}(\Psi_0)} - \frac{B\mu}{\mathcal{T}(\Psi_0)}\right], \tag{8}$$

where "temperature" $\mathcal{T}$ and "density" $\mathcal{N}$ are functions of the toroidal canonical momentum $\Psi_0 = \Psi + qF(\Psi)v_{\parallel}/mB$, where $F(\Psi) = RB_{\varphi}$ is the poloidal current flux function. At low collisionality, the full distribution f is still expected to be close to f_{CM}. Further considering f_{CM} as a background in this case thus appears advantageous as it enables to minimize the deviation δf and associated noise in the PIC representation. The self-collision operator is however naturally linearized with respect to a local Maxwellian (LM) background:

$$f_{LM} = \frac{n_0(\Psi)}{(2\pi T_0(\Psi)/m)^{3/2}} \exp\left[-\frac{mv_{\parallel}^2}{2T_0(\Psi)} - \frac{B\mu}{T_0(\Psi)}\right], \tag{9}$$

where temperature and density are functions of the poloidal magnetic flux Ψ. In the frame of the two-weight scheme, it is possible to take advantage of the collisionless equilibrium function even in collisional runs, by performing a transformation from a canonical to a local background for carrying out the collisional step [3]. This transformation is based on the conservation of the total weight for each marker in both δf-representations (CM and LM):

$$f = f_{CM} + \delta f_{CM} = f_{LM} + \delta f_{LM} \tag{10}$$

$$\Longrightarrow \quad w_{CM} + p_{CM} = w_{LM} + p_{LM}. \tag{11}$$

A second equation required for the transformation is provided by the ratio σ of the p_{CM} and p_{LM} weights, which can be related to the ratio of the corresponding backgrounds:

$$\sigma = \frac{p_{CM}}{p_{LM}} = \left.\frac{f_{CM}}{f_{LM}}\right|_z, \tag{12}$$

where z is the position of the marker in phase-space. Before the self-collision algorithm, the local weights (w_{LM}, p_{LM}) are thus computed from the canonical weights (w_{CM}, p_{CM}) using Eqs. (11) and (12):

$$p_{LM} = \frac{p_{CM}}{\sigma}, \tag{13}$$

$$w_{LM} = w_{CM} + \left(1 - \frac{1}{\sigma}\right) p_{CM}. \tag{14}$$

The self-collisions are then carried out in the frame of the LM representation, computing in particular new positions in phase space z^* as well as modifications of the weights w_{LM}^* and p_{LM}^* as described in [1]. A new ratio $\sigma^* = f_{CM}/f_{LM}|_{z*}$ is then computed. Transforming back to the CM representation, using the inverse of relations (11) and (12), one obtains the collisional modified weights (w_{CM}^*, p_{CM}^*) in the CM representation and the code is ready for the next collisionless step, which uses again a canonical background. The scheme of the algorithm is described in Figure 3.

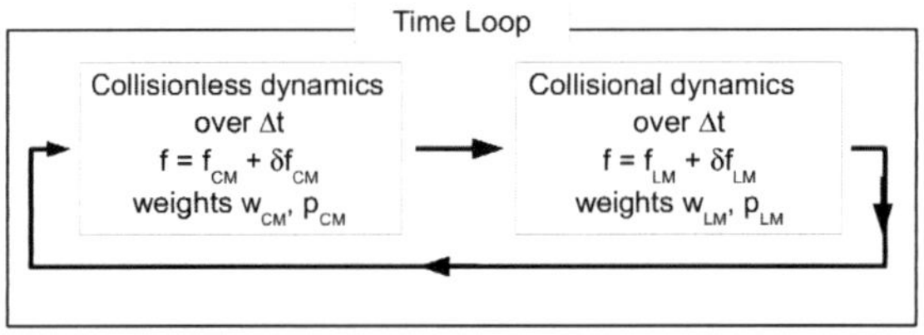

Figure 3. Mixed background collisional δf algorithm: Time loop for switching between a canonical Maxwellian background for stepping the collisionless dynamics and a local Maxwellian background for stepping the collisional dynamics.

6. Global collisional gyrokinetic simulations

The CYCLONE base case [13] is chosen in order to study the effects of the ion-ion self-collisions on ITG microturbulence, considering adiabatic electrons. The total initial distribution is a canonical Maxwellian, $f(t=0) = f_{CM}$, which is a stationary state of the collisionless gyrokinetic equation. Without collisions, simulations must start with a small perturbation to that equilibrium, in order for the turbulence to grow. With collisions, and at first enforcing axisymmetry by retaining only the toroidal Fourier mode $n = 0$, the full distribution will evolve towards a different distribution, i.e. a true neoclassical equilibrium together with the corresponding self-consistent axisymmetric electric field. After the neoclassical equilibrium has been reached, turbulence is switched on by letting the non-axisymmetric modes ($n \neq 0$) evolve freely. For the turbulent simulations, only every fourth toroidal Fourier mode is retained, namely $n = 0, 4, 8, \ldots, 56$, corresponding in real space to 1/4 toroidal wedge. 500 million markers are used along with the following grid in configuration space for solving the quasi-neutrality relation: 128 points in the radial direction, 512 points in the poloidal direction and 256 points in the toroidal direction. These numerical parameters yield an average of ~ 30 markers per cell. However, since the quasi-neutrality equation is solved in poloidal and toroidal Fourier spaces, the important ratio is the number of markers per Fourier mode [14]. Invoking the alignment of microturbulence with the magnetic field lines, only the poloidal modes $m \in [nq_s - 5, nq_s + 5]$ are retained for each toroidal mode n. This filter leads to ~ 3 million markers per Fourier mode. It has been verified that even with collisions this resolution ensures converged results.

For the sake of clarity, the turbulent contribution to the energy flux Q_{turb} is defined as follows:

$$Q_{turb} = \left\langle \frac{\vec{\nabla} r}{|\vec{\nabla} r|} \cdot \int \mathrm{d}^3 v \frac{Mv^2}{2} (f_0 + \delta f) \vec{v}_{E\times B} \right\rangle_S , \tag{15}$$

even though this term may contain a small neoclassical contribution, through the poloidally asymmetric modes ($n = 0, m \neq 0$). $\langle\rangle_S$ is the flux surface average operator and M the ion mass. The neoclassical contribution is defined as:

$$Q_{neo} = \left\langle \frac{\vec{\nabla} r}{|\vec{\nabla} r|} \cdot \int \mathrm{d}^3 v \frac{Mv^2}{2} \delta f (\vec{v}_{\nabla B} + \vec{v}_c) \right\rangle_S , \tag{16}$$

where $\vec{v}_{\nabla B}$ is the ∇B drift and $\vec{v}_c$ the curvature drift. The simulation system considered has no heat sources, so that the temperature profile is free to relax towards marginally stable gradients with respect to the instabilities underlying the microturbulence. Figure 4.a shows the initial and relaxed temperature profiles for both a collisionless case and for $\nu^*(r/a = 0.5) = 0.18$, which corresponds to a collisionality about 4.5 times larger than the one derived from the actual DIII-D parameters underlying the CYCLONE case. Let us point out that, in the collisional case, the system never fully relaxes, since the neoclassical heat transport persists even after the turbulent transport has vanished. As shown below and as expected, this neoclassical transport is however small compared to the fully developed turbulent transport in the first phase of the simulation, so that it is therefore appropriate to define the system as having reached a quasi-equilibrium state once the turbulent transport falls to the level of the neoclassical one. Figure 4.b shows the evolution of the ion heat diffusivity χ_{Hi} with respect to the effective ion temperature gradient R_0/L_{Ti}, averaged between $r = 0.3a$ and $r = 0.45a$. The critical gradient $R_0/L_{T\text{crit}}$ is defined as the temperature gradient of the system once it has reached the quasi-equilibrium state as just defined. As clearly seen from Figures 4.a and 4.b, the critical gradient of the collisional case is lower than the critical gradient of the collisionless case. Indeed, collisions yield a critical gradient close to the linear stability of the most unstable ITG modes, $R_0/L_{Ti} \approx 5$. This is consistent with the fact that collisions damp the zonal flows and thus reduce the so-called Dimits shift, defined as the difference between the critical gradient for linear stability and the critical gradient observed in nonlinear collisionless gyrokinetic turbulent simulations. The latter is larger due to the stabilizing effect of zonal flows. By damping the zonal flows, ion-ion collisions thus increase the level of ITG turbulent heat transport, as has already been observed and discussed in Ref. [15].

The time evolution of the energy flux at $r = 0.5a$ is plotted in Figure 5.a, for the same cases. For the collisional simulation, the neoclassical phase (keeping only $n = 0$ modes) is carried out up to time $t = 2\tau_{ii}$, after which the turbulent phase is initiated (keeping modes $n \neq 0$ as well). The small discrepancy between the total flux and the $\vec{v}_{\nabla B} + \vec{v}_c$ contribution in the neoclassical phase is due to the small inward neoclassical $\vec{v}_{E\times B}$ contribution through the poloidally asymmetric modes, as explained above. The fully developed turbulent transport, corresponding to the bursty phase, is clearly much larger than the neoclassical transport. The neoclassical flux nevertheless becomes dominant once the temperature profile has relaxed. Figure 5.b presents the time evolution of the signal/noise ratio. The standard way of evaluating the noise in ORB5 has been described in Ref. [16]. The basic idea is to define the signal as the modes in a field-aligned Fourier filter, while the noise is made by the modes outside the Fourier filter. The considered runs made use of the coarse graining procedure [2]. Note that even in the late phase of the collisional simulation, the ratio remains above the threshold (considered to be of order ~ 10, see Ref. [16]) under which the predicted transport level is not relevant. As expected, with the same numerical parameters the collisionless simulation remains significantly less noisy.

7. Conclusions

Collisional effects have been added to the global gyrokinetic δf Particle-In-Cell code ORB5. Some benchmarks against analytical predictions and results from other codes have been performed, showing very good agreement and validating the implementation of the collision operators. A novel δf-algorithm, which switches between a canonical and a local Maxwellian background for respectively carrying out the collisionless and collisional dynamics in the frame of a time splitting scheme, has been presented. Starting from global neoclassical equilibria, preliminary results concerning collisional ITG microturbulence have been obtained and compared to collisionless runs. The nonlinear upshifted critical gradient appears to be reduced by the effect of ion-ion collisions. The crucial issue of numerical noise for PIC codes has

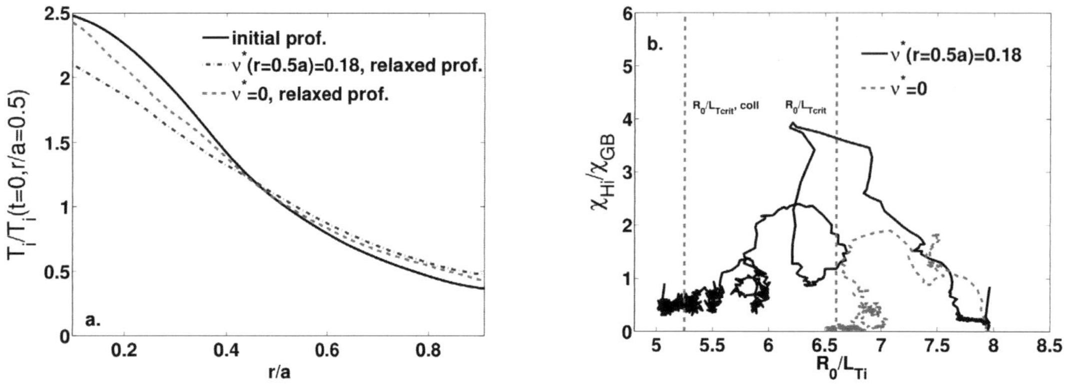

Figure 4. Collisionless and collisional turbulent CYCLONE case: a) Relaxation of the temperature profile towards the quasi-stationary state for both a collisionless and a collisional simulation. b) Evolution of the ion heat diffusivity $\chi_{Hi} = (Q_{neo} + Q_{turb})/(n_i|\nabla T_i|)$ in Gyro-Bohm units versus the temperature gradient. Collisions yield a lower nonlinear critical gradient. The quasi-stationary state is obtained when the heat diffusivity reaches the neoclassical level.

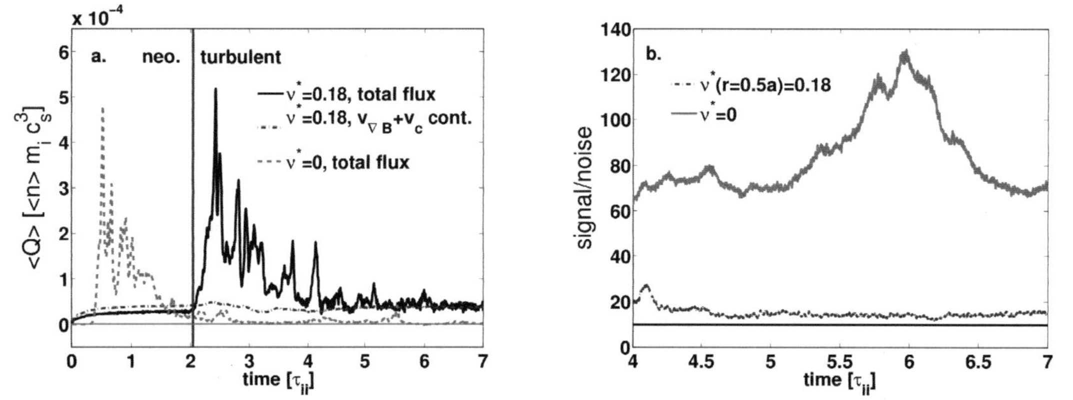

Figure 5. a) Time evolution of the total energy flux Q at $r/a = 0.5$ for both collisional (full line) and collisionless (dashed) simulations. Also shown is the neoclassical contribution for the collisional case (dash-dotted). In the bursty phase, the turbulent transport is much larger than the neoclassical transport. b) Signal/noise ratio for simulations using 500 million markers and coarse graining. The collisionless case is naturally less noisy, but the coarse graining enables to keep the collisional simulation above the relevance threshold ~ 10 even at later simulation times.

been adressed through the use of a coarse graining procedure, which enables high signal/noise ratios to be reached even in the turbulent phase of the simulation. These preliminary results open interesting perspectives for further work, especially the possibility to study collisional ITG microturbulence in presence of heat sources. The role of the FLR effects in ion-ion collision operator has also to be clarified. Finally, the study of collisional TEM microturbulence using kinetic electrons would be a challenging issue, taking account of the massive computer ressources it requires for getting physically relevant results.

Acknowledgements
Simulations were performed on the Monte Rosa CRAY XT-5 supercomputer of the Swiss National Supercomputing Center and the HPC-FF cluster of the Jülich Forschungszentrum. This work was partly supported by the Swiss National Science Foundation. The authors would like to thank S. Satake and Y. Idomura for supporting the benchmarking efforts by providing their own neoclassical data.

References

[1] Brunner S, Valeo E and Krommes J A 1999 *Physics of Plasmas* **6** 4504–4521
[2] Chen Y and Parker S E 2007 *Physics of Plasmas* **14** 082301
[3] Vernay T, Brunner S, Villard L, McMillan B F, Jolliet S, Tran T M, Bottino A and Graves J P 2010 *Physics of Plasmas* **17** 122301
[4] Jolliet S, Bottino A, Angelino P, Hatzky R, Tran T, McMillan B, Sauter O, Appert K, Idomura Y and Villard L 2007 *Computer Physics Communications* **177** 409 – 425
[5] Hahm T S 1988 *Physics of Fluids* **31** 2670–2673
[6] Hu G and Krommes J A 1994 *Physics of Plasmas* **1** 863–874
[7] Lin Z, Tang W M and Lee W W 1995 *Physics of Plasmas* **2** 2975–2988
[8] McMillan B, Jolliet S, Bottino A, Angelino P, Tran T and Villard L 2010 *Computer Physics Communications* **181** 715 – 719
[9] Satake S, Idomura Y, Sugama H and Watanabe T H 2010 *Computer Physics Communications* **181** 1069–1076
[10] Chang C S and Hinton F L 1982 *Physics of Fluids* **25** 1493–1494
[11] Hirshman S P and Sigmar D J 1981 *Nuclear Fusion* **21** 001079
[12] Hinton F L and Hazeltine R D 1976 *Rev. Mod. Phys.* **48** 239–308
[13] Dimits A M, Bateman G, Beer M A, Cohen B I, Dorland W, Hammett G W, Kim C, Kinsey J E, Kotschenreuther M, Kritz A H, Lao L L, Mandrekas J, Nevins W M, Parker S E, Redd A J, Shumaker D E, Sydora R and Weiland J 2000 *Physics of Plasmas* **7** 969–983
[14] Jolliet S 2009 Ph.D. thesis Ecole polytechnique fédérale de Lausanne
[15] Lin Z, Hahm T S, Lee W W, Tang W M and Diamond P H 1999 *Phys. Rev. Lett.* **83** 3645–3648
[16] Bottino A, Peeters A G, Hatzky R, Jolliet S, McMillan B F, Tran T M and Villard L 2007 *Physics of Plasmas* **14** 010701

Journal of Physics: Conference Series **260** (2010) 012022 doi:10.1088/1742-6596/260/1/012022

Kinetic structures of shear Alfvén and acoustic wave spectra in burning plasmas

F Zonca[1,2], A Biancalani[3,4], I Chavdarovski[5], L Chen[2,6], C Di Troia[1] and X Wang[2]

[1]Associazione Euratom-ENEA sulla Fusione, C.R. Frascati, C.P. 65, I-00044 Frascati, Italy
[2]Institute for Fusion Theory and Simulation, Zhejiang University, Hangzhou 310027, P.R.C.
[3]Max-Planck-Institut für Plasmaphysik, Euratom Association, D-85748 Garching, Germany
[4]Max-Planck-Institut für Sonnensystemforschung, Katlenburg-Lindau, Germany
[5]National Fusion Research Institute, Gwahangno 113, Yuseong-gu, Daejeon 305-333, Korea
[6]Department of Physics and Astronomy, University of California, Irvine, CA 92697-2575

E-mail: `fulvio.zonca@enea.it`

Abstract. We present a general theoretical framework for discussing the physics of low frequency fluctuation spectra of shear Alfvén and acoustic waves in toroidal plasmas of fusion interest. This framework helps identifying the relevant dynamics and, thus, interpreting experimental observations. We also discuss the roles of such general theoretical framework for verification and validation of numerical simulation codes vs. analytic predictions and experimental results.

1. Introduction

The low frequency fluctuation spectra of shear Alfvén and acoustic waves are important to burning plasmas of fusion interest, for they can be excited by both thermal as well as supra-thermal particles in various parameter regimes. There is an increasing consensus that accurate treatments of key physics processes of these spectra do require a kinetic approach [1, 2, 3, 4, 5, 6, 7, 8, 9, 10]; although whether a fully kinetic approach is really necessary is still being debated [11].

Here, we briefly summarize recently derived analytic theories, which demonstrate the crucial importance of employing kinetic analyses for fundamental processes in collisionless plasmas of fusion interest in the low frequency regime, i.e., where geometry effects, magnetically trapped particles [9, 10], wave particle interactions and finite (kinetic) plasma compressibility play major roles. Analytic expressions show that the acoustic branch is generally more strongly damped than the shear Alfvén branch, which is the one of practical interest for interpreting experimental observations. Meanwhile, analytic calculations can be used to benchmark results of numerical codes in the relevant limits [10, 12]; thus helping their verification processes. In this work, we also discuss possible further extensions of present analytic theories, which are already underway.

2. The general fishbone-like dispersion relation

Beta induced Alfvén Eigenmodes (BAE), observed originally in DIII-D [13], are particularly important in the study of low frequency fluctuations of the shear Alfvén wave (SAW) spectrum, since they can be excited by both fast ions (at long wavelengths) as well by thermal ions (at short wavelengths), as in the case of Alfvénic Ion Temperature Gradient driven modes (AITG) [14]. This was predicted theoretically [4] and then observed experimentally [15].

The BAE dispersion relation is yet another special case of the general fishbone-like dispersion relation (GFLDR) [1, 2, 3, 4, 5, 6], which can be written in the form

$$-i\Lambda + \delta W_f + \delta W_k = 0 \ , \tag{1}$$

where δW_f and δW_k play the role of fluid (core plasma, frequency independent) and kinetic (e.g. fast ion) contribution to the potential energy, respectively, while Λ represents a generalized inertia term. In ideal MHD, for frequencies much lower than the ion sound transit frequency, $\Lambda^2 = (\omega^2/\omega_A^2)(1+2q^2)$, where $\omega_A = v_A/(qR_0)$, v_A is the Alfvén speed and R_0 is the torus major radius, while q is the safety factor. In this case, Eq. (1) reduces to the well known fishbone dispersion relation [1]. More generally, Λ reflects finite thermal plasma diamagnetic responses [16, 17] as well as kinetic wave-particle interactions with the thermal plasma [4, 18].

Note that Eq. (1) is applicable both for low mode-number MHD fluctuations as well as for higher mode numbers [2, 3, 4]. In fact, the GFLDR can be generally derived by asymptotic matching the regular (ideal MHD) mode structure with the general (known) form of the SAW field in the singular (inertial layer) region. Following the idea that ballooning formalism can be generally used for computing the layer response of resistive MHD modes [19], the generic expression of Λ [4] can be used for application to low mode-number fluctuations as well [8, 10]. Actually, a more general Mode Structure Decomposition (MSD) approach can be used [20], which is based on the Poisson summation formula and reduces to the well-known ballooning formalism in the limit where spatial scale separation applies [20]. The MSD approach naturally yields [20, 21] the twisted field line coordinates discussed in [22], which have been successfully implemented as coordinate representation in numerical simulations, e.g. [23, 24]. It also suggests a straightforward way to handle both finite as well as vanishing magnetic shear $s = r(d\ln q/dr)$ [25]. Here, for brevity, we have written the GFLDR for $s \neq 0$ only.

The GFLDR generally demonstrates the existence of two types of modes [5]: a discrete gap mode, or Alfvén Eigenmode (AE), for $\mathrm{IRe}\Lambda^2 < 0$; and an Energetic Particle continuum Mode (EPM [3]) for $\mathrm{IRe}\Lambda^2 > 0$. Meanwhile, the SAW continuous spectrum is given by

$$\Lambda^2 = k_\parallel^2 q^2 R_0^2 = (nq-m)^2 \ , \tag{2}$$

for a fluctuation with given parallel wave vector $k_\parallel$ and poloidal/toroidal mode numbers m/n. Equations (1) and (2) demonstrate that, for both AE and EPM, the SAW accumulation point, $\Lambda = 0$, is the natural *gateway* through which modes are born at marginal stability [6]. Meanwhile, the GFLDR consistently treats SAW continuum, for which waves with local support Δr decay by phase mixing on a time scale $\propto (\Delta r \partial\tilde{\omega}_A(r)/\partial r)^{-1}$ [26], where $\tilde{\omega}_A(r)$ is generally to be intended as the solution of Eq. (2). As a consequence of the wide range of scale lengths on which SAW are excited [4, 14, 15], AE form a *dense population of eigenmodes (lighthouses) with unique (equilibrium-dependent) frequencies and locations* [6], nearby positions where $\partial\tilde{\omega}_A(r)/\partial r = 0$, and with characteristic properties depending on the peculiar equilibrium that is being considered via δW_f and δW_k. Note that δW_f and δW_k can be computed numerically, using different types of codes and models and, thus, even realistic mode structures. So, even if the GFLDR is given in the form of an energy functional, its validity is very general and, thus, it can be used as generic theoretical framework, which helps identifying the underlying dynamics and nature of the mode, as well as guidance for interpreting experimental observations. In conclusion, the GFLDR provides the unified theoretical framework for discussing kinetic structures of SAW and acoustic wave spectra in burning plasmas [5, 6], and can be used for a variety of MHD modes as well [19]. Here, for brevity, we will discuss the general fluctuation features described by Eqs. (1) and (2), which can be deduced from the properties of the generalized inertia term Λ.

3. BAE dispersion relation

In order to calculate the generalized inertia term Λ using kinetic theory, we follow [4] and consider finite mode number [8, 10, 21], in order to be able to use Eqs. (1) and (2) for all mode numbers (m,n). We also consider mode structures that may have a kinetic singular (inertial) layer located away from a mode rational surface, so that $k_\parallel qR_0 = (nq-m)$ is generally non vanishing. This allows us to derive Λ expressions that apply also for $s=0$, although Eq. (1) in such a case needs to be modified [25, 27][1], while Eq. (2) remains unaltered. For BAE one typically has $\omega \approx \omega_{ti} \approx \omega_{*pi} \approx k_\parallel v_A$, where $\omega_{ti} = (2T_i/m_i)^{1/2}/qR_0$ is the thermal ion transit frequency and $\omega_{*pi} = \omega_{*ni} + \omega_{*Ti}$ is the thermal ion diamagnetic frequency, separated as density and temperature gradient components, $\omega_{*ni} = (T_i c/eB)(\boldsymbol{k}\times\boldsymbol{b})\cdot\nabla(n_i)/n_i$ and $\omega_{*Ti} = (T_i c/eB)(\boldsymbol{k}\times\boldsymbol{b})\cdot\nabla(T_i)/T_i$, with $\boldsymbol{b}=\boldsymbol{B}/B$. Thus, $k_\parallel qR_0 \approx \beta^{1/2}$, with β the ratio of kinetic to magnetic pressures. In the case analyzed here, we have $k_\parallel = k_{\parallel 0} + \Delta k_\parallel$, with $k_{\parallel 0} = (nq-m)/(qR_0)$ computed at the singular layer and $\Delta k_\parallel qR_0 \approx \beta^{1/2}$.

As in [4], the problem of coupled SAW and acoustic fluctuations is defined by the coupled quasi-neutrality and vorticity equations, i.e.,

$$\left(1+\tau^{-1}\right)(\delta\phi - \delta\psi) = (T_i/ne)\langle \delta K_i\rangle \ , \tag{3}$$

$$B\mathbf{b}\cdot\nabla\left[\frac{1}{B}\frac{k_\perp^2}{k_\vartheta^2}\mathbf{b}\cdot\nabla\delta\psi\right] + \frac{\omega^2}{v_A^2}\left(1-\frac{\omega_{*pi}}{\omega}\right)\frac{k_\perp^2}{k_\vartheta^2}\delta\phi + \frac{\alpha}{q^2R^2}g(\theta)\delta\psi = \left\langle \frac{4\pi e}{k_\vartheta^2 c^2}\omega\omega_{di}\delta K_i\right\rangle . \tag{4}$$

Here, $\tau = T_e/T_i$, angular brackets denote velocity space integration and the scalar fields $\delta\phi$ and $\delta\psi$ are related with the vector potential fluctuation $\delta A_\parallel$ and parallel electric field $\delta E_\parallel$ by $\delta A_\parallel \equiv -i\,(c/\omega)\,\mathbf{b}\cdot\nabla\delta\psi$ and $\delta E_\parallel \equiv -\mathbf{b}\cdot\nabla(\delta\phi-\delta\psi)$. The fluctuating parallel magnetic field $\delta B_\parallel$ is given by the condition of perpendicular pressure balance [5]. Meanwhile, for the sake of simplicity, we have assumed massless electrons and one thermal ion species with unit electric charge e. The drift-kinetic thermal ion distribution function response is decomposed as

$$\delta f_i = \left(\frac{e}{m}\right)_i\left[\frac{\partial F_{0i}}{\partial\mathcal{E}}\delta\phi - \frac{QF_{0i}}{\omega}\delta\psi\right] + \delta K_i \ , \tag{5}$$

where F_{0i} is the equilibrium distribution function, $\mathcal{E} = v^2/2$ the energy per unit mass, $QF_{0i} = (\omega\partial_\mathcal{E} + \hat{\omega}_*)_i F_{0i}$, $\hat{\omega}_{*i}F_{0i} = (m_i c/eB)(\mathbf{k}\times\mathbf{b})\cdot\nabla F_{0i}$. We also adopted straight field line toroidal flux coordinates (r,θ,ζ) [20, 23, 24], as in the MSD approach, assuming a simple $(s, \alpha = -R_0 q^2 d\beta/dr)$ model equilibrium with shifted circular magnetic surfaces, for which the perpendicular wave vector is $k_\perp^2 = k_\vartheta^2[1+(s\theta-\alpha\sin\theta)^2]$ and the magnetic drift frequency ω_{di} is $\omega_{di}(\theta) = g(\theta)k_\vartheta m_i c(v_\perp^2/2+v_\parallel^2)/eBR$, $g(\theta) = \cos\theta + [s\theta - \alpha\sin\theta]\sin\theta$. Finally, the drift-kinetic thermal ion compressibility response is obtained from

$$\left[\omega_{tr}\partial_\theta - i\,(\omega-\omega_d)\right]_i \delta K_i = i\,(e/m)_i\, QF_{0i}\left[(\delta\phi-\delta\psi) + (\omega_d/\omega)_i\,\delta\psi\right] \ , \tag{6}$$

where $\omega_{tr} = v_\parallel/qR$ is the transit frequency. The kinetic expression of the generalized inertia term Λ is obtained from the formal manipulation of the layer Eqs. (3), (4) and (6) via asymptotic expansion in $\beta^{1/2}$ [4, 14, 30]. From the lowest order solution of Eq. (6), substituted back into Eq. (3), we have the ideal MHD response $\delta\psi^{(0)} = \delta\phi^{(0)}$ for the mode numbers m/n.

At the next order, $\delta\phi^{(1)} = e^{i\theta}\delta\phi^{(1+)} + e^{-i\theta}\delta\phi^{(1-)}$ with a corresponding $\delta K_i^{(1)} = e^{i\theta}\delta K_i^{(1+)} + e^{-i\theta}\delta K_i^{(1-)}$. The $\delta\psi^{(1)}$ component is negligible, as it is readily verified from Eq. (4). Let $\omega_{tr}^{(\pm)} = [1\pm(nq-m)]\,(v_\parallel/qR_0)$. Then

$$i\left[\pm\omega_{tr}^{(\pm)} - \omega\right]\delta K_i^{(1\pm)} = i\left(\frac{e}{m}\right)_i QF_{0i}^{(\mp)}\delta\phi^{(1\pm)} \mp i\left(\frac{e}{m}\right)_i QF_{0i}\frac{v_\perp^2/2+v_\parallel^2}{R_0\omega_{ci}\omega}\frac{k_r}{2i}\delta\phi^{(0)} \ . \tag{7}$$

[1] Equation (1) needs also to be modified for $k_\parallel$ values satisfying the Bragg reflection condition [3, 28, 29].

Note that the dominant effect comes from the geodesic curvature (large $k_r \propto \theta$), causing radial magnetic drifts. Sideband generation is evident in the wave-particle interaction as well as in the diamagnetic effect in $QF_{0i}^{(\mp)}$, which is computed at $m \mp 1$ [8, 10]. Substituting back into the quasi-neutrality condition and letting $\omega_{ti}^{(\pm)} = (2T_i/m_i)^{1/2}\left[1 \pm (nq-m)\right]/qR_0$, one obtains

$$\delta\phi^{(1\pm)} = \mp i \frac{cT_i}{eB_0} \frac{N_m(\omega/\omega_{ti}^{(\pm)})}{D_{m\mp 1}(\omega/\omega_{ti}^{(\pm)})} \frac{k_r}{\omega R_0} \delta\phi^{(0)} \; . \tag{8}$$

Here, $Z(x) = \pi^{-1/2} \int_{-\infty}^{\infty} e^{-y^2}/(y-x)dy$ and subscripts m and $m \mp 1$ indicate poloidal mode numbers to compute the functions N_m and D_m [4]

$$N_m(x) = \left(1 - \frac{\omega_{*ni}}{\omega}\right)\left[x + (1/2 + x^2)\,Z(x)\right] - \frac{\omega_{*Ti}}{\omega}\left[x\,(1/2+x^2) + (1/4 + x^4)\,Z(x)\right] \; , \tag{9}$$

$$D_m(x) = \left(\frac{1}{x}\right)\left(1 + \frac{1}{\tau}\right) + \left(1 - \frac{\omega_{*ni}}{\omega}\right) Z(x) - \frac{\omega_{*Ti}}{\omega}\left[x + (x^2 - 1/2)\,Z(x)\right] \; . \tag{10}$$

Corresponding solutions of the drift-kinetic equation are

$$\delta K_i^{(1\pm)} = \mp i \frac{cT_i}{eB_0}\frac{e/m_i}{\omega \mp \omega_{tr}^{(\pm)}}\left[\frac{m_i}{2T_i}\left(\frac{v_\perp^2}{2} + v_\parallel^2\right) QF_{0i} - \frac{N_m(\omega/\omega_{ti}^{(\pm)})}{D_{m\mp1}(\omega/\omega_{ti}^{(\pm)})} QF_{0i}^{(\mp)}\right]\frac{k_r}{\omega R_0}\delta\phi^{(0)} \; . \tag{11}$$

When substituted back into the second order vorticity equation for the singular layer (large $|\theta|$), reducing to the canonical form yielding the GFLDR [4, 14, 30],

$$\left(\partial_\theta^2 + \Lambda^2\right)\delta\psi^{(0)} = 0 \; , \tag{12}$$

these results give the following expression of Λ^2 [8, 10], which accounts for charge separation effects due to geodesic curvature, but neglects higher order terms due to finite Larmor radius (FLR) and finite magnetic drift orbit widths (FOW) [30], which will be considered elsewhere (see final section). Accounting for FLR and FOW introduces finer scales in the problem and modifies the structure of the vorticity equation in the singular layer [30], similar to considering small but finite electron inertia [31]; the GFLDR is readily recovered from such treatments in the appropriate (long) wavelength limit [7, 30, 31].

$$\begin{aligned}\Lambda^2 = {} & \frac{\omega^2}{\omega_A^2}\left(1 - \frac{\omega_{*pi}}{\omega}\right) + q^2\frac{\omega_{ti}^2}{2\omega_A^2}\left[\left(1 - \frac{\omega_{*ni}}{\omega}\right)\left((\omega/\omega_{ti}^{(+)})F(\omega/\omega_{ti}^{(+)}) + (\omega/\omega_{ti}^{(-)})F(\omega/\omega_{ti}^{(-)})\right)\right. \\ & - \frac{\omega_{*Ti}}{\omega}\left((\omega/\omega_{ti}^{(+)})G(\omega/\omega_{ti}^{(+)}) + (\omega/\omega_{ti}^{(-)})G(\omega/\omega_{ti}^{(-)})\right) \\ & \left. - \left((\omega/\omega_{ti}^{(+)})N_m(\omega/\omega_{ti}^{(+)})\frac{N_{m-1}(\omega/\omega_{ti}^{(+)})}{D_{m-1}(\omega/\omega_{ti}^{(+)})} + (\omega/\omega_{ti}^{(-)})N_m(\omega/\omega_{ti}^{(-)})\frac{N_{m+1}(\omega/\omega_{ti}^{(-)})}{D_{m+1}(\omega/\omega_{ti}^{(-)})}\right)\right] \; .\end{aligned} \tag{13}$$

Here, the functions $F(x)$ and $G(x)$ are defined as [4]

$$\begin{aligned} F(x) &= x\,(x^2 + 3/2) + (x^4 + x^2 + 1/2)\,Z(x) \; , \\ G(x) &= x\,(x^4 + x^2 + 2) + (x^6 + x^4/2 + x^2 + 3/4)\,Z(x) \; . \end{aligned} \tag{14}$$

Details of diamagnetic sideband generation are unnecessary for drift Alfvén waves (DAW) at short wavelength; they are needed for moderate wavelength MHD-type modes. The importance of these effects, which are missing in Ref. [11], has been lately emphasized in Refs. [8, 10].

The structures of the low-frequency SAW continuum can be studied with the given Λ^2 expression and solving Eq. (2). Such analyses, however, must be limited to exploring the frequency range above the thermal ion bounce frequency, since the expression of Λ^2, given above, does not describe trapped particle dynamics (cf. later discussions). The importance (dominance at low frequency) of trapped particle dynamics has been recently discussed both analytically [9] and numerically [10].

4. Acoustic wave couplings

The notion of Beta induced Alfvén Acoustic Eigenmode (BAAE) was originally formulated [32, 33] on the basis of a fluid approach, which may not be valid in collisionless plasmas of fusion interest. This approach was extended later on to kinetic analyses accounting for circulating thermal particle effects [11]. A simple pictorial view of the BAAE gap formation is given in Fig. 1 of Ref. [11]. The BAAE gap is formed essentially because of the coupling of two (sideband, $m \pm 1/n$) acoustic waves both among themselves as well as with the reference (flute-like, m/n) SAW. That figure, however, neglects crucially important kinetic effects, which completely alter that picture, since Landau damping essentially rules out the possibility of a significant plasma response to the acoustic and Alfvén acoustic polarizations.

In order to show this, we give a simple derivation of the kinetic BAAE response, following the analogous procedure used for BAE in the previous section. For Alfvénic polarization, the dominant component is flute-like and satisfies $\delta\psi^{(0)} = \delta\phi^{(0)}$ (ideal Ohm's law), as shown above. For acoustic wave coupling to become important, as in the BAAE picture, we must as well consider an $O(1)$ sideband $\delta\phi_s = e^{i\theta}\delta\phi^{(+)} + e^{-i\theta}\delta\phi^{(-)}$; i.e., the sideband does not enter as first order modulation as for BAE/KBM (kinetic ballooning mode [17]). Note that the $\delta\psi_s$ sideband component is negligible at low-β, due to the vorticity equation constraints.

At the lowest order of the $\beta^{1/2}$ asymptotic expansion, solution of the quasi-neutrality condition implies $\delta\psi^{(0)} = \delta\phi^{(0)}$ (as for the BAE/KBM problem) and

$$D_{m\mp1}(\omega/\omega_{ti}^{(\pm)})\delta\phi^{(\pm)} = 0 \ . \tag{15}$$

So, no connection is provided at the lowest order between $\delta\phi^{(\pm)}$ and $\delta\phi^{(0)}$. This result shows that, at the lowest order, the "true" acoustic mode (no SAW coupling) is merely the usual (sideband) electrostatic drift wave (e.s. DW), given by $D_{m\mp1}(\omega/\omega_{ti}^{(\pm)}) = 0$, which is known to be generally affected by strong Landau damping unless, e.g., $\tau \gg 1$.

At the next $O(\beta^{1/2})$ order, the solution of the quasi-neutrality condition gives back Eq. (8) with $\delta\phi^{(1\pm)}$ replaced by $\delta\phi^{(\pm)}$, i.e. the same link between sidebands and flute-like component that was written for the BAE/KBM case. The only difference is that, in the present case, we may generally have $\delta\phi^{(\pm)} \approx \delta\phi^{(0)}$ for $D_{m\mp1}(\omega/\omega_{ti}^{(\pm)}) \approx \beta^{1/2}$, or even $|\delta\phi^{(\pm)}| \gg |\delta\phi^{(0)}|$ for a purely electrostatic polarization [4]. The final solubility condition is obtained from the vorticity equation at $O(\beta)$, which yields the same expression for Λ^2, given above in Eq. (13).

Thus, the kinetic BAAE dispersion relation reduces trivially to the GFLDR in the limit where trapped particle dynamics are neglected, which are crucially important at low-frequencies [9, 10] (see next section on trapped particle effects). This result is not unexpected, since the first derivation of BAE/KBM kinetic responses solved the coupled vorticity and quasi-neutrality equations [4], thus completely accounting for SAW coupling to acoustic waves.

It was also noted [4] that the low-frequency kinetic SAW and acoustic fluctuation response is analytically expressed in terms of transcendental functions, which in general have an infinite number of (stable) roots. Only the least stable must be considered when computing the plasma behavior. In fact, when solving the initial value problem, the contribution of the more stable branch points and cuts becomes rapidly negligible.

The solution of Eq. (2), with Λ^2 given by Eq. (13), is shown in Figs. 1 and 2 for both DAW (BAE) as well as e.s. DW (BAAE) branches. Due to the intrinsic difficulties in comparing experimental observations with theoretical predictions of mode frequencies (see Sec. III.D of [11]), we have chosen $\beta_i = 0.01$, $\omega_{*ni}/\omega_{ti} = 0.1$, $\omega_{*Ti}/\omega_{ti} = 0.2$ and $\tau = 2$ as fixed parameters, reasonably close to experimental values, and considered fluctuations with $m/n = 6/3$ localized near $q = 2$. The close resemblance of the present Fig. 1 with Fig. 1 of Ref. [11] is evident for the two higher frequency branches. Lower frequency modes of [11] are not considered here, since their correct treatment requires the inclusion of trapped particle dynamics (see below).

Journal of Physics: Conference Series **260** (2010) 012022 doi:10.1088/1742-6596/260/1/012022

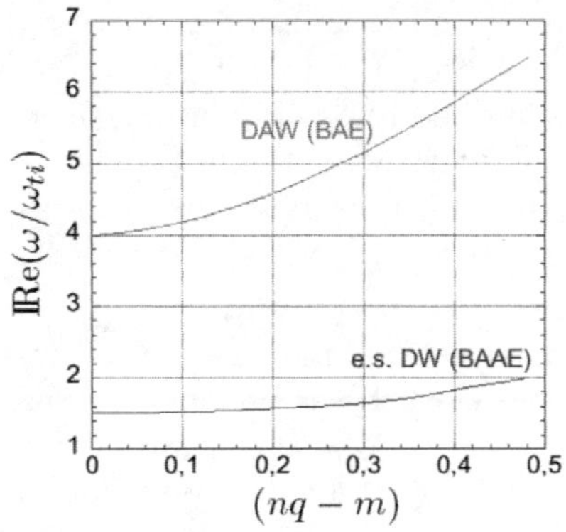

Figure 1. $\mathbb{R}\mathrm{e}(\omega/\omega_{ti})$ is plotted vs. $(nq - m)$ for DAW (BAE) and e.s. DW (BAAE).

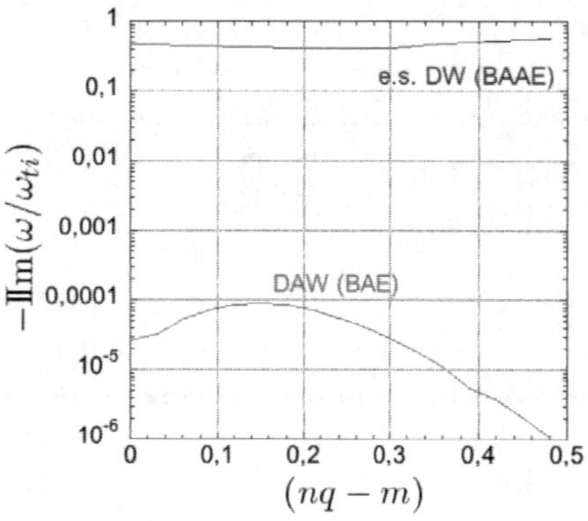

Figure 2. $-\mathbb{I}\mathrm{m}(\omega/\omega_{ti})$ is plotted vs. $(nq-m)$ for DAW (BAE) and e.s. DW (BAAE).

The acoustic polarization has stronger Landau damping due to the typically lower fluctuation frequency and the stronger a.c. electric field component. Thus, most unstable modes tend to have Alfvénic polarization and a general flute-like structure. Meanwhile, the various kinetic interactions with the thermal plasma are properly described by the generalized inertia term Λ^2 and account for the different fluctuation types in the low frequency Kinetic Thermal Ion (KTI) gap [6], which are connected with different physics; e.g., diamagnetic drift for KBM [17], thermal ion compressibility for BAE [13], kinetic thermal ion compressibility and wave-particle resonances for AITG [14, 30]. Note that fluctuation frequencies below the SAW accumulation points are not indicative of the existence of BAAE; they simply show that AE/EPM can be shifted down in frequency with respect to the accumulation point, as prescribed by the GFLDR (see section 2). In addition, SAW accumulation points, computed with Λ^2 from Eq. (13), tend to overestimate the actual frequency, due to the absence of trapped particle responses [9, 10]. Plasma shaping, such as elongation, may also lower the SAW accumulation point frequency [10].

5. Trapped particle effects

Detailed descriptions of trapped particle effects [9, 10], including finite $k_\parallel qR_0$ and diamagnetic sideband generation, are beyond the scope of the present work. Here, we briefly remark two important elements. First, the low frequency ($|\omega| < \omega_{bi}$, with ω_{bi} the thermal ion bounce frequency) renormalized plasma inertia is dominated by trapped particle dynamics. Second, trapped particle compressions generally lower the BAE accumulation point frequency and are needed for a correct description of experimental observations [10]. At low frequency, the SAW dynamics in toroidal system reflects the compression effects of trapped thermal ions [34]:

$$\Lambda^2 = \left(\omega^2/\omega_A^2\right)\left(1 - \omega_{*pi}/\omega\right)\left(1 + 1.6q^2\epsilon^{-1/2} + 0.5q^2\right) . \tag{16}$$

It is possible to show that the renormalized inertia expression in Eq. (16) must be equal [27] to the zonal flow (ZF) polarizability [35]. This is a consequence of the same physics analogy argument that is used to demonstrate BAE accumulation point degeneracy with the Geodesic acoustic mode (GAM) for long wavelengths [6, 36] and has implications on cross-scale couplings mediated by zonal structures and long time-scale behaviors in burning plasmas [37, 38]. Note the difference of this expression with that for the *modified Alfvén branch, "shielded" by acoustic sidebands*, given in [11] when ignoring trapped particles. Besides not including trapped particle responses, the expression given in [11] fails, for it assumes isotropic thermal plasma compressibility, whereas at low-frequency the plasma behavior is strongly anisotropic. The low frequency limit of Eq. (13), which accounts for anisotropic compressibility but not for trapped particles, correctly reproduces the $\propto (1 + 0.5q^2)$ real part of Eq. (16) [27]; however, it must be corrected for the imaginary part, in order to account for the modified circulating ion response due to finite trapped particle fraction [9]. A general expression of Λ^2, smoothly connecting that of Eq. (16) with that of Eq. (13), has been derived for the case of deeply trapped particles [9].

6. Interpretation of experimental data, conclusions and discussions

One of the key experimental evidences [15] is the observation of low frequency SAW in a wide range of mode numbers, driven by both energetic (long wavelength) as well as thermal (short wavelength) ions, confirming theoretical predictions [4, 14]. New interest was attracted by observations of BAAE [32, 33] at frequencies below the BAE accumulation point of the SAW continuum, by the more recent evidence of "Sierpes modes" in ASDEX Upgrade [39], interpreted as BAE excited by energetic ions generated by Ion Cyclotron Resonance Heating (ICRH) [40], and also by measurements of ICRH driven BAE in Tore Supra [41]. For interpretation of these observations, as shown above, it is necessary to use kinetic theories for the proper treatment of wave-particle interactions with circulating as well as trapped thermal plasma particles [9, 10]. The features of the kinetic dispersion relations and the intrinsic non-linear (non additive) interplay of trapped and circulating particle dynamics are complex [9, 10] and make simplified extensions and/or renormalizations of fluid theories inappropriate [11]. Meanwhile, for sufficiently strong energetic particle drive (ions and electrons), experimental observations show evidence of a continuous transition between various SAW and MHD fluctuation branches [8, 42, 43, 44, 45], consistent with theoretical predictions [1, 2, 3, 28] based on the GFLDR.

The present theoretical framework suggests that low-frequency SAW and acoustic oscillations should be seen as either EPM or AE, localized within the Kinetic Ion Thermal (KTI) gap [6]. The notion of BAAE, which was lately emphasized [11, 32, 33], seems instead to be based on a paradigmatic approach dictated by improper analogies with fluid theories. In fact, the kinetic BAAE dispersion relation reduces trivially to the GFLDR in the limit where trapped particle dynamics are neglected, which are crucially important at low-frequencies [9, 10]. The intrinsic limits of MHD/fluid analyses at low frequency in collisionless tokamak plasmas of fusion interest suggest developing new numerical investigation tools, with aim at bridging simulations of meso- and micro-scale phenomena [6, 36, 37, 38, 46]. The linear gyrokinetic code LIGKA is capable to address these issues [10, 46]. Meanwhile, an eXtended version of the HMGC code (XHMGC [47], nonlinear hybrid MHD-gyrokinetic code), has been developed to handle kinetic thermal ion compressibility effects [48]. Successful comparisons of XHMGC numerical studies of kinetic BAE excited by energetic particles vs. analytic predictions have been already carried out [12]. Global non-perturbative (full-f) electromagnetic gyrokinetic simulations are also becoming available, such as for the GTC [49] and GYRO [50] codes. Work is ongoing to complete the present theoretical-analytical framework; e.g., extending the Λ^2 expression including (deeply) trapped particle dynamics [9] with finite $k_{\|}qR_0$ and diamagnetic sideband generation, as well as finite Larmor radius and finite magnetic-orbit width effects [14, 30] in the whole frequency range. With these results, the further generalization of the GFLDR at short scales [14, 30] will complete the present theoretical framework and allow us to verify local analytic theories vs. local and global numerical simulations, consolidating the basis for a complete toolbox for numerical computation and application of the GFLDR in situations of practical interest.

Acknowledgments

This work is supported by the Euratom Communities under the contract of Association between EURATOM/ENEA, US DOE Grants and the National Basic Research Program of China.

References

[1] Chen L, White R B and Rosenbluth M N 1984 *Phys. Rev. Lett.* **52** 1122
[2] Tsai S T and Chen L 1993 *Phys. Fluids B* **5** 3284
[3] Chen L 1994 *Phys. Plasmas* **1** 1519
[4] Zonca F and Chen L 1996 *Plasma Phys. Control. Fusion* **38** 2011
[5] Zonca F and Chen L 2006 *Plasma Phys. Controlled Fusion* **48** 537
[6] Chen L and Zonca F 2007 *Nucl. Fusion* **47** S727
[7] Nguyen C, Garbet X and Smolyakov A I 2008 *Phys. Plasmas* **15** 112502
[8] Zonca F, Chen L, Botrugno A et al. 2009 *Nucl. Fusion* **49** 085009
[9] Chavdarovski I and Zonca F 2009 *Plasma Phys. Controlled Fusion* **51** 115001
[10] Lauber Ph et al. 2009 *Plasma Phys. Controlled Fusion* **51** 124009
[11] Gorelenkov N N, Van Zeeland M A, Berk H L et al. 2009 *Phys. Plasmas* **16** 056107
[12] Wang X, Zonca F and Chen L 2010 *Plasma Phys. Control. Fusion* **52** 115005
[13] Heidbrink W W, Strait E J, Chu M S and Turnbull A D 1993 *Phys. Rev. Lett.* **71** 855
[14] Zonca F, Chen L, Dong J Q and Santoro R A 1999 *Phys. Plasmas* **6** 1917
[15] Nazikian R et al. 2006 *Phys. Rev. Lett* **96** 105006
[16] Coppi B and Porcelli F 1986 *Phys. Rev. Lett.* **57** 2272
[17] Biglari H and Chen L 1991 *Phys. Rev. Lett.* **67** 3681
[18] Romanelli F and Chen L 1991 *Phys. Fluids B* **3** 329
[19] Pegoraro F and Schep T J 1986 *Plasma Phys. Control. Fusion* **28** 647
[20] Zonca F, Chen L and White R B 2004 *Theory of Fusion Plasmas* (Società Italiana di Fisica, Bologna, 2004) J.W. Connor, O. Sauter and E. Sindoni (Eds.) pp. 3-12
[21] Zonca F 2008 *Lecture Notes of the "Gyrokinetic Theory and Numerics" high level course of the "Goal Oriented Training in Theory" EFDA project* B.D. Scott (Ed.) (IPP, Garching, Germany, 3 - 14 November, 2008)
[22] Roberts K V and Taylor J B 1965 *Phys. Fluids* **8** 315
[23] Scott B 1998 *Phys. Plasmas* **5** 2334
[24] Scott B 2001 *Phys. Plasmas* **8** 447
[25] Zonca F, Briguglio S, Chen L et al. 2002 *Phys. Plasmas* **9** 4939
[26] Chen L and Hasegawa A. 1974 *Phys. Fluids* **17** 1399
[27] Zonca F et al. 2007 *Nucl. Fusion* **47** 1588
[28] Chen L and Zonca F 1995 *Physica Scripta* **T60** 81
[29] Zonca F and Chen L 1992 *Phys. Rev. Lett.* **68** 592
[30] Zonca F, Chen L, Santoro R A and Dong J Q 1998 *Plasma Phys. Control. Fusion* **40** 2009
[31] Smolyakov A I, Nguyen C and Garbet X 2008 *Plasma Phys. Control. Fusion* **50** 115008
[32] Gorelenkov N N, Berk H L, Fredrickson E and Sharapov S E 2007 *Phys. Lett. A* **370** 70
[33] Gorelenkov N N et al. 2007 *Plasma Phys. Control. Fusion* **49** B371
[34] Graves J P, Hastie R J and Hopcraft K I 2000 *Plasma Phys. Control. Fusion* **42** 1049
[35] Rosenbluth M N and Hinton F L 1998 *Phys. Rev. Lett.* **80** 724
[36] Zonca F et al. 2006 *Plasma Phys. Control. Fusion* **48** B15
[37] Zonca F and Chen L 2008 "Nonlinear Dynamics and Complex Behaviors in Magnetized Plasmas of Fusion Interest" Frontiers in Modern Plasma Physics *AIP Conf. Proc.* **1061** 34
[38] Zonca F and Chen L 2008 "Structures of the low frequency Alfvén continuous spectrum and their consequences on MHD and micro-turbulence" Theory of Fusion Plasmas *AIP Conf. Proc.* **1069** 355
[39] Garcia-Munoz M et al 2008 *Phys. Rev. Lett.* **100** 055005
[40] Lauber Ph et al 2008 *35th EPS Conf. on Plasma Phys. (Hersonissos, 9 - 13 June 2008)* ECA **32** O-4.030
[41] Sabot R et al. 2009 *Nucl. Fusion* **49** 085033
[42] Nabais F et al. 2005 *Phys. Plasmas* **12** 102509
[43] Fredrickson E D et al. 2006 *Phys. Plasmas* **13** 056109
[44] Darrow D S et al. 2008 *Nucl. Fusion* **48** 084004
[45] Fredrickson E D et al. 2009 *Phys. Plasmas* **16** 122505
[46] Lauber Ph and Günter S 2008 *Nucl. Fusion* **48** 084002
[47] Wang X, Briguglio S, Chen L, Fogaccia G, Vlad G and Zonca F 2010 "An extended hybrid magnetohydrodynamics gyrokinetic model for numerical simulation of shear Alfvén waves in burning plasmas" *submitted to Phys. Plasmas*
[48] Wang X, Bierwage A, Briguglio S, Chen L, Fogaccia G, Vlad G, Di Troia C, Zonca F, Zhang H and Lin Z 2010 "Kinetic thermal ions effects on Alfvénic fluctuations in tokamak plasmas" submitted to *IAEA Fusion Energy Conference (Daejon, Korea)*
[49] Lin Z, Hahm T S, Lee W W, Tang W M and White R B 1998 *Science* **281** 1835
[50] Candy J and Waltz R E 2003 *Phys. Rev. Lett.* **91** 045001

AUTHOR INDEX

Abiteboul, J. 72, 126
Akers, R. 94
Allfrey, S. 72, 126
Amicucci, L. 58
Angioni, C. 40, 65
Bagnoli, F. 32
Balakin, A. A. 9
Belmondo, V. 1
Bertelli, N. 9
Biancalani, A. 168
Bilato, R. 1
Bonfiglio, D. 16
Borchardt, M. 102
Borgogno, D. 87, 109
Bottino, A. 159
Brambilla, M. 1
Braun, S. 24
Brunner, S. 79, 159
Bufferand, H. 32
Burckel, A. 40
Candy, J. 40
Cappello, S. 16
Cardinali, A. 50
Cesario, R. 58
Chacón, L. 16
Chavdarovski, I. 168
Chen, L. 168
Chowdhury, J. 79
Ciraolo, G. 32
Dannert, T. 79
De Arcangelis, D. 58
Dennis, G. 94
Di Troia, C. 168
Dif-Pradalier, G. 126
Fable, E. 40, 65
Ferrari, M. 58
Fietz, S. 65
Fitzgerald, M. 94
Fusco, V. 50
Garbet, X. 72, 126, 135
Garcia, O. E. 9
Ghendrih, P. 32, 72, 126
Görler, T. 79
Grandgirard, V. 72, 126
Grasso, D. 87, 109, 150
Guo, S. C. 135
Hatzky, R. 102
Helander, P. 24
Hobirk, J. 65
Hole, M. 94
Hornsby, W. A. 143
Jenko, F. 79
Jolliet, S. 159
Kauffmann, K. 102
Kleiber, R. 102
Lapillonne, X. 40, 79
Latu, G. 72, 126
Lepri, S. 32
Liu, S. F. 135
Livi, R. 32
Maj, O. 1
McClements, K. G. 94
McMillan, B. F. 79, 159
Merz, F. 79
Napoli, F. 58
Naulin, V. 9
Nielsen, A. H. 9
Peeters, A. G. 143
Pegoraro, F. 87
Pereverzev, G. 65
Perona, A. 109
Poli, E. 143
Pometescu, N. I. 120
Predebon, I. 135
Pullara, E. 58
Sarazin, Y. 72, 126
Sattin, F. 135
Sauter, O. 40, 159
Schep, T. J. 87
Siccinio, M. 143
Smolyakov, A. 72
Spizzo, G. 16
Strugarek, A. 72, 126
Tamain, P. 32
Tassi, E. 150
Told, D. 79
Tran, T. M. 159
Veranda, M. 16, 135
Vernay, T. 159
Villard, L. 79, 159
Waelbroeck, F. L. 150
Wang, X. 168
Westerhof, E. 9
Zonca, F. 168